Communications in Computer and Information Science 590

Commenced Publication in 2007
Founding and Former Series Editors:
Alfredo Cuzzocrea, Dominik Ślęzak, and Xiaokang Yang

More information about this series at http://www.springer.com/series/7899

Wenguang Chen · Guisheng Yin · Gansen Zhao
Qilong Han · Weipeng Jing · Guanglu Sun
Zeguang Lu (Eds.)

Big Data Technology and Applications

First National Conference, BDTA 2015
Harbin, China, December 25–26, 2015
Proceedings

 Springer

Editors
Wenguang Chen
Department of Computer Science and
   Technology
Tsinghua University
Beijing
China

Guisheng Yin
Harbin Engineering University
Harbin
China

Gansen Zhao
South China Normal University
Guangzhou
China

Qilong Han
Harbin Engineering University
Harbin
China

Weipeng Jing
Northeast Forestry University
Harbin
China

Guanglu Sun
Harbin University of Science and
   Technology
Harbin
China

Zeguang Lu
Harbin Sea of Clouds and Computer
   Technology Services Ltd.
Harbin
China

ISSN 1865-0929   ISSN 1865-0937 (electronic)
Communications in Computer and Information Science
ISBN 978-981-10-0456-8   ISBN 978-981-10-0457-5 (eBook)
DOI 10.1007/978-981-10-0457-5

Library of Congress Control Number: 2015960775

This Springer imprint is published by SpringerNature
The registered company is Springer Science+Business Media Singapore Pte Ltd.

Preface

The Conference on the Forum of Big Data Technology and Applications (BDTA) 2015 took place in Harbin, Heilongjiang Province, China, on December 25, 2015, hosted by CCF and co-hosted by CCF TCAPP.

The development and application of big data and cloud computing have made great progress. Big data are expected to make a breakthrough in key industries such as communications, Internet, finance, medical treatment, and smart cities. Although big data has brought about changes and opportunities, China's big data industry is still in the initial stage of development. There are many challenges to be faced, for example, overall strategic planning is still to be deployed, the core technology of big data and the development of data resources need to be improved, the collaboration of industry partners and the bottlenecks in business models still need to be resolved, while the situation of data security and privacy protection is grim.

BDTA 2015 provided a high-end interactive platform bringing together key figures in the area. They reported on and discussed the problems faced by China's researchers working on big data core technology and different application areas.

This year, BDTA received 116 submissions. After a thorough reviewing process, 26 English papers were selected for presentation as full papers, with an acceptance rate of 22.4%. This volume contains the 26 English full papers presented at BDTA 2015 and one short papers. The high-quality program would not have been possible without the authors who chose BDTA 2015 as a venue for their publications. We are also very grateful to the Program Committee members and Organizing Committee members, who put a tremendous amount of effort into soliciting and selecting research papers with a balance of high quality and new ideas and new applications.

We hope that you enjoy reading and benefit from the proceedings of BDTA 2015.

November 2015

Weipeng Jing
Wenguang Chen

Organization

Conference Chair

Wenguang Chen   Tsinghua University, China

Conference Co-chair

Guisheng Yin   Harbin Engineering University, China

Program Chair

Gansen Zhao   South China Normal University, China

Program Co-chairs

Qilong Han   Harbin Engineering University, China
Weipeng Jing   Northeast Forestry University, China

Local Chair

Zeguang Lu   Harbin Sea of Clouds and Computer Technology
Services Ltd., China

Committee Members

Aibin Chen   Central South University of Forestry and Technology, China
Xuebin Chen   North China University of Science and Technology, China
Ziyun Deng   Hunan Vocational College of Modern Logistics, China
Youxiang Duan   China University of Petroleum, China
Yuan Gao   Tsinghua University, China
Guangjie Han   Hohai University, China
Xin Jin   Central University of Finance and Economics, China
Junyu Lin   Harbin Engineering University, China
Peng Liu   North China Electric Power University, China
Yuekun Ma   North China University of Science and Technology, China
Jun Niu   Ningbo University, China
Jianquan Ouyang   Xiangtan University, China
Zhiyou Ouyang   Nanjing University of Posts and Telecommunications, China

Zhaowen Qiu   Northeast Forestry University, China
Jianjing Shen   The PLA Information Engineering University, China
Lei Shu   Guangdong University of Petrochemical Technology, China
Yuxia Sun   Jinan University, China
Ming Tao   Dongguan University of Technology, China
Zumin Wang   Dalian University, China
Hongzhi Wang   Harbin Institute of Technology, China
Kun Wang   Nanjing University of Posts and Telecommunications, China
Mingjun Wei   North China University of Science and Technology, China
Xiaoling Wu   Guangzhou Institute of advanced Technology, China
Ke Zhang   University of Electronic Science and Technology of China,
                     China
Gansen Zhao   South China Normal University, China
Junhui Zhao   Beijing Jiaotong University, China

Contents

A General MHSS Iteration Method for a Class of Complex Symmetric Linear Systems

Yan-Ping Wang$^{(\boxtimes)}$ and Li-Tao Zhang$^{(\boxtimes)}$

Department of Mathematics and Physics, Zhengzhou University of Aeronautics,
Zhengzhou, Henan 450015, People's Republic of China
wangyanping68@126.com, litaozhang@163.com

Abstract. Recently, Bai et al. [Modified HSS iteration methods for a class of complex symmetric linear systems, computing, 87 (2010), 93–111] introduced and analyzed a modification of the Hermitian and skew-Hermitian splitting iteration method for solving a broad class of complex symmetric linear systems. In this paper, based on the Modified HSS iteration methods (MHSS) designed by Bai et al., we present a general MHSS iteration method for a class of complex symmetric linear systems. Moreover, we analyze the convergence of general MHSS method.

Keywords: Complex symmetric matrix · Hermitian and skew-Hermitian splitting · Convergence · Preconditioning

MSC: 65F10 · 65F15

1 Introduction

Consider the linear equations of the form

$$Ax = b, \tag{1}$$

where $x, b \in C^n$ and $A \in C^{n \times n}$ is a complex symmetric matrix, whose form is

$$A = W + iT, \tag{2}$$

and $W, T \in R^{n \times n}$ are real symmetric matrices, with W being positive definite and T positive semidefinite. Here and in the sequel we use $i = \sqrt{-1}$ to denote the imaginary unit. We assume $T \neq 0$ which implies that A is non-Hermitian.

Complex symmetric linear systems of this kind arise in many problems in scientific computing and engineering applications, including diffuse optical tomography, FFT-based solution of certain time-dependent PDEs, quantum mechanics, molecular scattering, structural dynamics, and lattice quantum chromodynamics.

The Hermitian and skew-Hermitian parts of the complex symmetric matrix $A \in C^{n \times n}$ are given by

© Springer Science+Business Media Singapore 2016
W. Chen et al. (Eds.): BDTA 2015, CCIS 590, pp. 1–6, 2016.
DOI: 10.1007/978-981-10-0457-5_1

$$H = \frac{1}{2}(A + A^*) = W \text{ and } S = \frac{1}{2}(A - A^*) = iT$$

respectively, hence, $A \in C^{n \times n}$ is non-Hermitian, but positive definite matrix. Here A^* is used to denote the conjugate transpose of the matrix A. Based on the Hermitian and skew-Hermitian splitting (HSS)

$$A = H + S$$

of the matrix $A \in C^{n \times n}$, Bai et al. [1] gave HSS iteration method, which is as follows:

The HSS Iteration Method [1]. Let $x^{(0)} \in C^n$ be arbitrary initial guess. For $k = 0, 1, 2, \ldots$ until the sequence of iterates $\{x^{(k)}\}_{k=0}^{\infty} \subset C^n$ converges, compute the next iterate $x^{(k+1)}$ according to the following procedure:

$$\begin{cases} (\alpha I + W)x^{(k+\frac{1}{2})} = (\alpha I - iT)x^{(k)} + b, \\ (\alpha I + iT)x^{(k+1)} = (\alpha I - W)x^{(k+\frac{1}{2})} + b. \end{cases} \tag{3}$$

where α is a given positive constant and I is the identity matrix.

However, a potential difficulty with the HSS iteration method is the need to solve the shifted skew-Hermitian sub-system of linear equations at each iteration step, which is as difficult as that of the original problem. Recently, by making use of the special structure of the coefficient matrix $A \in C^{n \times n}$, Bai et al. established the following modified HSS iteration (MHSS) method for solving the complex symmetric linear system (1–2) in an analogous fashion to the HSS iteration scheme in [7].

The modified HSS Iteration Method [7]. Let $x^{(0)} \in C^n$ be arbitrary initial guess. For $k = 0, 1, 2, \ldots$ until the sequence of iterates $\{x^{(k)}\}_{k=0}^{\infty} \subset C^n$ converges, compute the next iterate $x^{(k+1)}$ according to the following procedure:

$$\begin{cases} (\alpha I + W)x^{(k+\frac{1}{2})} = (\alpha I - iT)x^{(k)} + b, \\ (\alpha I + T)x^{(k+1)} = (\alpha I + iW)x^{(k+\frac{1}{2})} - ib. \end{cases} \tag{4}$$

where α is a given positive constant and I is the identity matrix.

Concerning the convergence of the stationary MHSS iteration method, Bai et al. [7] the convergence properties, including convergence condition, spectral radius of the iterative matrix and the choices of iterative parameter etc., of the MHSS iteration for complex symmetric linear system (1–2). Based the modified HSS (MHSS) method, we design the general MHSS iteration method and analyze the convergence of the general MHSS iteration method similar to the proving process of Sect. 2 in [7].

2   The General MHSS Method $\times$ MSSOR

In this section, based modified HSS iteration method, we present the general MHSS iteration method for a class of complex symmetric linear system, which is as follows:

The reneral MHSS Iteration Method. Let $x^{(0)} \in C^n$ be arbitrary initial guess. For $k = 0, 1, 2, \ldots$ until the sequence of iterates $\{x^{(k)}\}_{k=0}^{\infty} \subset C^n$ converges, compute the next iterate $x^{(k+1)}$ according to the following procedure:

$$\begin{cases} (\alpha I + W)x^{(k+\frac{1}{2})} = (\alpha I - iT)x^{(k)} + b, \\ (\beta I + T)x^{(k+1)} = (\beta I + iW)x^{(k+\frac{1}{2})} - ib. \end{cases} \tag{5}$$

where α is a given positive constant and I is the identity matrix.

Remark 2.1. When $\beta = \alpha$, then the general MHSS iteration method reduces to the modified HSS iteration method. So, the general MHSS method is the generalization of the modified HSS method.

Remark 2.2. Evidently, each iterate of the MHSS iteration alternates between the two symmetric matrices W and T. As $W \in R^{n \times n}$ is symmetric positive definite, $T \in R^{n \times n}$ is symmetric positive semidefinite and $\alpha \in R$ is positive, we see that both matrices $\alpha I + W$ and $\alpha I + T$ are symmetric positive definite.

After straightforward derivations we can reformulate the general MHSS iteration scheme into the standard form

$$x^{k+1} = \mathcal{M}(\alpha, \beta)x^{(k)} + \mathcal{G}(\alpha, \beta)b, k = 0, 1, 2, \ldots,$$

where

$$\mathcal{M}(\alpha, \beta) = (\beta I + T)^{-1}(\beta I + iW)(\alpha I + W)^{-1}(\alpha I - iT)$$

and

$$\mathcal{G}(\alpha, \beta) = (\beta - \alpha i)(\beta I + T)^{-1}(\alpha I + W)^{-1}.$$

In addition, if we introduce matrices

$$\mathcal{B}(\alpha, \beta) = \frac{\beta + \alpha i}{\alpha^2 + \beta^2}(\alpha I + W)(\beta I + T) \text{ and } \mathcal{C}(\alpha, \beta) = \frac{\beta + \alpha i}{\alpha^2 + \beta^2}(\beta I + iW)(\alpha I - iT).$$

Then it holds that

$$A = \mathcal{B}(\alpha, \beta) - \mathcal{C}(\alpha, \beta) \text{ and } \mathcal{M}(\alpha, \beta) = \mathcal{B}(\alpha, \beta)^{-1}\mathcal{C}(\alpha, \beta).$$

Concerning the convergence of the general MHSS iteration method, we have the following theorem.

Theorem 2.1. Let $A = W + iT \in C^{n \times n}$ with $W \in R^{n \times n}$ and $T \in R^{n \times n}$ symmetric positive definite and symmetric positive semidefinite, respectively, and let α and β be two positive constant. If the parameters satisfy one of the following conditions

(i) $\beta = \alpha$,
(ii) $\beta > \alpha$ and $\beta^2 < \alpha^2 + 2\alpha\lambda_j$,
(iii) $\beta < \alpha$ and $\alpha^2 \leq \beta^2 + 2\beta\mu_j$.
Then the spectral radius $\rho(\mathcal{M}(\alpha, \beta))$ of the general MHSS iteration matrix $\mathcal{M}(\alpha, \beta) = (\beta I + T)^{-1}(\beta I + iW)(\alpha I + W)^{-1}(\alpha I - iT)$ holds that

$$\rho(\mathcal{M}(\alpha, \beta)) \leq \sigma(\alpha, \beta) < 1, \ \forall \alpha, \beta > 0,$$

where

$$\sigma(\alpha, \beta) \equiv \max_{\lambda_j \in \mathrm{sp}(W)} \frac{\sqrt{\beta^2 + \lambda_j^2}}{\alpha + \lambda_j}$$

where $sp(W)$ denotes the spectrum of the matrix W. i.e., the general MHSS iteration converges to the unique solution $x_* \in C^n$ of the complex symmetric linear system (1–2) for any initial guess.

证明. By direct computations we obtain

$$
\begin{aligned}
\rho(\mathcal{M}(\alpha, \beta)) &= \rho(\beta I + T)^{-1}(\beta I + iW)(\alpha I + W)^{-1}(\alpha I - iT) \\
&= \rho((\beta I + iW)(\alpha I + W)^{-1}(\alpha I - iT)(\beta I + T)^{-1}) \\
&\leq \|(\beta I + iW)(\alpha I + W)^{-1}(\alpha I - iT)(\beta I + T)^{-1}\|_2 \\
&\leq \|(\beta I + iW)(\alpha I + W)^{-1}\|_2 \|(\alpha I - iT)(\beta I + T)^{-1}\|_2.
\end{aligned}
$$

Because $W \in R^{n \times n}$ and $T \in R^{n \times n}$ are symmetric, there exist orthogonal matrices $U, V \in R^{n \times n}$ such that

$$U^T W U = \Lambda_W, V^T T V = \Lambda_T,$$

where

$$\Lambda_W = \mathrm{diag}(\lambda_1, \lambda_2, \ldots, \lambda_n)$$

and

$$\Lambda_T = \mathrm{diag}(\mu_1, \mu_2, \ldots, \mu_n),$$

with $\lambda_j (1 \leq j \leq n)$ and $\mu_j (1 \leq j \leq n)$ being the eigenvalues of the matrices W and T, respectively. By assumption, it holds that $\lambda_j > 0$ and $\mu_j > 0, j = 1, 2, \ldots, n$. Now, based on the orthogonal invariance of the Euclidean norm $\|\bullet\|_2$, we can further obtain the following upper bound on $\rho(\bar{M}(\alpha, \beta))$:

$$\rho(\mathcal{M}(\alpha,\beta)) \le \|(\beta I + \Lambda_W)(\alpha I + \Lambda_W)^{-1}\|_2 \|(\alpha I - i\Lambda_T)^{-1}\|_2$$

$$= \max_{\lambda_j \in sp(W)} |\frac{\beta + i\lambda_j}{\alpha + \lambda_j}| \bullet \max_{\mu_j \in sp(T)} |\frac{\alpha - i\mu_j}{\beta + \mu_j}|$$

$$= \max_{\lambda_j \in sp(w)} \frac{\sqrt{\beta^2 + \lambda_j^2}}{\alpha + \lambda_j} \max_{\mu_j \in sp(T)} \frac{\sqrt{\alpha^2 + \mu_j^2}}{\beta + \mu_j}.$$

Case 1: When $\beta = \alpha$

Since $\mu_j \ge 0$ holds for all $\mu_j \in sp(T)(1 \le j \le n)$, we find that $\sqrt{\alpha^2 + \mu_j^2} \le \alpha + \mu_j = \beta + \mu_j$.

Then, we can get

$$\rho(\mathcal{M}(\alpha,\beta)) \le \max_{\lambda_j \in sp(w)} \frac{\sqrt{\beta^2 + \lambda_j^2}}{\alpha + \lambda_j} = \sigma(\alpha,\beta).$$

Obviously, $\sigma(\alpha,\beta) < 1$ holds true for any $\alpha = \beta > 0$, therefore the general MHSS iteration converges to the unique solution of the complex symmetric linear system (1–2).

Case 2: When $\beta > \alpha$ and $\beta^2 < \alpha^2 + 2\alpha\lambda_j$

Since $\beta > \alpha$, then $\sqrt{\alpha^2 + \mu_j^2} < \sqrt{\beta^2 + \mu_j^2} \le \beta + \mu_j$. So, we can immediately obtain

$$\rho(\mathcal{M}(\alpha,\beta)) \le \max_{\lambda_j \in sp(w)} \frac{\sqrt{\beta^2 + \lambda_j^2}}{\alpha + \lambda_j} = \sigma(\alpha,\beta).$$

Obviously, $\sigma(\alpha,\beta) < 1$ holds true for $\beta^2 < \alpha^2 + 2\alpha\lambda_j$, therefore the general MHSS iteration converges to the unique solution of the complex symmetric linear system (1–2).

Case 3: When $\beta < \alpha$ and $\alpha^2 \le \beta^2 + 2\beta\mu_j$

Since $\alpha^2 \le \beta^2 + 2\beta\mu_j$, then $\sqrt{\alpha^2 + \mu_j^2} \le \beta + \mu_j$. We can immediately obtain

$$\rho(\mathcal{M}(\alpha,\beta)) \le \max_{\lambda_j \in sp(w)} \frac{\sqrt{\beta^2 + \lambda_j^2}}{\alpha + \lambda_j} = \sigma(\alpha,\beta).$$

Obviously, $\sigma(\alpha,\beta) < 1$ holds true for $\beta < \alpha \Rightarrow \sqrt{\beta^2 + \lambda_j^2} < \sqrt{\alpha^2 + \lambda_j^2} < \alpha + \lambda_j$, therefore the general MHSS iteration converges to the unique solution of the complex symmetric linear system (1–2).

Remark 2.3. Theorem 2.1 shows that the convergence rate of the general MHSS iteration method is bounded by $\sigma(\alpha, \beta)$ in the parameters satisfying certain conditions, which depends on the eigenvalues of the symmetric positive definite matrix W.

Remark 2.4. Obviously, Theorem 2.1 in this paper includes Theorem 2.1 in [7].

Funds. This research of this author is supported by NSFC Tianyuan Mathematics Youth Fund (11226337), NSFC (11501525, 11471098, 61203179, 61202098, 61170309, 91130024, 61272544, 61472462 and 11171039), Aeronautical Science Foundation of China (2013ZD55006), Project of Youth Backbone Teachers of Colleges and Universities in Henan Province (2013GGJS-142, 2015GGJS-179), ZZIA Innovation team fund (2014TD02), Major project of development foundation of science and technology of CAEP (2012A0202008), Defense Industrial Technology Development Program, China Postdoctoral Science Foundation (2014M552001), Basic and Advanced Technological Research Project of Henan Province (152300410126), Henan Province Postdoctoral Science Foundation (2013031), Natural Science Foundation of Zhengzhou City (141PQYJS560).

References

1. Bai, Z.Z., Golub, G.H., Ng, M.K.: Hermitian and skew-Hermitian splitting methods for non-Hermitian positive definite linear systems. SIAM J. Matrix Anal. Appl. **24**, 603–626 (2003)
2. Bai, Z.Z., Golub, G.H., Ng, M.K.: On inexact Hermitian and skew-Hermitian splitting methods for non-Hermitian positive definite linear systems. Linear Algebra Appl. **428**, 413–440 (2008)
3. Bai, Z.Z., Benzi, M., Chen, F., Wang, Z.Q.: Preconditioned MHSS iteration methods for a class of block two-by-two linear systems with applications to distributed control problems. IMA J. Numer. Anal **33**(1), 343–369 (2013)
4. Bai, Z.Z.: On Hermitian and skew-Hermitian splitting iteration methods for continuous Sylvester equations. J. Comput. Math. **29**(2), 185–198 (2011)
5. Bai, Z.Z., Benzi, M., Chen, F.: On preconditioned MHSS iteration methods for complex symmetric linear systems. Numer. Algorithms **56**(2), 297–317 (2011)
6. Bai, Z.Z.: On semi-convergence of Hermitian and skew-Hermitian splitting methods for singular linear systems. Computing **89**(3–4), 171–197 (2010)
7. Bai, Z.Z., Benzi, M., Chen, F.: Modified HSS iteration methods for a class of complex symmetric linear systems. Computing **87**(3–4), 93–111 (2010)
8. Bai, Z.Z., Guo, X.P.: On Newton-HSS methods for systems of nonlinear equations with positive-definite Jacobian matrices. J. Comput. Math. **28**(2), 235–260 (2010)
9. Bai, Z.Z.: Optimal parameters in the HSS-like methods for saddle-point problems. Numer. Linear Algebra Appl. **16**(6), 447–479 (2009)
10. Bai, Z.Z., Yang, X.: On HSS-based iteration methods for weakly nonlinear systems. Appl. Numer. Math. **59**(12), 2923–2936 (2009)
11. Benzi, M., Bertaccini, D.: Block preconditioning of real-valued iterative algorithms for complex linear systems. IMA J. Numer. Anal. **28**, 598–618 (2008)

Big Data Storage Architecture Design in Cloud Computing

Xuebin Chen[1], Shi Wang[1(✉)], Yanyan Dong[1], and Xu Wang[2]

[1] College of Science, North China University of Science and Technology,
Tangshan, Hebei, China
ws10121@126.com
[2] College of Information Engineering,
North China University of Science and Technology, Tangshan, Hebei, China

Abstract. To solve the lag problem of the traditional storage technology in mass data storage and management, the application platform is designed and built for big data on Hadoop and data warehouse integration platform, which ensured the convenience for the management and usage of data. In order to break through the master node system bottlenecks, a storage system with better performance is designed through introduction of cloud computing technology, which adopts the design of master-slave distribution patterns by the network access according to the recent principle. Thus the burden of single access the master node is reduced. Also file block update strategy and fault recovery mechanism are provided to solve the management bottleneck problem of traditional storage system on the data update and fault recovery and offer feasible technical solutions to storage management for big data.

Keywords: Big data · Cloud computing · Hadoop · Data warehouse · Storage architecture

1 Introduction

With the advent of the era of the cloud, the data format and size is growing at an unprecedented speed. Reasonable storage and management on growing huge amounts of data will be beneficial to provide support to industry analysis which is applied to forecast, as well as to effectively take advantage of big data under the background of opportunities and challenges, to the integrate traditional decision method and the concept of the decision method of big data, to build a platform of innovative management of computer technology for intelligent analysis prediction and evaluation of prediction. Eventually the ability of enterprise to apply new technology on management and decision making will be improved [1].

Big data is put forward for the first time in 2009, and then have found application in the field of multiple business and development, especially the mature usage on medical field. In the era of big bang data, using data service in the industry is the inevitable result of the activation of era. To use data efficiently and accurately, the premise is the efficient storage and management of data, to take appropriate data storage model according to different application requirements, so as to more efficiently real-time process and

© Springer Science+Business Media Singapore 2016
W. Chen et al. (Eds.): BDTA 2015, CCIS 590, pp. 7–14, 2016.
DOI: 10.1007/978-981-10-0457-5_2

analyze data. Relevant data is the collected and under storage management for analysis and process so as to dig out the potential value of the information provided to the senior leadership for decision making judgment. In big data environment, users put forward higher request to the storage service on the availability, reliability and durability of data. In order to prevent the data from being lost or damaged and ensure data privacy, the environment of users' storage system is essential [2]. Google's programming model can effectively parallel-process mass data, which adopt the GFS and BigTable model for data management, the big data platform based on Hadoop implements Google's big data storage system with source opened.

Early in the information economy, the enterprise only plays as a resource to collect and store the data with simple statistical analysis at most, while the intrinsic value of the data is usually ignored. Along with the progress of the storage and analysis technology, the enterprise further mined and processed the data collected with growing awareness of the importance of active mastery on data. Ability to develop potential value of data becomes one of the core competitiveness of enterprises. The value of data shows its important position in the era of intelligence science and technology.

Hysteresis phenomenon exists in the application of traditional data storage technology on mass data storage management and security. At present, in the continuous development of industry data, there are vast amounts of unstructured data and semi-structured data. Reasonable process and analysis on these data to unearth valuable information complied with the requirements for big data policy decisions and service provision. Data model is the precondition and foundation of big data analysis and forecast. This paper builds the big data application platform and studies the big data storage system architecture based on the Hadoop data platform.

Application platform provides industry analysis, reporting, prediction and decision making, etc. At the same time feasible technical solutions are provided in storage management for big data.

2 Big Data Platform Design

2.1 Platform Environment

How to collect and store huge amounts of data, how to integrate heterogeneous data, how to mining and process large data sets, which are of concern to the third party service provider. Data collection, data storage, data process, data analysis, data application will be the basic task of the enterprise performance in wisdom economy era. Judgment and decision based on data will become the skills and means of enterprises for development. A platform, which conform to the big data management and support the development in the field of application, is designed for the analysis of relevant industry development, integration of heterogeneous data and customized user requirements, etc. The platform is based on Hadoop big data platform architecture, relying on HDFS, MapReduce and MongoDB, etc. distributed framework which are deployed into more cheap hardware equipment, for the application with high throughput data access mechanism. HDFS as open source distributed file system supports data storage and management of high fault tolerance. HDFS is open source implementation of GFS, which can deploy to cheap PC devices and is a suitable application for big data [3]. HDFS takes a master-slave mode structure. In the cluster, there is a NameNode and multiple DataNodes, which is in charge of data storage option and namespace, database storage and optimal strategy choice [4]. MapReduce is proposed by Google to concurrent processing mass data parallel programming model. It has the characteristics of simple and efficient for the shielding mechanism of the underlying implementation details, effectively reduction on programming difficulty. Cheap machine deployment clusters can be used to achieve high performance with good scalability and provide high-efficient scheduling interface, implementing task scheduling, load balancing and fault tolerance and consistency management etc. [5–8]. MongoDB supports multiple binaries and two kinds of storage ways as storage subsystem automatically divided and user-defined divided. In order to realize load balance, MongoDB implements distributed storage in multiple servers for the same data file directed into a multiple block. There will be no necessity for the user to know where the data is stored for the servers will record the shard [9, 10].

2.2 Architecture Design

Business application requirements include data acquisition, data gathering, and data analysis and data application. A unified data application platform will be built for data real-time loading, storage and processing different types of data. Data processing tools and services are integrated for the management of heterogeneous data. Structured and unstructured data warehouse analysis tool are also integrated. The platform can implement the concentration of large data sharing and collaborative access at anytime, anywhere, by any terminal equipment; Application platform can support the modeling of new business development and business strategy, and promote the development of the industry insight, real-time early warning analysis.

Big data application platform can satisfy the processing requirements of data of large amount, multiple style and fast flow. It also possess the ability to implement huge amounts of data collection, storage, processing and analysis, meet the basic requirements of high reliability, easy extensibility, strong fault tolerance, high availability, high security and high secrecy of the enterprise application, and ensure the compatibility of existing technology with the platform and realize data storage and processing.

Big data application platform conforms to two standard systems, the system safety standard system and service management standard system. Big data platform in Hadoop and data integration platform in data warehouse implement a lot of data storage, analysis, processing and usage, including the front layer, core layer, manager layer, data layer, and application layer, etc. Office automation, risk assessment, data acquisition, smart analysis, real-time processing is integrated to create a data integration and management platform. With integration of data warehouse and analysis tools, this paper puts forward a real-time forecast analysis solution.

A connection is realized among core layer, big data platform and data warehouse integration platform, in order to realize intelligent warning and real-time analysis as well as integration of data warehouse and intelligent analysis system, using analytical tools for analyzing visualization and form the electronic report and analysis report. As shown in Fig. 1 as the big data general architecture.

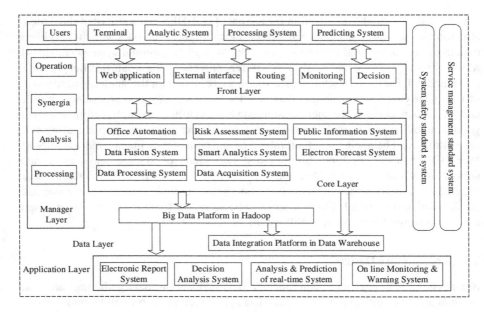

Fig. 1. General architecture

3 Key Technologies

3.1 Storage System Design Based on Cloud Computing

Cloud computing virtualization technology is used in the design of storage system to achieve high concurrency and high fault tolerance under the condition of the consistency. The master-slave distribution pattern is used in the design of data storage to avoid the data loss and damage caused by the outage under the traditional storage technology which adopts a single mode of storage system. System uses partitioned storage in different physical and backup storage device, So as to improve the security and integrity of data. Virtualized physical resources are integrated into the master node as the system management node, which is responsible for the management and monitoring the daily operation of the slave nodes as well as to ensure the normal state of nodes. The master node network virtualization and distributed management are put forward as the design idea to solve the problems of master node management bottleneck of the traditional technology. The master node is visited by the principle of recent-visit, to a certain extent, alleviate the access and management of the master node burden, avoid server failures caused by centralized access to the master node which will lead to the collapse of whole system, as well as to solve the problem of the bottleneck of the whole system operation, thus improve the overall efficiency. Multiple slave nodes as the data storage, achieves load balanced distribution of stored data. The difficulty of storage of different data types under the traditional relational database is solved, and multiple of backup storage of data is distributed into data nodes. The loss of data is avoided on the premise of superior system operation performance, the storage architecture design is shown in Fig. 2.

Storage system access flow begins with access request sent by client, then the message accesses near master node server through network. After received and responded to users request to read and write, the node will locate block of data in specified slave node according to the address to implement specified operation. Data is stored by partition distributed storage, convenient in management. With the use of analysis of visualization tools, electronic report is generated in the users' terminal display, which provides convenience in analysis and decision. As long as connecting to the Internet client terminal can access the system. The terminal can be hardware, fixed and mobile devices and embedded devices, etc.

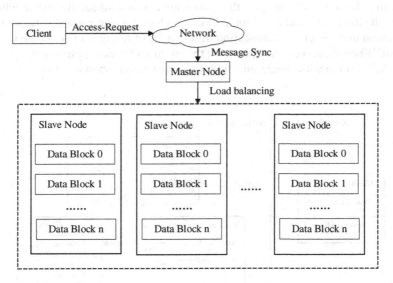

Fig. 2. Storage architecture

3.2   Updating Algorithm Design Based on File Block

Store system adopts distributed master-slave mode, which automatically implement load balanced block storage. Data may be stored in different data nodes and blocks. In order to ensure transactional consistency operation when read and write updates, necessary coordination mechanism need to be taken to realize the feasible operation. Traditional technology solutions using the famous paxos algorithm, while this paper uses the Chubby coarse-grained lock service to solve consistency problem when the file block update. On one hand, the design is consistent to data set partitioned storage design patterns; on the other hand, ensure normal network traffic during system data update, to avoid network congestion phenomena caused by data update in the system. Update process is the communication between the client and the Server by RPC. Redundant technology strategy with lock service is used to ensure the consistency of data update. The server is made of five machines. The single server consists of fault tolerant database and fault tolerant log file. Communication between servers is through protocols.

Snapshots are stored locally. Data in the form of file blocks is under parallel transmission and synchronization update, thus update efficiency is improved.

When data update normally, data block is updated in server, while update operation is written into the fault tolerant database and fault-tolerant log files, at the same time, the client is notified to update data. When data update abnormally, an inconsistency occurs during data block update, recovery on error data in the data block will be implemented according to the copy of the rest data block and fault tolerant database and log file, and notify the client to update the abnormal data or to ignore the update of the data. When data failure occurs, the server will revoke to normal transactions according to the log file. In order to achieve the goal that data is always up-to-date, the system will real-time test the data state, and avoid unnecessary updates take up normal network traffic, reduce client and server communication flow. Cache of log files of sever are saved by the client. When the server data update, on the premise of lock the client will update the data in the local file with synchronization, basic structure is shown in Fig. 3.

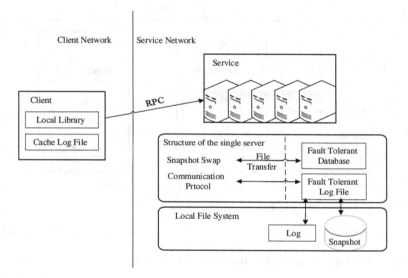

Fig. 3. Basic structure

3.3 Fault Recovery Mechanism Design Based on Cloud Storage

When system failure occurs during the running of application, without timely recovery on failure site, huge economic and customer losses will be caused to the enterprise. Storage service based on cloud is a kind of network storage service, which hold a high efficient and safe design of data and fault recovery, on the one hand, the efficiency of data computing, communication and storage is improved, on the other hand, the efficiency of detection is improved as high precision in fault detection of data storage and in data recovery. Master-slave distributed design method is used for the mass data storage system, as well as master-slave mode design on fault recovery mechanism based on cloud storage. According to the global and local fault detection and recovery mechanism, the operational efficiency of the system and the accuracy of fault positioning recovery are improved and unnecessary waste of resources reduced. Data error occurrence, security and reliability are real-time detected by the integration of idle resources of network through the network communication mechanism of the cloud storage system. For data resources are in cloud storage, detect movement can be implemented while user is offline, which makes the system always in a state of the data accuracy, safety and reliability.

In this paper, the fault recovery mechanism integrates global action, distributed management and local action at an organic whole, which makes a balance of the independent operation of the master node and slave nodes. A middle tier of the proxy server is added between the master node and slave nodes so as to reduce the workload of the master node and avoid the bottleneck of system management. When data inconsistency occurs, cloud storage server will lessen the fault detected to the management master node. After receiving the fault message, through the cloud storage servers, the node send the fault report back to the proxy server to process. This move is a global action. The fault message through the external interface service is sent to a proxy server. Proxy server implements the distributed management and share part of the master node work, which is responsible for the records and addressing the fault point. After received fault message, according to the error log file, slave node starts to recover the data to normal status and update and respond to proxy server the current status of storage systems. The process is a local action. Fault recovery mechanism is shown in Fig. 4.

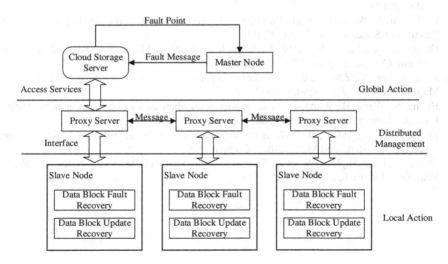

Fig. 4. Fault recovery mechanism

4   Conclusions

The importance of the application of data for the enterprise is analyzed in this paper in the first step. To satisfy the demand of big data processing platform, data application platform in Hadoop and data integration platform in data warehouse are put forward. Cloud computing technology is adopted in the design of storage system and improves the traditional technology in the master node server management in distributed storage. Through analyzing and realizing key technologies of storage system, such as file block update strategy and fault recovery mechanisms, feasible technical solutions in storage management for big data are provided.

Acknowledgment.   This work is partially supported by The National College Student Innovative Program of China (Grant No. 201410081014). The authors also gratefully acknowledge the helpful comments and suggestions of the reviewers, which have improved the presentation.

References

1. Zhu, D., Zhang, Y., Wang, X., et al.: Research on the methodology of technology innovation management with big data. Sci. Sci. Manage. S. & T. **4**, 172–180 (2013)
2. Feng, D., Zhang, M., Li, H.: Big data security and privacy protection. Chin. J. Comput. **1**, 246–258 (2014)
3. Qi, Yu., Jie, Ling: Research of cloud storage security technology based on HDFS. Comput. Eng. Des. **8**, 2700–2705 (2013)
4. Wang, Y., Mao, S.: A blocks placement strategy in HDFS. Comput. Technol. Dev. **5**, 90–92+96 (2013)
5. Li, J., Cui, J., Wang, D., et al.: Survey of MapReduce parallel programming model. Acta Electronica Sinica **11**, 2635–2642 (2011)
6. Huang, B., Xu, S., Pu, W.: Design and implementation of MapReduce based data mining platform. Comput. Eng. Des. **2**, 495–501 (2013)
7. Song, J., Liu, X., Zhu, Z., et al.: An energy-efficiency optimized resource ratio model for MapReduce. Chin. J. Comput. **1**, 59–73 (2015)
8. Huang, S., Wang, B., Wang, G., et al.: A survey on MapReduce optimization technologies. J. Front. Comput. Sci. Technol. **10**, 885–905 (2013)
9. Zheng, J., Ye, Y., Tai, T., et al.: Design of live video streaming, recording and storage system based on Flex, Red5 and MongoDB. J. Comput. Appl. **2**, 589–592 (2014)
10. Zhang, Y., Feng, H., Hao, W., et al.: Research on the storage of file big data based on NoSQL. Manufact. Autom. **6**, 27–30 (2014)

Visual Analysis for Civil Aviation Passenger Reservation Data Characteristics Based on Uncertainty Measurement

Huaiqing He, Hongrui Du[✉], and Haohan Liu

Department of Computer Science and Technology, Civil Aviation University of China,
Tianjin, China
dhrui2009@163.com

Abstract. Aviation data analysis can help airlines to understand passenger needs, so as to provide passengers with more sophisticated and better services. How to explore the implicit message and analyze contained features from large amounts of data has become an important issue in the civil aviation passenger data analysis process. The uncertainty analysis and visualization methods of data record and property measurement are offered in this paper, based on the visual analysis and uncertainty measure theory combined with parallel coordinates, radar chart, histogram, pixel chart and good interaction. At the same time, the data source expression clearly shows the uncertainty and hidden information as an information base for passengers' service recommendations.

Keywords: Civil aviation passenger data · Uncertainty · Visual analysis

1 Introduction

With China's social progress and economic development, its civil aviation passenger constitute in the domestic market has undergone great changes, which have an important impact on the demand characteristics of travelers and are gradually changing the pattern of the future society transport market. Currently, the scale of civil aviation passenger data records becomes increasing, which can provide good suggestions for improving the business by passenger travel behavior-depth analysis. At the same time it needs to be aware that the travel choice randomness exists in passenger travel due to the impact of various events, such as weather, travel location, personal preferences, etc., namely uncertainty. The degree of personal uncertainty largely determines the future evolution of travel. The uncertainty analysis applied to the passenger data in recommender system can help researchers analyze the transfer degree of historical preference for users, which can provide passengers with personalized service and create value.

Aviation passenger reservation data is a typical of high-dimensional dataset, including the user's age, gender, travel records and other properties. A simple and fast way to reflect the characteristics of these properties is required, and visualization techniques can display multi-dimensional data in a low-dimensional space intuitively, effectively and interactively, which also explore data patterns and models to uncover the relationship between each property [1]. At present, the basic methods of multi-dimensional data

© Springer Science+Business Media Singapore 2016
W. Chen et al. (Eds.): BDTA 2015, CCIS 590, pp. 15–25, 2016.
DOI: 10.1007/978-981-10-0457-5_3

visualization: space mapping, icons and pixel-based visualization methods, such as parallel coordinates, scatter, Chernoff Faces, radar chart, pixel chart. Parallel coordinates is a common method for multidimensional data, but the overlapping will affect the judgement for how much the value of the shaft, therefore the histograms is introduced in parallel coordinates to display statistical distribution in every property. Radar chart is suitable for the comparative analysis of multivariate data and pixel chart for multivariate data visualization. The data record and property uncertainty are measured and visualized in this paper based on the existed methods. Parallel coordinates with histogram are used to explore the relationship between different properties and the distribution features of each property. Radar chart shows the uncertainty of the properties of different groups. Pixel chart is used for travel sites Analysis. All of these lay a foundation for future research recommender system.

2   Uncertainty Measurement of Multidimensional Data

2.1   Measurement Method

Uncertainty refers to the variation in objective world or entity itself which acts as imprecision, randomness and fuzziness. Currently, the calculation methods have been proposed from probability theory, fuzzy set theory, evidence theory and information theory and so on [2]. The randomness is an effective measurement method for data uncertainty, information theory measurement is suitable for uncertainty of random data. Shannon is introduced in information theory as a measurement of the uncertainty of a random event in 1948, and the information measurement method is established [3]. The degree of uncertainty of an event can be described by the probability of its occurrence: the smaller the probability of the event is, the bigger the amount of information contained in the event is, so there is a lot of uncertainty; if the event is inevitable, the probability of occurrence information is 1, it should be zero pass, there is no uncertainty; if the event is not possible (with probability 0), then it will have endless information.

If U is the domain, $X = \{X_1, X_2, ..., X_n\}$ is a division, there is a probability distribution $p_i = p(X_i)$, then:

$$H(X) = -\sum_{i=1}^{n} p_i \ln p_i \qquad (1)$$

is the Information entropy for information source X. where i represents the i-th state (total n states); p_i represents the probability that the i-th state appears; H(X) indicates the amount of needed information to eliminate the uncertainty of things. If there is a Pi which is equal to 1, while the rest are 0, then H(X) = 0, there is no uncertainty at this time, the result is certain. If all events are all equally, where $p_1 = p_2 = \cdots p_n = 1/n$. For such case, it is difficult to predict the occurrence of an event, the uncertainty of this event system is greatest.

Because of air passenger data contains several properties, and randomness exists due to passengers' travel under varies of influence. Therefore, information entropy method is applied for calculating the uncertainty of passengers' properties.

2.2   Uncertainty Measurement of Properties and Data Record

It's analyzed from the passenger's age, sex, airline, departure, arrival, the departure time, aircraft cabin, discounts altogether eight major properties. Put a property as a discrete random variable, then calculate the entropy of the variable according to the formula (1). Fewer the states of the variable are, the smaller entropy is, which means that this variable uncertainty is smaller. If the variable has only one state, then the uncertainty disappears, and its entropy is zero.

Passenger data of more than three times of travel are selected, total of 4180 travel records, 682 users. For the passenger's uncertainty, firstly, because each passenger has multiple travel records, the information entropy (Eq. (1)) is used to measure the uncertainty of each property; Secondly, calculate the mean value of each property uncertainty (Eq. (2)). The greater the result is, indicating that passenger's travel without regulation, the greater its uncertainty is, and vice versa.

$$U_i = \sum_{j=1}^{m} H(X)_j/m \quad i <= n \tag{2}$$

Where i represents the i-th passenger, $H(X)_j$ represents the j-th property of passenger i (Eq. 1), m represents the number of properties.

For a group of passenger's uncertainty, just use formula (1) directly to calculate every property's uncertainty of the group.

3   Visual Analytic

The implied passenger travel characteristics have been analyzed from different perspectives through three kinds of visual analysis methods. In this paper, parallel coordinates with histogram are used to explore the relationship between different properties associated and the distribution features of each property, radar chart shows the uncertainty of the properties of different groups, pixel chart of the travel site.

3.1   Parallel Coordinates with Histogram

3.1.1   Parallel Coordinates
Parallel coordinates is an important visualization technique for representation of multi-dimensional data analysis and mutual relationships. By n equidistant parallel shaft mapped onto a two-dimensional plane, every axis represents an attribute dimension ranging from a minimum value to a maximum evenly distributed. So each property value for each record may mark a point in the corresponding attribute axis, these points are connected as a line indicating a data record. It enables the user to see the whole picture of the data set directly, and analyze the distribution in the same axis for every object and the relationships between properties [4]. There are nine axes representing the passengers' id, age, gender, airline, departure, arrival, the departure time, space, discounts. For id axis, id are pre-sorted in the data processing stage by travel times which have been marked in the left of axis; age (age axis) is in the range of 18 to 73 old years; gender

(gender axis) has two attributes: male (m) and female (w); airlines (airline axis) is total of 24; departure (departure axis) and destination (destination axis) all have more than 100 values; the departure time (time axis) is discrete into 24 time zones; accommodation (space axis) has four kinds: First Class (F), business class (C), Premium Economy (W) and economy class (Y), and finally the discount (discount axis) has 11 species, where 11 represents discount fares higher than all the original price. Every trip record can be marked in the corresponding attribute axis, and the dots are connected into a polyline. It should be noted that there are some travel records per passenger. The travel records of all passengers are showed in Fig. 1, in addition to using parallel coordinates display the original data, the uncertainty of each passenger is normalized and divided into five intervals, using blue - green - yellow - orange - red five colors to represent the uncertainty ranges (0~0.2, 0.2~0.4, 0.4~0.6, 0.6~0.8, 0.8~1.0).

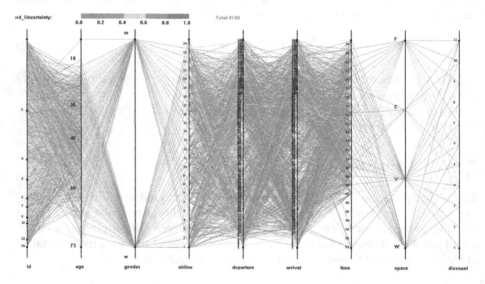

Fig. 1. Uncertainty expression of civil aviation passenger data (Color figure online)

In Fig. 1, it's clear to see gathered beneath the id axis is red, orange and yellow line, indicating the passengers with more travel times usually have greater uncertainty. They are more than 20 times of trips, such users may belong to business travelers due to the more opportunity to travel with richer choices and larger differences. Travel more than 10 times the frequency ratio is about 20 %, indicating that the promotion of the compa-ny's frequent flyer program has great space for development. For space, major users of high uncertainty selected First (F), Business Class (C) and high-end economy class (W), the lower uncertainty of the passengers choose economy class which are the popular trend. In addition to observation of one axis, the association between the adjacent prop-erties can be discovered, such as departure time and space, the lines are across, indicating the same class has different discounts, which may be caused by the airline's pricing policy or aviation distribution network caused.

3.1.2 Histogram

Although the parallel coordinate plots can show the multidimensional data better, each attribute axis has intersect point which causes lines overlap, and mutual interference between lines will further affect access to information. Despite some literatures use hierarchy [5], clustering [6] to make improvements in visual aspect, it did not make a specific quantitative analysis. Therefore, in order to know how many data records intersect at the same point of each attribute axis, histogram as a supplement for statistical analysis is introduced which can describe the points distribution trends and patterns of every axis more clearly [7]. In Fig. 2, the gender (gender axis) data for male is 2 times higher than female, which suggests that male travels more frequently. For space (space axis), economy class has a majority of passengers, followed by business class and first class passengers with fewer passengers, which may be increased in recent years, travelers and travel expense ratio increased passenger concerned. Such a simple classification on one dimension of data, it is easy to see the data distribution of the property axis. Of course, too much histogram embedded in parallel coordinates affects the overall visual effect, thus adding some transparency so as not to block lines behind the trend, and age is divided into several partitions to reduce the number of histograms.

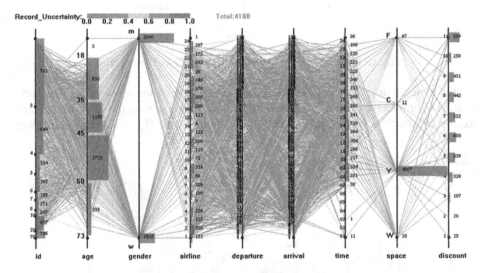

Fig. 2. Histogram embedded into parallel coordinates

3.2 Radar Chart

Since the parallel coordinates plots emphasis on the links between dimensions, and lack of longitudinal research capabilities. In order to explore the characteristics of air passenger group, radar chart [8] is applied for the uncertainty comparison of each attribute of the group, and determines its development trend. This paper focuses on five kinds of uncertainty to analyze user groups. Radar chart has eight rays, where each ray represents a property. The value is in ascending order from the center to the circumference, then the value of each attribute axis are joined to form a closed poly-line after calculating the uncertainty of various attributes of a group by using the formula (1). It's easy to observe the uncertainty degree of various attributes of the group and the comparison for different groups. In Fig. 3, the uncertainty ranged from 0.2–0.4 (green), and 0.8–1.0 (red) of two passenger groups attribute uncertainty are displayed.

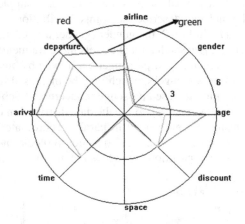

Fig. 3. Radar chart of property uncertainty (Color figure online)

For a single line, such as the red line, the uncertainty of some properties is signifi-cantly larger than the other attributes from the radar chart, such as the value of age (age axis) is bigger than space (space axis), and combined with Fig. 4 where age attribute

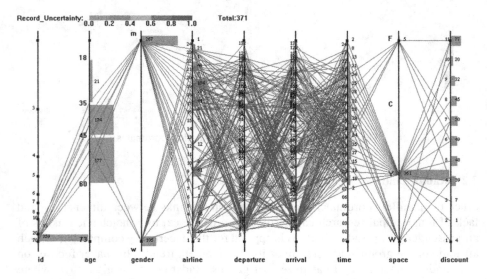

Fig. 4. Parallel coordinate plots of passenger travel of uncertainty from 0.8 to 1.0

histogram distribution is more dispersed, space attribute axis histogram focused on the Y class. It can be concluded: for a property, if its value is concentrated on one or a few property values, namely probability distributions of the value for these properties are significantly greater, the corresponding entropy is small. The information source provides less information, and the attribute uncertainty is lower. For two lines, the age attribute for green line has greater uncertainty than the red, so the age distribution of this traveler group has a larger span. Then combined with parallel coordinate plots for further analysis, although such groups each small passenger uncertainty, but greater number, so the overall value of the property is wide. By the method of radar chart, to some extent, can make up for the deficiencies of relying solely on parallel coordinates and histogram, showing attribute uncertainty which can effectively help analyze the overall changes.

3.3 Pixel Chart

Hot cities and hot routes can be found in the study on passenger travel sites, which helps provide a basis for the recommendation of airline passenger service. When the number of travel sites is large, parallel coordinate plots can't show the relationship of departure and destination clearly. Therefore pixel-oriented graphical visualization technique which supports the large-scale data output is introduced [9]. Each value has a corresponding screen pixel with color, and different data properties are expressed by various of windows. It can display data as much as possible. Departure-destination form a matrix (123 * 130). In this paper, the number of trips is mapped into five colors, respectively: (0), (1, 2), (3–10), (10–20) (20–). Using blue - green - yellow - orange - red color gradient pattern to represent frequency changes from low to high, which is in line with people's

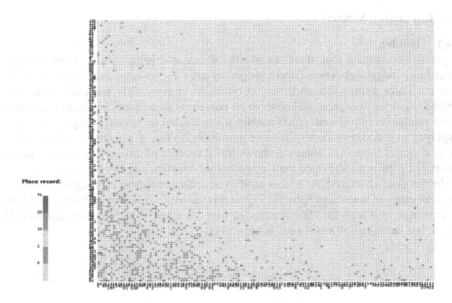

Fig. 5. Pix chart for different cities

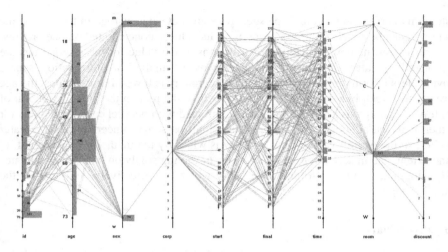

Fig. 6. Parallel coordinate plot using hidden

visual habits and obtains better visualization effect. Since there are a lot of nodes and not all of them have a relationship, thus a sparse matrix generated. In order to reduce the consumption of matrix calculation, show the regulation of structure, and enhance readability of the visualized result, the nodes are ordered based on the degree firstly and re-order the same degrees according to values [1]. In Fig. 5, the longitudinal represents departure city, the lateral direction indicates arrival city, which facilitates the travel times more concentrated in the lower-left point angle, horizontal empathy.

3.4 Interaction Method

3.4.1 Hidden

In parallel coordinate plot, there are overlap situation existing in records of different uncertainty ranges between adjacent properties axes due to large dataset, even though different color expression uncertain information entropy data recording has been applied. In order to express and observe air passenger data more clearly, hidden inter-active method is introduced which enables user-selected uncertainty range records are displayed so that shows the results more intuitively. In Fig. 4, it's the records of uncertainty range of 0.8 to 1.0. Figure 6 shows travel records of airline 9, where the larger number of trips is middle-aged male and other information can be observed.

In the pixel chart, hidden is also used to in order to analyze routes preferences of different range of uncertainty groups. Figure 6 is the user's travel records of uncertainty in the range of 0.8 to 1.0. The number of users is small, which can be observed in Fig. 4, but the range of travel sites is large (Fig. 7).

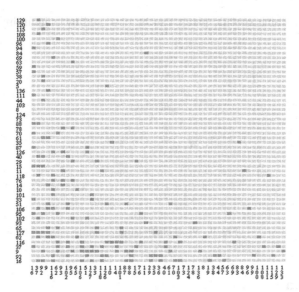

Fig. 7. Pixel chart using hidden

3.4.2 Zooming and Moving

In parallel coordinate plot, when the dimension is higher, it can't show the lines of adjacent dimensions due to close distance of dimensions clearly. Zooming can solve that problems. Use the mouse to drag the axis to the center of the scene, the attribute distribution characteristics can be easier to observe and the effect is more accurate through the amplification (Fig. 8).

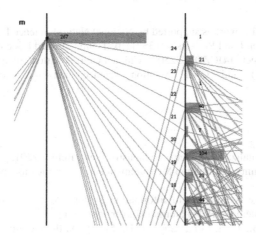

Fig. 8. Parallel coordinate plot using zooming and moving

In pixel chart, some specific pixels can be observed by magnifying, as shown in Fig. 9.

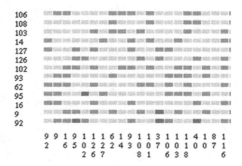

Fig. 9. Pixel chart using zooming

4   Conclusion

Several kinds of multidimensional data visualization technologies are combined together in this paper, through the interactive technology realized by parallel coordinates, histogram, and pixel chart visual analysis method. On the one hand, make full use of the advantages of various visualization technologies from a different perspective to reveal the multidimensional data; On the other hand, users are provided with convenient and flexible interactions which are easier for users to analyze and understand the data.

Next focus is puts forward a kind of additional service recommender with visual and good interactive functions based on the uncertainty measurement and visual analysis, combined with the civil aviation recommender additional service.

Acknowledgment. This work is supported by National Natural Science Foundation of China - Civil Aviation Research Fund Projects (U1333110), Tianjin applied basic research and frontier technologies Key Project (14JCZDJC32500), China's Civil Aviation college preparatory major projects(3122013P003) and Civil Aviation science and technology fund projects (MHRDZ201207).

References

1. Chen, W.: Data Visualization. Electronic Industry Press, Beijing (2013)
2. Mei, X.-F.: Uncertainty Information Measurement and its Application. Northwest University (2006)
3. Xie, B.: Knowledge Acquisition in Information Systems With Uncertainty Measurement Research. The Problems of Hebei Normal University (2011)
4. Sun, Y., Feng, X.-S., Tang, J.-Y., Xiao, W.-D.: Survey on the research of multidimensional and multivariate data visualization. J. Comput. Sci. **35**(11), 1–7 (2008)

5. Fua, Y.-H., Ward, M.O.: Hierarchical parallel coordinates for exploration of large datasets. In: Proceedings of IEEE Visualization, pp. 43–50. IEEE USA (1999)
6. Zhou, H., Yuan, X., Qu, H., Cui, W., Chen, B.: Visual clustering in parallel coordinates. In: Eurographics IEEE-VGTC Symposium on Visualization 2008, vol. 27 (2008)
7. Chen, Y., Cai, J.-F., Shi, Y.-B., Chen, H.-Q.: Coordinated visual analytics method based on multiple views with parallel coordinates. J. Syst. Simul. 25(1), 81–86 (2013)
8. Ding, J.-M., Li, W.-B.: Visualization representation and feature analysis of radar-map-based multi-dunebsuibal data. Mod. Electron. Tech. 33(23), 24–26 (2010)
9. Janetzko, H., Simon, S., Neuhaus, K., et al.: Visual boosting in pixel-based visualizations. Comput. Graph. Forum 30(3), 871–880 (2011)

The Mobile Personalized Recommendation Model Containing Implicit Intention

Jing Liu[1,2(✉)], Jun Zhang[1,2], Yan Li[1,2], Shuqun He[1], and Caixue Zheng[3]

[1] School of Computer Science and Engineering, Hebei University of Technology,
Tianjin 300401, China
88451500@qq.com
[2] Hebei Province Key Laboratory of Big Data Calculation, Tianjin 300401, China
[3] School of Economic Management, Hebei University of Technology, Tianjin 300401, China

Abstract. Because mobile e-commerce is limited by the mobile terminal, network environment and other factors, accurate personalized recommendations become more and more important. We establish a large data intelligence platform, aiming at the characteristics of mobile e-commerce; we put forward a personalized recommendation model with implicit intention further.

Firstly, create an intelligence unit with the virtual individual association set, virtual demand association set and virtual behavior associated set; Secondly, calculate the complex buying behavior prediction engine; Finally, give the predictive value of complex buying behavior.

This method takes full account of factors such as hidden wishes perturbations that affect the predict of complex buying behavior, which to some extent solve a long-span composite purchasing behavior prediction. It shows that this method improves the purchasing behavior prediction accuracy effectively through experiments.

Keywords: Mobile e-commerce · Personalized recommendations · Hidden wishes · Big data intelligence platform

1 Introduction

With the rapid development of mobile communication technology, more and more users use the mobile terminal to access e-commerce platform for shopping. Due to the difference of mobile terminals, the network environment and shopping scenarios, the mobile users' information requirement and shopping ways have a big difference with PC users. Therefore, optimizing the personality recommendation for mobile e-commerce by modeling and analysing the mobile terminal data will greatly enhance consumers satisfaction degree, which ultimately has a good role in improving the sales.

This paper is a research project of the mobile commerce association recommendation model and Algorithm Research (project number: 71271186) and Hebei Province Natural Science Foundation Project "the research on the G2013203237 based on cloud computing for mobile business users" (item number: YB20152204).

W. Chen et al. (Eds.): BDTA 2015, CCIS 590, pp. 26–33, 2016.
DOI: 10.1007/978-981-10-0457-5_4

In recent years, it has paid more and more attention on the researches on the relevance of mobile e-commerce at home and abroad. The current mobile search personalized recommendation model can be divided into three categories: filtering technology based on coordination [1]; mining technology based on association rules [2, 3] and technology based on situational awareness [4, 5]. The above three kinds of technology take the consumers' personal attributes, shopping records and other information in to account, which achieves good results. However, online shopping is a more complex behavior, which can be regarded as a result of the implicit willingness with a small probability and the dominant demand influenced by the implicit will. This form of purchase behavior is called a composite purchase behavior. It is difficult to find the association between implicit intention and purchasing behavior data. The traditional personalized recommendation method does not take the influence of implicit will on the model in to account. In this paper, a large data intelligent platform is proposed, which is based on the large data analysis technique and the distributed parallel computing capability [6]. A new model of mobile personalized recommendation is proposed and the validity and superiority of the model are proved by experiments.

2   Mobile Personalized Recommendation Model with Implicit Intention

The concealment and potential of the implicit intention determines that is difficult to extract the characteristic value of the purchase behavior and rely on historical data or experience in direct response and processing. In order to overcome the implicit intention prediction based on empirical knowledge and human-oriented view in the existing techniques that causes the problem of poor prediction effect, the Mobile Personalized Recommendation Model with implicit intention is proposed. By modeling the relationship between the elements of purchasing behavior with time variation, the dynamic behavior of purchasing behavior, feedback and iterative structure are rapidly identified and understood from the overall forecasting method and the technical framework. As shown in Fig. 1, the main steps of the personalized recommendation model with implicit intention are as follows:

1. It determines the status of the purchase behavior of dominant demand signs by taking advantage of association rules algorithm to establish the dominant demand knowledge database, so that we can carry out the next step;
2. Match the purchase behavior monitoring data and the dominant demand pattern in the knowledge database. If there is no match results, it explains that the purchase behavior is not a feature of the dominant demand, and continue to the next data matching;
3. It indicates that the purchase behavior has the potential dominant demand if the matching succeed in step 2. Start the intelligent unit, we will get the implicit intention and form the composite purchase behavior forecast to return to the system at the end. Details are as follows:

- Calculate the dominant demand matching result set including the matching results and the probability of the potential dominant demand;
- Start the intelligent unit which is based on the virtual individual association set, the virtual demand association set and the virtual behavior association set to construct the implicit willingness to predict engine, and calculate the influence of the implicit intention of the time T;
- The composite purchase behavior forecast value was obtained by the comprehensive calculation of the dominant demand matching result set and the implicit intention influence value using the composite purchase behavior prediction engine;
- Return the composite purchase behavior forecast to system at the end;
4. If the composite purchase behavior is predicted to exceed the preset threshold, the result is indicated, or the next set of matching is performed.

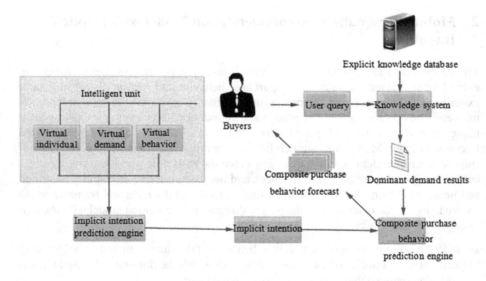

Fig. 1. The structure of a personalized recommendation model with implicit intent

3 Research on Key Technologies Based on MPRI Method

MPRI method is mainly divided into three steps: First, construct the dominant demand knowledge database, and match the purchase behavior monitoring data and knowledge database in the dominant demand historical data. If there is no match results, the system is running and continue to match the next data; On the other hand, if the matching successes, it shows that there is potential dominant demand in the system, and return the dominant demand matching result set;Secondly, start the intelligent unit and calculate the influence of the implicit intention of the time T basing on the virtual individual association set, the virtual demand association set and the virtual behavior association set and using the prediction engine of the implicit intention. Finally, the composite purchase behavior forecast value was obtained by the comprehensive calculation of the dominant demand matching result set and the implicit intention influence value using the composite purchase behavior prediction engine, and ultimately return to the system. If more than the threshold set in advance, start the next set of matching.

3.1 Constructing Explicit Demand Knowledge Database

First, clean and remove the inconsistent data;The data is normalized in order to prevent the the attribute weights with large values are larger than those of the smaller values; Scale the data in the normalized processing, so that they are on the [0,1]; The data is clustered and discretized, so that we can obtain the generalized results of the data. Set the correlation degree is 40 % and the reliability is 60 %. Make sure the input is the historical data after cleaning, while the output is the dominant demand pattern. According to the Apriori algorithm to carry on the judgment, obtain the explicit demand knowledge database.

Apriori algorithm specific steps are as follows:

① Scan a database to generate frequent 1 item sets L1;
② In the K cycle, firstly by the frequent item sets from the k− 1 connection and pruning the candidate frequent item sets generated K Ck;
③ Scan the database and monitoring each data T to calculate the monitoring data T and figure out which candidate purchase behavior is characterized by frequent K item sets. And the number of support for these candidate purchase behaviors of frequent K item sets plus 1. If the number of support for a candidate purchase behavior is greater than or equal to the minimum number of support, the candidate purchase behavior is characterized by frequent K item sets;
④ Cycle until no longer produce candidate purchasing behavior and the frequent K item sets end;

Algorithm description is as follows:

L1 = find_frequent_1-itemsets(D);

for (k=2;Lk-1 $\neq \Phi$;k++)

{

Ck = apriori_gen(Lk-1 ,min_sup);

for each transaction t $\in$ D

{   //scan D for counts

Ct = subset(Ck,t);//get the subsets of t that are candidates

for each candidate c $\in$ Ct

c.count++;

}

Lk ={c $\in$ Cklc.count$\geq$min_sup}

}

Return L= $\cup$ k Lk;

Among which, the specific calculation of the dominant demand matching result set is as follows:

Input: monitor data values
Output: potential dominant demand, probability values

The monitoring data set is the precondition of the association rule, and the explicit requirement knowledge base is generated by the Apriori algorithm.

Rank the successful matching of the N bar by high to low according to the degree of support, match the purchase behavior D with the support degree s, and calculate the probability values P whose computational formula is for $p = s - [I/(n + I)]I/N$, among them I is an integer, and $0 < I$ is less than or equal to n.

3.2 Calculate Implicit Will Influence Value

Establish the intelligent unit with virtual individual association set, the virtual demand association set and the virtual behavior related set. The details are as follows:

(1) Virtual individual association set: construct a set of object oriented virtual individual association using memory i1, cognitioni2, adaption i3, behavior i4, in order to achieve their autonomy;

(2) Virtual demand association set: comprehensively considering multiple factors such as location S1, time S2, stateS3, demand S4 to realize multisource and dynamic evolution;

(3) Virtual behavior associated set: behavior rules are established and the rule of decision behavior is described with the previous purchase behavior time difference B1, purchase behavior similarity B2 and trend of B3.

The influence of implicit intention is calculated by using the self organization competition model of neural network in the MATLAB: Specific calculation is as follows:

Input: The 11 main attributes that set up the purchase behavior information: the input of neural network are virtual individual association (i1–i4), virtual requirement association (S1–S4) and virtual behavior association (b1–b3).
Output: Implicit will influence value.

Using the function NEWC to create a self organizing competitive network. The number of neurons is set to 11 and the learning rate is 0.1, which is the net = newc (minmax (P), 11, 0.1).

- Initialize the self organization competing neural network model initialization, namely net = init (net).
- Set the training parameters, which is "net trainparam. epochs = 300"
- Determine the connection weights of the neural network with the training of the sample.
- Use the trained network to determine the value of the implicit will.

3.3 Calculate the Predictive Value of the Composite Purchase Behavior

To construct a composite purchasing behavior prediction engine model, the relationship between implicit intention and explicit demand in mass dynamic information is established in accordance with agile methods. Establish a comprehensive evaluation function of R = H*D, whose index are the degree of importance and similarity, comprehensive influence and demand factor, which are based on purchasing behaviors. R is a comprehensive probability value, H is a combination of various factors of the purchase behavior of the weighted function collection, and D is a combination of various factors that affect the relationship set. The process is shown in Fig. 3. Firstly, treat the dominant demand matching result set and the implicit will influence the value weighting as the weighted function set H of the final composite purchasing behavior risk source. Use the BP neural network to further predict its corresponding probability of D, and so that get the forecast value with the comprehensive evaluation function R. The main steps include choosing the standard purchase behavior sample, learning from each standard purchase behavior sample and entering the test sample into the BP neural network model that has been trained well, which can obtain the corresponding set.

4 Experimental Analysis

The purchase behavior data of a large domestic electricity business platform is predicted in order to verify the validity of the personalized recommendation model based on large data technology. There are 21 samples of the purchase behavior to be cleaned and removed inconsistent data; The data is normalized in order to prevent the attribute weights with large values are larger than those of the smaller value; Scale the data in the normalized processing so that they are on the [0,1], and establishes purchasing behavior monitoring data stream. The generalization of the clustering range of various types of data is obtained by making clustering analysis and discretization of continuous historical data that belongs to the purchasing behavior monitoring data set, and then the monitoring data flow is obtained.

The experts and data engineers determine the threshold of each index, and convert the continuous real number to a discrete value; Build 3 test libraries S1, S2, S3, which have the following characteristics:

1. Test library S1 contains 100 monitoring data, which contains 80 data browsing and purchase data 20;
2. Test library S2 contains 300 monitoring data, which contains 220 data browsing and purchase data 80;
3. Test library S3 contains 800 monitoring data, which contains 650 of the browse data and purchase of line data 150.

Do experiments on the windows7 operating system testing machine with CPU 3.4 GHz and memory 2G, and use the VC++6.0 to write test procedures. The test results are shown in Table 1. It also lists the most common method for the purchase of behavior prediction (Apriori algorithm) which has the same set of samples to determine the results in Table 1. It can be seen in the S1, S2, S3 test library that the correct rate of MPRI is greater than Apriori algorithm, and with the increase of data, the correct rate is relatively stable. It can obviously improve the accuracy of the judgment of the MPRI method in the implicit intention of virtual computation.

Table 1. Two methods in S1, S2, S3 test results

Data library	Total sample number	Purchase data	Predict the correct number of times		Correct rate	
			Method A	Method B	Method A	Method B
S1	100	20	4	8	20 %	40 %
S2	300	80	32	33	40 %	41 %
S3	800	150	48	65	32 %	43 %

A – method for Apriori method. B – method MPRI method

5   Conclusion

Due to the mobile terminal screen, network and other restrictions, accurate personalized recommendation becomes more and more important. In this paper, according to the characteristics of mobile electronic commerce, a large data intelligent platform is established. Based on this platform, a personalized recommendation model with implicit intention is put forward, which sets up the virtual individual connection set, the virtual demand association set and the intelligent unit of the virtual behavior association. Calculate the composite purchase behavior prediction engine and finally figure out the predictive value of composite purchase behavior. This method replaces the traditional method of "prediction and judgment", which has a strict structure and the dependence on historical data. It Fully understand the current "implicit state", and make full assessments based on the future of the "dominant state". In order to improve the accuracy of purchasing behavior prediction, we adequately considerate the influence of implicit intention disturbance and other factors on the prediction of composite purchase behavior, therefore we solve the problem of large span composite purchase behavior prediction in a certain extent.

References

1. Lu, G., Li, F.: Research on mobile personalized recommendation service based on improved collaborative filtering. Inf. Explor. **02**, 101–105+110 (2014)
2. Zhang, D., Xu, F.: Research and implementation of a personalized integrated framework for mobile applications android platform. Comput. Sci. **11**, 63–68 (2014)
3. Li, Y., Liu, J., Hou, H.: Multi source association of mobile e-commerce personalized recommendation framework. Intell. Theory Pract. **04**, 98–100 (2014)
4. Zhang, S., Guo, S.: Research on personalized recommendation model of university mobile library based on situation awareness. Inf. Explor. **10**, 6–11 (2014)
5. Zhou, L., Longzhen: Digital library alliance in context aware based personalized recommendation service. Libr. Theory Pract. **07**, 67–69+87 (2014)

Personalized Recommendation System
on Hadoop and HBase

Shufen Zhang[✉], Yanyan Dong, Xuebin Chen, and Shi Wang

College of Science, North China University of Science and Technology, Tangshan, Hebei, China
{hblgzhsf,453705197,chxb}@qq.com, ws10121@126.com

Abstract. In view of the existing recommendation system in the Big Data have two insufficiencies: poor scalability of the data storage and poor expansibility of the recommendation algorithm, research and analysis the IBCF algorithm and the working principle of Hadoop and HBase platform, a scheme for optimizing the design of personalized recommendation system based on Hadoop and HBase platform is proposed. The experimental results show that, using the HBase database can effectively solve the problem of mass data storage, using the MapReduce programming model of Hadoop platform parallel processing recommendation problem, can significantly improve the efficiency of the algorithm, so as to further improve the performance of personalized recommendation system.

Keywords: Hadoop · HBase · MapReduce · Personalized recommendation

1 Introduction

With the innovation of Internet technology, information resources on the network has become more and more abundant, it makes people step into the ocean of data - the Era of Big Data. In the era of big data, it becomes extremely difficult for people to find the information they need from the vast amounts of data quickly and efficiently. This problem is also known as Information Overload. Personalized recommendation technology using the existing historical data and real-time data, analysis of each user's behavior intention, and establish a corresponding interest model for each user. According to users' interest model push potential items, realize the personalized recommendations.

The traditional personalized recommendation system has two insufficiencies: poor scalability of data storage and poor expansibility of recommendation algorithm. This's because traditional personalized recommendation system underlying data storage using a Relational Database Management System, aka RDBMS. Over the system time running, the number of users and projects in the system will continue increasing; eventually leading RDBMS systems is difficult to load mass data storage. Moreover, the execution speed of recommendation algorithm is the core part of the personalized recommendation system; it directly affects the accuracy and efficiency of the recommendation. However, traditional recommendation system using a single node serial operation, it is hard to meet user demand for real-time recommendation result in the huge amounts of data environment.

© Springer Science+Business Media Singapore 2016
W. Chen et al. (Eds.): BDTA 2015, CCIS 590, pp. 34–45, 2016.
DOI: 10.1007/978-981-10-0457-5_5

In view of two aspects issue above-described in the traditional personalized recommendation systems, this paper proposes a personalized recommendation system based on Hadoop and HBase. The underlying system used HBase database manages mass and sparse user behavior data, use of HBase column oriented storage and storage scalability characteristics to solve the poor scalability problems that exist in the traditional recommendation system. Data processing using Hadoop distributed computing advantage change every stage of the recommendation algorithm to parallel processing. Improve the execution efficiency of algorithm, reduce recommended time consuming process, enhance scalability of data processing, make recommendation system can better adapt to the practical application of huge amounts of data.

2   Personalized Recommendation System and Hadoop, HBase

2.1   Personalized Recommendation System

In recent years, with the constant innovation of new network technology, personalized recommendation system is becoming more and more widely used in our daily life, resulting in personalized recommendation technology become a challenge and opportunity coexist academic research field. A complete personalized recommendation system should include three parts: system show page, algorithm calculation engine, and data storage module. Recommendation system architecture diagram is shown in Fig. 1.

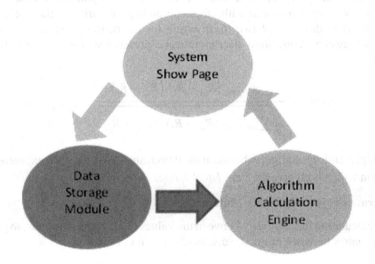

Fig. 1. Recommended system architecture

Algorithm calculation engine is the core part of the personalized recommendation system; better recommendation algorithm can calculate the user's potential interest information timely and accurately, provide good experience for users. The recommendation algorithm most widely used in actual production environment is Item-Base

Collaborative Filtering, aka IBCF. The main idea of IBCF is based on history rates of user for item, and generate item similarity matrix. Search out the item which is highest similarity with other items as the result of recommendations. From the perspective of computing, IBCF recommendation algorithm implementation process is divided into the following five steps:

1. **Item Rating Item Quantization**

When using IBCF algorithm to calculate the two items similarity, first, collecting the existing historical data from the database, quantify the rate of two items, Dimension is the number of users who has comment rating on an item, the value of dimensions is users' ratings for an item. Assuming that recommendation system contains m users, collection is expressed as $U = \{U_1, U_2, \cdots, U_m\}$ and n items, collection are expressed as $I = \{I_1, I_2, \cdots, I_n\}$. So the item I_i corresponding rate vector of items can be represented as $R_i = \{R_{1i}, R_{2i}, \cdots, R_{ui}, \cdots, R_{mi}\}$. Among them, R_{ui} expressions the rating of user U_u to item I_i. The rating value can be viewed as the preference degree of user U_u to item I_i.

2. **Similarity between Items**

The commonly used algorithm which to calculate similarity of two items include Cosine index similarity or Pearson Correlation Coefficient, etc. The following use Pearson Correlation Coefficient as an example, the Coefficient defined the Similarity through the linear relationship between two vector which corresponding to two rating for items, from the perspective of linear algebra is to compute the cosine value of between two vectors. Compared with other computing similarity method, it also take into account the difference of two mean rating for items, to eliminate the effects of differences between items, show user's authentic preferences for item. As shown in formula (1).

$$Sim_{ij} = \frac{\sum_{u \in N_i \cap N_j} (R_{ui} - \bar{R}_i)(R_{uj} - \bar{R}_j)}{\sqrt{\sum_{u \in N_i \cap N_j} (R_{ui} - \bar{R}_i)^2 (R_{uj} - \bar{R}_j)^2}} \tag{1}$$

Sim_{ij} represents the similarity between the item I_i and item I_j. $N_i \cap N_j$ expresses users intersection which both give the item I_i and I_j rate.

3. **Generate Project Similarity Matrix**

After calculating the similarity between the values of all project systems, to generate similarity matrix between projects, such as (2) shown in Equation.

$$\begin{bmatrix} Sim_{11} & Sim_{12} & \cdots & Sim_{1n} \\ Sim_{21} & Sim_{22} & \cdots & Sim_{2n} \\ \vdots & \vdots & \ddots & \vdots \\ Sim_{n1} & Sim_{n2} & \cdots & Sim_{nn} \end{bmatrix} \tag{2}$$

Which Sim_{ij} represents the similarity value between item I_i and I_j (where I_i and I_j belong to I), the value can be seen as the degree of similarity between items.

4. Generate nearest neighbor set

After obtaining the similarity between items, typically use two methods to select the nearest neighbor set for target item: one is a pre-set threshold similarity; the similarity between all the target items exceeds the threshold composition project nearest neighbor set V [11]. Another approach is to select the maximum K similarity value as a neighbor set items $V = \{I_1, I_2, \cdots, I_k\}$. Wherein $I_i \notin V$, and satisfies $Sim_{ik} > Sim_{ik+1} > \cdots Sim_{in}$, n as the total number of items.

5. Generate the final recommendation list

According to the target items' nearest neighbor set of rates to predict the current user's score to the target item, the highest score before the election predicted a number of items as a result of recommendation to the current user. User rating prediction target item target item by nearest neighbor set score obtained, calculated as shown in Eq. (3).

$$R_{uj} = \overline{R_i} + \frac{\sum\limits_{i \in V} Sim(R_{uj} - \overline{R_j})}{\sum\limits_{i \in V} Sim_{ij}} \tag{3}$$

Which R_{uj} represents the system predicts user U_u to item I_i rate. $R_{uj} - \overline{R_j}$ represents the average score of the project user ratings minus the project, it is to project the center received an average rating of (Mean-centering) treatment [12], to eliminate differences in scores between projects.

2.2 Hadoop Platform

Hadoop platform is one of the most popular solutions for big data problem. The Hadoop distributed file system and MapReduce programming model is the core of the distributed computing framework. It is provided by the Apache Software Foundation, aka ASF. Users can in the case of don't need to understand the underlying system details, use Hadoop organize computing resources to build their own distributed computing platform, Develop a distributed application, to make full use of cluster performance distributed and parallel processing massive data problem.

2.3 HBase Platform

HBase is a column oriented, scalable, and distributed No-SQL database, it can manage mass data which stored on thousands of server nodes efficiently and reliably. HBase is the open source implementation for Google's paper named Big-Table, and it has become the top project of the Apache Software Foundation. Compared with RDBMS, HBase's column oriented storage mechanism is suitable for storing sparse data. RDBMS is mainly suitable for transactional demanding situations, According to the theory of CAP, in order to achieve the strong consistency, need to synchronize by strict ACID transactions, this makes the performance of the system is reduced greatly in availability and scalability. HBase database is an eventual consistency mechanism, in the process of storage for transparent data segmentation, make store itself has a horizontal scalability and extensibility.

3   Personalized Recommendation System Algorithm Optimization

In the actual production environment, the number of users and projects in the personalized recommendation system often will be more than ten million levels, and the number of daily magnitude is also very large. Such large computing tasks, online real-time computing systems simply not load it in many cases, and use of off-line calculation also takes a long time to complete the update of similarity between the items, but user's demand for the system is more biased in favor of the real-time feedback. Therefore, to study and solve the problem of poor scalability recommendation system exists, it is very valuable.

Hadoop's distributed computing implementation is based on the MapReduce programming model, the basic idea is to split the entire data file into a number of blocks with a certain size, and store the data block in each storage node in the Hadoop cluster. When performing recommendation algorithm, split the entire calculation job into plurality Map subtask accordance with the number of data blocks, mapped the subtask to each computing nodes in the cluster for parallel computing. Map stage reads the data block file, generates intermediate key/value pairs, namely <Key, Value>, and writes the output to a local disk for persistent storage. After the completion of all Map stages, each computing node starts Reduce stage, read the output results of Map, after the reduction process outputs the final result. It is implementation shown in Fig. 2.

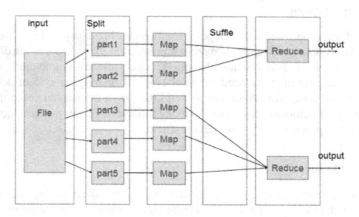

Fig. 2. MapReduce implementation process

According to the Sect. 2.1 implementation IBCF algorithm, now use the Hadoop MapReduce programming model to improve it. The process of calculating the predicted ratings of user for item will be divided into four MapReduce task is completed.

The first MapReduce task: Map stage reads the data block file with key-value pair <Key, Value>. After Map function processing, data UserID as Key, (ItemID, Rating) for the Value, i.e. <UserID, (ItemID, Rating)>. Reduce stage to perform a reduction process User ID as Key, the resulting generate User-Item scoring matrix. It is implementation shown in Fig. 3.

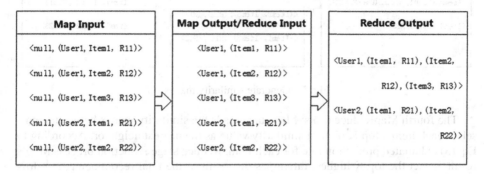

Fig. 3. Generate User-Item rating matrix

The second MapReduce task: Map stage reads the data block file, output to ItemID as the Key, (UserID, Rating) as the Value, i.e. <ItemID, (UserID, Rating)>. Reduce stage perform reduction processing ItemID as the Key to generate Item-User rating matrix. It is implementation shown in Fig. 4.

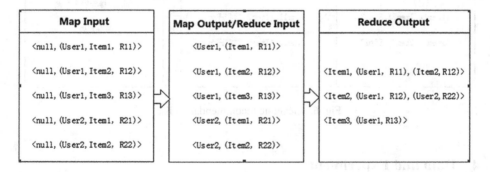

Fig. 4. Generate Item-User rating matrix

The third MapReduce tasks: Map stage reads the Item-User Matrix produced by the last process, and output result use Item pair (I_i, I_j) as the Key, the set of Rating as the Value. Reduce stage to calculate the similarity between items according to Eq. (1), the resulting similarity matrix as Item-Item (2) formula. It is implementation shown in Fig. 5.

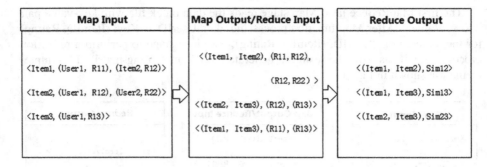

Fig. 5. Generate similarity matrix

The fourth MapReduce tasks: Map stage reads the similarity matrix between items, select each item's top K highest similarity value as the nearest neighbor. According to Eq. (3) calculated prediction rate for each item. Reduce stage reduction process Map's result; select the top K highest rateing as the result of the final recommendation. It is implementation shown in Fig. 6.

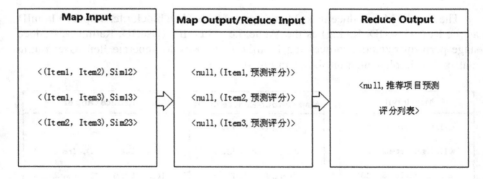

Fig. 6. Generate recommendation list

4   Data and Experiment

4.1   Experimental Data

The experimental data Movie-Lens is provided by the Group Lens Research project team is a free movie recommendation data, contents include: about ten million lines, data size is 252 MB, number of users is 71567, and number of film is 10681. The scope of user's ratings is [1, 5], the value of the rating on behalf of the evaluation of users.

4.2   Experiment and Result Analysis

This experiment to configure Hadoop cluster on 7 nodes, configure the Name-Node Server on the NO.01 and NO.02 machines, and configure Resource-Manager Server on the NO.03 machine. On the remaining 4 machines configure Data-Node Server and Node-Manager Server as storage nodes of HDFS and compute nodes of MapReduce. The configuration of each machine as follows: the CPU Model is Intel(R) Core(TM) i5-3470, the Memory is 4 GB, Hadoop version is 2.4.1, Linux version is CentOS-6.6.

4.2.1   Data Storage Optimization Experiment

According to the characteristics of the data set, design the data storage model for RDBMS and HBase. The data model based on RDBMS storage, need to design three tables: the User table, Movie table and Rating table. Respectively, used to storing User's basic information, movies' basic information and the ratings of user for Movie. For example as the following User table, shown in Table 1.

Table 1. User table design

UID	UName	UAge	UGender	...
ID1	Name1	Age1	Gender1	...
...	...	...	...	...
IDn	Namen	Agen	Gendern	...

Among them, the field UID is the primary key of User table and can uniquely identify each row in the table. The field UName, UAge, UGender meaning the basic attribute of user, the data Namen, Agen, Gendern meaning the value of users' each attribute.

Use HBase to store the experimental data, just need to design one table only. This experiment based on HBase design the data storage model, as shown in Table 2.

Table 2. HBase data model design

Row key	Time stamp	Column family:c1		Column family:c2	
		Info	Value	Rating	Value
U_1	t6	c1:Age	Value1		
	t5	c1:Name	Value2		
	t4	c1:Gender	Value3		
	t3			c2:MovieID	R_{ui}
U_2	t2	c1:name	Value1		
	t1			c2:MovieID	R_{2i}

Among them, the U1 and U2 is the HBase database's RowKey, is equivalent to the primary Key of the RDBMS. The field c1 and c2 respectively HBase database's Column Family, and the Age, Name, MovieID in the equivalent of the columns for an RDBMS. The default value of Timestamp is the system time stamp, Can also be customized according to the business needs, each record has a corresponding timestamp.

During the experiment, selected 10M, 20M, 50M, 100M and 200M five groups of data sets of different sizes as the experimental data, using the above-described two types data model, the experimental data are being stored in the RDBMS database and HBase database, And measured different data models corresponding data storage file size. The results are shown in Fig. 7.

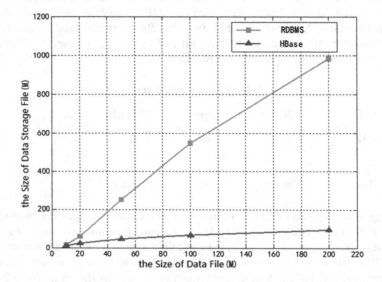

Fig. 7. Storage file size corresponding to different data models

Among them, Abscissa represents the different sizes of selected experimental data sets, in megabytes (MB) of units. Ordinate respectively represent the storage file size of RDBMS-based storage and HBase-based, in megabytes (MB) of units.

The above results show that, with the increasing size of the data set, RDBMS-based storage file size increased significantly, while HBase-based storage file size increase is relatively flat. This is because there are a lot of sparse data in the recommended system, namely a large number of meaningless null (Null value) exists. For these Null values, based RDBMS storage takes a lot of storage space, largely reducing the use of performance RDBMS database. The HBase column-oriented storage features for null values (Null value) does not take up any storage space, thus saving storage space and improved read performance, it is suitable for storing sparse data.

4.2.2   Recommended Algorithm Optimization Experiments

The first set of experiments to verify the performance of the Hadoop platform recommendation algorithm execution, gradually increase the number of nodes in the cluster, and record the response time of recommendation algorithm. First, use a single node test the time of calculate a user's predicted rate for all items, and then gradually increase the number of nodes from 2 to 6, and recorded all the time of calculate a user's predicted rate for all items. The results are shown in Fig. 8.

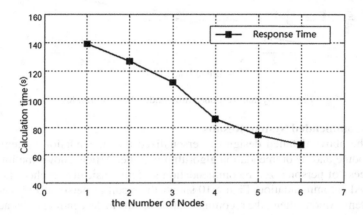

Fig. 8.   System response time corresponding to different nodes

Wherein, Abscissa represents the number of nodes. Ordinate represents the corresponding system response time for different number of nodes, in seconds (s) as a unit.

The above results show that as the number of nodes increases, the computing time continue to decrease, indicating that the system performance continues to improve, but also found time decreasing amplitude is also declining because of the increasing number of nodes, the system do Map/Reduce operating system overhead required is also increasing.

The second set of experiments selected 10M, 20M, 50M, 100M and 200M five groups of data sets of different sizes as experimental data. Computing a user's predicted rate for all items, and record the recommendation algorithm's response time under different environmental. The results are shown in Fig. 9.

Among them, Abscissa represents the different sizes of selected experimental data sets, in megabytes (MB) as a unit. Ordinate represents the recommendation algorithm's response time for different modes, in seconds (s) as a unit.

The above results show that with the increase of the amount of data input, serial environment recommendation algorithm for memory resource consumption increases, resulting in decreased performance of the algorithm is not complete computing tasks. In a multi-node Hadoop cluster recommendation algorithm can complete large-scale data computing tasks, which fully shows that the use of based Hadoop parallel computing method can solve recommendation algorithm computing scalability issues.

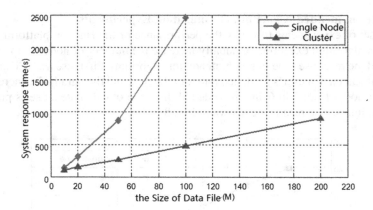

Fig. 9. System response time under different modes

4.2.3 Recommended System Results

Through the above detailed design of personalized recommendation system in data storage module and recommendation algorithm module. Article shows the implementation process of personalized recommendation system, basically reached the goal of personalized recommendation. Figure 10 shows the result of personalized recommendation system. Among them, the recommendation results in descending order according to the size of the recommended values, recommended values represent the recommended system predicted value for the target users' interests.

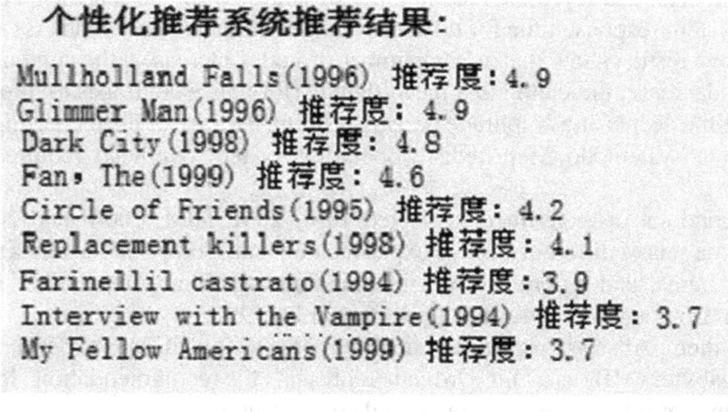

个性化推荐系统推荐结果：

Mullholland Falls(1996) 推荐度：4.9
Glimmer Man(1996) 推荐度：4.9
Dark City(1998) 推荐度：4.8
Fan, The(1999) 推荐度：4.6
Circle of Friends(1995) 推荐度：4.2
Replacement killers(1998) 推荐度：4.1
Farinellil castrato(1994) 推荐度：3.9
Interview with the Vampire(1994) 推荐度：3.7
My Fellow Americans(1999) 推荐度：3.7

Fig. 10. Personalized recommendation system recommended results

5 Conclusion

Depending on the needs of users to quickly and accurately push the required information, provide personalized recommendation service is currently a hot research topic. In this paper, the current recommendation system in the data storage scalability and recommendation algorithm expansibility of both the problems, research and analysis of the IBCF algorithm, combined with Hadoop, HBase technology, design a suitable for large data environment data storage and processing solutions. Proposed based on Hadoop and HBase based personalized recommendation system, and use Group Lens Research project offers movies datasets, programming analysis and testing of the different size of the data file. Experimental results show that the proposed personalized recommendation system not only improves data storage scalability, and enhanced recommendation algorithm scalability.

References

1. Cao, H., Fu, K.: Clustering collaborative filtering recommendation system search method. Comput. Eng. Appl. **50**(5), 16–20 (2014)
2. Fu, S.: Personalized information retrieval technology review. Inf. Theory Pract. **32**(5), 107–113 (2012)
3. Sarwar, B.: Sparsely. Scalability and distribute in recommender systems. University of Minnesota (2011)
4. Guopxia, W., Heping, L.: Personalized recommendation system overview. Comput. Eng. Appl. **48**(7), 66–76 (2012)
5. Ying, C., Wang, Z.: SVD feature: a tool kit for feature-based collaborative filtering. J. Mach. Learn. Res. **13**(1), 3619–3622 (2012)
6. Scarab, Katipos, G., Konstanz, J.: Incremental singular value decomposition algorithms for highly. In: Scalable Recommender Systems Fifth International Conference on Computer and Information Science, pp. 27–28(2012)
7. Dean, J., Ghemawat, S.: MapReduce: simplified data processing on large clusters. Commun. ACM **51**(1), 107–113 (2011)
8. Braes, J., Heckerman, D.: Empirical analysis of predictive algorithms for collaborative filtering. In: Proceedings of the 14th Conference on Uncertainty in Artificial Intelligence (UAI 1998), pp. 43–52 (2012)
9. Miller, B.N., Albert, I., Lam, K., et al.: Movie-Lens unplugged: experiences with an occasionally connected recommender system. In: Proceedings of the Conference on Human Factors in Computing Systems, pp. 210–217 (2011)
10. Linden, G., Smith, B., York, J.: Amazon.com recommendations: item to Item collaborative filtering. IEEE Internet Comput. **7**(1), 76–80 (2013)
11. Zeng, C., Xing, C.X., Zhou, L.Z., et al.: Similarity measure and instance selection for collaborative filtering international. J. Electron. Commer. **4**(8), 115–129 (2011)
12. Schafer, J., Konstan, J., Riedl, J.: Recommender systems in ecommerce. In: Proceedings of ACM E-Commerce, pp. 158–166. ACM Press, New York (2013)

A Social Stability Analysis System Based on Web Sensitive Information Mining

Wei Wang[(⊠)]

Department of Electronic Technology,
Engineering University of CAPF, Xi'an, China
wjwangwei@pku.edu.cn

Abstract. Researches on domestic social stability analysis mainly focus on construction of social stability theory, architecture and index, while few pay attention on quantitative analysis. In this paper, a social stability supervising framework is proposed based on sensitive Web information mining, semantic pattern matching and quantitative calculating. A sensitive information knowledge base is constructed by analyzing sensitive information about social environment, national harmonious and happy index of people's live in natural language online news texts from Internet, and recognizing hot keywords as well as the event trends led by the keywords. A social stability index theoretic model and a quantitative calculating model are proposed to evaluate social stability quantitatively. Parameters of the calculating model are determined by employing social investigations and an iterative feedback learning method. A prototype system is built on proposed framework and experiments are conducted on six frontier provinces, e.g., Xinjiang and Tibet. The result of an average accurate of 73.29 % shows the effectiveness of the proposed model.

Keywords: Sensitive information · Social stability index · Web text mining

1 Introduction

There are many kinds of information on the Web, e.g., information about gaps between the rich and the poor, bad social security and unemployment which related to the social environment; information about religious convictions, different lifestyles and penetrations of foreign culture which involved in ethnic harmony; as well as information about living environment, social insurance and disposable incomes which related to people's livelihood. With the continuous improving of the popularity of the Internet, the virtual online space has a growing influence on the real world. Facts proved that some real affairs such as parades, meetings and associations in the real world came into being by discussion in online community at first. So, employing some information technologies to perform a comprehensive, accurate and timely supervision on the sensitive information on the Internet, and then to issue early warnings for fast responses in the real word, the social stability and unity, the vigorously development of the economy can be effectively maintained and protected.

At present, the supervision of network information is mainly done by public opinion monitoring systems. These systems are able to monitor web information, trace

© Springer Science+Business Media Singapore 2016
W. Chen et al. (Eds.): BDTA 2015, CCIS 590, pp. 46–58, 2016.
DOI: 10.1007/978-981-10-0457-5_6

hot events and carry out correlation analysis and trend analysis, but they cannot give apparent results on the social stability [1, 2]. The existing domestic work on social stability analysis mainly focus on the theory, system, index construction [3–11], few aims to achieve real-time assessment on the situation of social stability using Web information [12]. The deficiency of existing work lies in two aspects. On one hand, a large number of studies have been carried out only for qualitative analysis, but not for more meaningful quantitative results. On the other hand, some work is limited to a single factor, for example, research only focuses on happiness index without taking into account the combined effects of multiple factors.

In this paper, a social stability index theoretic model and a quantitative calculation model are proposed to evaluate social stability quantitatively. The theoretic model comprehensively considered three kinds of factors, the social environment factor, the national harmonious factors and the happiness index factors. Parameters of the calculating model are determined by employing social investigations and an iterative feedback learning on massive natural language text on the Web. Based on these models, a social stability supervising framework is proposed by sensitive Web information mining, semantic pattern matching and quantitative calculating. A prototype system is built on proposed framework. By analyzing sensitive information about social environment, national harmonious and happy index of people live in natural language news of six frontier provinces, e.g., Xinjiang and Tibet on the Internet, and recognizing hot keywords as well as the event trends led by the keywords, the system is able to monitor the social stability in time. The experiment result of an average accurate of 73.29 % shows the effectiveness of the system.

The rest of the paper is organized as follows. We first review related work in Sect. 2. The proposed technical framework is introduced in Sect. 3. Implementation of the prototype system and a case study are described in Sect. 4. In Sect. 5, we evaluate the system and demonstrate its feasibility and applicability. In Sect. 6, we conclude this paper.

2 Related Work

Some social science researchers in China have worked on the social stability situation analysis, index system construction and management system development. Li [2] gives an empirical analysis on the influence factors of social stability in the frontier minority areas from two aspects: the fact evaluation index and the stable confidence index on the basis of structural survey statistics. In documents [4, 6], the economic significance of happiness index, the construction of the index system, as well as the collection and the empirical analysis of happiness index were studied. The psychological factors of social group events were discussed in [7, 8]. The index system of harmonious development of economy society, which is composed of 38 important indicators, is constructed in [9]. From the perspective of economics, documents [10, 11] make a tentative analysis of the social and political stability of our country respectively. In [12], construction of a social stability early warning and management system is described, while the system can only use information input manually but not information grabbed automatically from the Web. So far, we can see that the research mainly focus on the theory, system and index, and Web information mining technology has not been used to analyze the social stability.

The public opinion monitoring applications are directly related to this paper. A public opinion monitoring system is able to discover and extract useful information from semi-structure or non-structure data in Web page content automatically, find and trace hot spots and focus events that newly happened and interested by people in the vast amounts of Web information, form a certain correlation analysis and trend analysis. There are some relatively good public opinion monitoring systems, for example, the Founder's Intellectual Thought public opinion early warning and assistant decision support system[1], the TRS Internet public opinion information monitoring system[2], the People's opinion[3], the Eagle micro blogging and emotion[4]. These systems are based on the information acquisition technology and employ information processing, content management, knowledge management and information classification technologies to achieve network public opinion monitoring, hot news tracking and supervision.

In this paper, we use text mining technology to realize the monitoring of social stability. Different from previous work, we pay more attention to quantitative social stability situation analysis. Namely, according to the proposed theoretical model and calculate model of social stability index, on the basis of acquisition, analysis and processing of social stability related network information, we perform quantitative calculation to get the situation of social stability.

3   Social Stability Analysis Technology Framework

This paper proposes a social stability situation analyzing framework based on sensitive information mining technology. In the framework, an exponential model of social stability is constructed, and an automatic quantitative social stability index calculation process is realized. The overall technical framework is shown in Fig. 1, it comprises three layers:

- *The Web Mining Layer:* This layer offers massive Web text mining services. By examining elements involved in the social stability model, this layer first crawls relevant Web pages, then uses TML (Text Mining Language) to extract keywords, understand semantic meaning, recognize associate relation and analyze sensitive information within page content. TML encapsulated complex web crawling and natural language processing technologies, it can easily maps the theoretical model and the extracting rules to the specific text mining process [13].
- *The Knowledge Discovery Layer:* This layer performs theoretical modeling, rule extraction and knowledge discovery. According to the social stability index theory model, it first recognizes the key words and the relationships of each factor to construct a sensitive information matching rule base. Then it uses an iterative feedback mechanism to determine the weight of each factor in the social stability index calculate model, to realize the quantitative calculation of social stability.

---

[1] http://www.founder.com/templates/T_Second/index.aspx.

[2] http://www.trs.com.cn/product/product-om.html.

[3] http://yuqing.people.com.cn.

[4] https://www.eagtek.com.

– **The Data Presentation Layer:** This layer supports data visualization and maneuverability. It takes the sensitive information extracted by the social stability index quantitative calculation model as input, represents them in forms of charts or other visualized methods to show the changes of social stability, and provides human-machine interface for further intelligent information analysis and decision making.

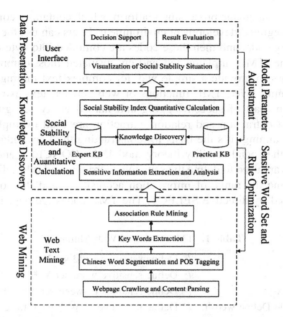

Fig. 1. A quantitative social stability analysis framework based on web sensitive information mining.

3.1   The Web Mining Layer

At the bottom of the framework, the text mining layer extracts sensitive words and matching rules from the massive network text under the guidance of the knowledge discovery layer. The extracted words and rules can be classified into three categories which have direct associations to the social stability, i.e., social environment, ethnic harmony and happiness index.

News is a kind of style that is widely used by newspapers, radio and other media to record facts, transfer information and reflect the times. The openness of the Internet enables the network news accounting the real society more directly and more quickly. So, many factors affecting the social stability situation can be found in online news. Two ways are employed to obtain sensitive information on the Web in this layer:

1. Extract sensitive information manually. This method is executed by reading news page artificially, selecting sensitive information as 'seed' according to some public

views on the current situation and the policy. It has a good effect in the initial state, but its efficiency is relatively low.

2. Using TML to capture online information automatically. Initially, some manually obtained sensitive words are feed to TML's web crawler as key words to carry out a directed crawl, and then the returned web pages are analyzed to obtain more sensitive words and all these words are put into a sensitive word set to support an iterative crawling process.

TML is a natural language processing platform, which contains a compiler, a virtual machine and an integrated development environment. Users can use the TML language to write text mining code, and then these codes are compiled into byte-codes to run on the virtual machine. TML implemented and encapsulated most commonly used text mining technologies to provide a simple way for complex text mining.

In the text mining layer, TML functions of web crawler, text extraction, Chinese word segmentation, part of speech tagging and named entity recognition, keyword extraction, concept extraction and relation extraction are used to implement the sensitive information mining; it's the basis for constructing the knowledge base.

According to the social stability theory model, the CONCEPT and PREDICATE directives of TML are used to define the sensitive word sets and rules, directive PAGES is used to determine the range of information acquisition, and the concept and relationship is extracted by directive SELECT. The TML codes are described in Table 1.

Table 1. TML Codes of Web Mining.

CONCEPT x;	/* Define sensitive word set X */
CONCEPT y;	/* Define sensitive word set Y */
PREDICATE x-y;	/* Define relations between sensitive word sets */
PAGES Sample Define website	/* Define the range of Web page for crawler */
SELECT x-y from Sample;	/* Extract the relations */
OUTPUT;	/* Output the results in XML */

For example, in analyzing the 'economic income' factor in 'social environment', we first manually recognize word set CONCEPT (income) as {"收入", "工资", "薪水", "生活费"}, after taking it as a seed to crawl the Web and expand it with synonyms, we get CONCEPT (income) = {"收入", "工资", "薪水", "生活费", "平均收入", "平均生活费", "经济", "物质", "生活必需品", "饮食质量", "伙食费", "平均工资", "平均薪水", "可支配收入", "可支配工资", "可支配薪水", "生活用品"}.

The semi-automatic learning process only completed the identification of sensitive words. In order to observe the changes of social stability, some verbs are needed to describe the trend of sensitive words. For example, in the 'social environment' factors, we need to analyze the changes related to the sensitive information 'economic income'. So the semi-automatic learning method is also used to construct a word set of verbs which can denote the status of 'economic income', namely CONCEPT (income-v) = {"低", "下降", "减少", "滑", "降", "回落", "低落", "低下", "没有", "不够", "拮据"}.

In order to realize the precise semantic matching, and avoid complex analysis of Chinese grammar, the 'co-occurrence' method is used to define the predicate modifier relations between the sensitive words and the trend verbs. Function *PREDICATE SE₁(income n₁, income-v v₁) {dist_15(n₁, v₁);}* means that at a distance of 15 words (the average length of a sentence), a word from the collection of *income* and a word from *income-v* formed a relation of the subject and predicate, which describes a kind of factor which will affect the social stability. This matching method based on the distance between two sets forms a mapping of |*income*| × |*income-v*|, which will improve the rule's coverage, and will also improve the recall rate just like synonym expansion.

3.2   The Knowledge Discovery Layer

The function of knowledge discovery layer is to construct the theoretic and the quantitative calculation models of social stability. When the models are built, they can be used to guide the text mining layer to execute the rule extraction and knowledge discovery tasks.

Theoretic Model of Social Stability Index. Through empirical analysis, we find the diversified characteristics of factors which have effect on social instability. That is to say, a bunch of factors such as economy, employment, social security, price, interest, political, ethnic, cultural, religious, hostile forces penetration, emergencies, land acquisition and so on are found undermining the social stability.

In this paper, by thorough analysis, discussion and investigation, we believe that the social stability index (*SI*) is a linear combination of the social environment (*SE*) factors, the national harmony (*NH*) factors and the happiness index (*HI*) factors. The definition of *SI* is shown as Eq. (1).

$$SI = \alpha SE + \beta NH + \gamma HI \tag{1}$$

Where $SE = \alpha_1 RP + \beta_1 SS + \gamma_1 ES + \cdots$, it means the social environment is defined as a combination of a variety of factors related to social environment. Here, RP, SP, EQ... respectively represents the element of the rich and the poor, the social security, employment situation and other elements.

$NH = \alpha_2 RC + \beta_2 FP + \gamma_2 LS + \cdots$, it means national harmony is defined as a combination of a variety of factors related to national unity. Here, RC, FP, LS, ... respectively represents the element of religion convictions, foreign penetration, lifestyle and so on.

$HI = \alpha_3 DI + \beta_3 SA + \gamma_3 EQ + \cdots$, the happiness index, is defined as a combination of a variety of factors related to happiness. Here, DI, SA, EQ, ... respectively represents the element of disposable income, social assurance, environmental quality and other factors.

Factors Affecting Social Stability. As mentioned above, the social stability index is influenced by many factors. In order to determine the importance of each factor, we designed a questionnaire to investigate the factors' influences on social stability. In 2013 March and April, we randomly issued a total of 600 questionnaires in universities, enterprises and streets to make a survey. The response rate was 91 % and 500 questionnaires were available, where 187 people were ethnic minority and 313 people were Han. The age distribution, occupation distribution and the education level distribution are found in Table 2.

Table 2. Statistics of questionnaire participants.

Age	Num.	%	Professional	Num.	%	Education	Num.	%
<20	119	23.8	Migrant worker	100	31.4	Middle school	54	10.8
20~30	178	35.6	Student	103	21.6	High school	89	17.8
30~40	103	20.6	Teacher	89	17.85	University	198	39.6
40~50	67	13.4	Doctor	39	10.2	Postgraduate	97	19.4
>50	33	6.6	Businessman	79	15.8	Ph.D	62	12.8
			Worker	90	24.2			

The statistics results of the survey showed the influence factors of social stability. We classified the detail factors into social environment, national harmony and happiness index. They are shown in Table 3.

Table 3. The factors that affect social stability.

Categories	Factors	Items
SE	Economic income	(1) per capita income (2) income growth and price growth ratio (3) is stable
	Employment status	(1) employment situation (2) attitude to current occupation (3) is stable
	Career promotion	(1) chance of promotion (2) self fulfillment
	Social status	(1) local or migrant (2) rural or urban (3) regional superiority
	Welfare support	(1) endowment insurance (2) medical insurance (3) city infrastructure (4) environment
	Family life	(1) housing and transportation (2) marriage (3) spouse (4) family relationship network
	Living condition	(1) pollution degree (2) city planning (3) public security level
	Group event	(1) social contradictions (2) illegal assembly activities (3) Riot (4) fury
NH	Economic development	(1) backward economy (2) the gap between rich and poor (3) price rise (4) unemployed persons
	Government duty	(1) unbalanced social development (2) social security (3) social injustice (4) increased crime rate
	Ethnic issues	(1) ethnic separatist activities (2) ethnic conflicts (3) religious issues

(*Continued*)

Table 3. (*Continued*)

Categories	Factors	Items
HI	Quality of life	(1) consumption level (2) environmental quality index (3) per capita disposable income of urban residents (4) Engel coefficient (5) per capita living space
	Social order	(1) incidence of mass incidents (2) duty crime rate of civil servants (3) incidence of major accidents (4) negative political rumors
	Social stability	(1) inflation rate (2) actual unemployment rate of urban (3) social security coverage (4) medical insurance coverage

3.3   Quantitative Calculation of Social Stability Index

In the proceed of questionnaire survey, the respondents were asked to sort the factors from big to small by the factors' influences on social stability according to their personal feelings. The same sorting was made on items in each kind of factors. For the sorting result of a certain class of factors, assuming the number in the first place is x_1, the number in the second place is x_2,..., the number in the m place is x_m, then according to the statistical results, the influence coefficient a_i of the factors on social stability was calculated by Eq. (2).

$$a_i = \frac{x_1 \times \theta_1 + x_2 \times \theta_2 + \cdots x_m \times \theta_m}{500 \times m} \tag{2}$$

Where $\theta_j = \frac{m-j+1}{m}, j = 1, 2, \ldots, m$.

According to Eq. (2), we get a subjective estimated parameters of the model. In order to refine these parameters, we need to select several websites from the frontier provinces to get actual experimental data. Six sites such as Xinjiang and Tibet are chosen as data sources to be crawled by comparing the capacity and the update frequency of the content on the websites, as shown in Table 4.

Table 4. Website list for data sampling and analyzing in the prototype system.

Province	Website URL
Tibet	http://www.chinatibetnews.com
Xinjiang	http://www.ts.cn
Guangxi	http://www.gxnews.com.cn
Inner Mongolia	http://www.nmg.xinhuanet.com
Jilin	http://www.jl.xinhuanet.com/
Yunnan	http://www.yn.xinhuanet.com

By analyzing empirical data, parameters of the model are verified and adjusted, and the calculation formula of the social stability index *SI* is eventually defined in Eq. (3):

$$SI = 0.45 \times SE + 0.35 \times NH + 0.2 \times HI \qquad (3)$$

In the equation, the social environment SE is:

$$\begin{aligned} SE = {} & 0.25 \times (\text{Economic income}) + 0.09 \times \\ & (\text{Employment status}) + 0.05 \times \\ & (\text{Career promotion}) + 0.12 \times (\text{Social status}) + 0.13 \times \\ & (\text{Welfare support}) + 0.15 \times (\text{Family life}) + 0.08 \times \\ & (\text{Living environment}) + 0.13 \times (\text{Group event}) \end{aligned} \qquad (4)$$

The national harmony factor NH is:

$$\begin{aligned} NH = {} & 0.5 \times (\text{Economic development}) + 0.3 \times (\text{Government duty}) \\ & + 0.2 \times (\text{Ethnic issues}) \end{aligned} \qquad (5)$$

The happiness index factor HI is:

$$\begin{aligned} HI = {} & 0.4 \times (\text{Quality of life}) + 0.4 \times (\text{Social order}) \\ & + 0.2 \times (\text{Social stability}) \end{aligned} \qquad (6)$$

3.4   The Data Presentation Layer

In the data presentation layer, graphs and tables are employed to display the social stability data. The graphic interface can provide a dynamic and visualized view for users to make better decisions. The optional data display modes include:

1. Line Chart: a social stability index linear chart is drawn according to the quantitative analysis result of stability index, and the line chart can intuitively shows the stability trend of the supervised provinces in a period of time.
2. Situation Map: a China Map is rendered everyday to dynamically monitor the stability index of different provinces in time. To display the stability status intuitively, the Map is colored into green, blue, yellow, orange and red according to the general international security level.

4   Prototype System Demonstration

4.1   System Construction

According to the proposed technical framework, a Browser/Server based frontier province social stability index analysis system is implemented in this paper. Apache, TML and JSP are used to develop the server side program; JavaSript and Ajax are employed to build friendly use interfaces in the client side. Figure 2 shows the architecture of the social stability index analysis system.

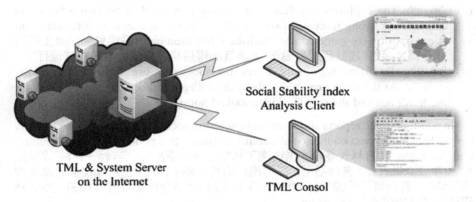

Fig. 2. Structure of the frontier province social stability analysis system.

4.2 Case Study

We ran the system on the Internet, and analyzed the social stabilities of six frontier provinces such as Tibet, Xinjiang and so on. The stability index line charts of these provinces from 2013/5/6/to 2013/9/6 are shown in Fig. 3.

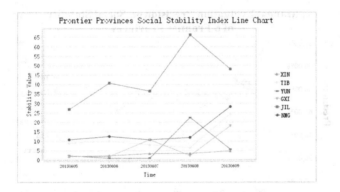

Fig. 3. A social stability line chart of 6 frontier provinces.

In the figure, the stability indexes of Jilin Province are high and the values change significantly. By manually analyzing the content in the grabbed web page, we found that a fire explosion accident had occurred in Jilin province on June 3rd, 2013. So, during those days, many reports appeared about the event. Some extracted sensitive information of the system are listed below:

(1) 3/6–5/6: the 6.3 serious fire explosion event happened, reports appeared to reveal the accident.
(2) 6/6–7/6: the death in the accident is announced one after another.

(3) 7/6–8/6: The explosion accident continuously fermented, it became a hot event and led to a strong reaction in society. News about the accountability and influence control were gathered and published. For example, "当地曾为出事工厂违规开路", "政府道歉后还需追责", "液氨高温后易造成流行病与疫病流行".

(4) 8/6–9/6: Problems were solved, reports about the fire explosion gradually reduced. At the same time, news about the college entrance examination became the headlines and the stability index looked normal.

In addition to the reports about fire and explosion accident, a large number of other news which had impact on the social stability of Jilin in that period were also extracted. For example, "吉林长春市一地铁施工处发生施工事故", "吉林一法院'温馨提示'引发公众批评", "吉林石化乙二醇出厂报价小幅上涨", "吉林榆树高考乱象娱乐了谁", "吉林男子行凶 见义勇为者身中多刀", "韩企白菜价进口中国人参暴利吉林千亿计划阻击", and so on.

5   Evaluation and Analysis

To verify the feasibility of the system and the proposed social stability index model, the system results were compared with the fact we manually extracted from the actual news. The statistics results are shown in Fig. 4.

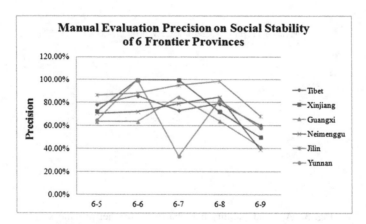

Fig. 4. The manual evaluation accuracy on social stability of 6 frontier provinces in 2013-6-5 ∼ 2013-6-9.

The precision is defined as:

$$percision = \frac{|sensitive\ words\ identified\ manually|}{|sensitive\ words\ identified\ automatically|} \times 100\%$$ (7)

The figure shows the precisions of the sensitive information extraction from 6 representative websites in the frontier provinces. In June 7th and June 9th, the precisions of Yunnan are low. By tracing the calculating procedure of social stability index, the problems in the expansion of sensitive words and the design of the words set structure were found. So we classified the sensitive word sets according to their semantics and optimized the cross correlation among the word sets. After correction and optimization, the average accuracy rate of the system raised to 73.72 %.

The experimental results show that the proposed model and technical framework can better monitor the social stability, and timely reflect the changing trend of social stability. To further improve the practicality of the system, more work should be done in two aspects: (1) When selecting sensitive information, refer to Baidu hot words list and other resources to enhance the authority of the constructed sensitive information knowledge base; (2) Employing text polarity analysis technology to grasp public opinion's trend in a finer granularity.

6 Conclusions and Future Work

Sensitive information extraction and social stability analysis technologies are studied in this paper. A social stability index theoretic model and a quantitative calculation model are proposed to evaluate social stability quantitatively. A B/S based prototype system is implemented based on TML's text mining and exact semantic matching technologies. Practical evaluation were conducted on six frontier provinces such as Xinjiang and Tibet, the results confirmed that the proposed model and the prototype system could better reflect the situation of social stability.

This work is able to provide useful information to the army, government and public security intelligence departments for making better decision, and eventually maintain the national stability and unity.

Acknowledgments. This work is supported by the Young Scientists Fund of the National Natural Science Foundation of China (Grant No. 61309022) and the Military Application Research Project of CAPF (Grant No. WXK2015-13).

References

1. Shou, L., Chen, G., Hu, T., Chen, K., Wang, Y.: A relevance mining method of Internet hot spot topic. Invention patent CN101158957 (2008)
2. Li, Y., Sun, L.: Hot-word detection for Internet public sentiment. J. Chin. Inf. Proc. 25(1), 48–59 (2011)
3. Li, Y.: Analysis of social stability influence factors in frontier ethnic areas. Heilongjiang Nat. Periodicals 2010(1), 36–43 (2010)
4. Tang, X., Yang, P.: On evaluation model for Chinese citizens happiness index. J. Anhui Sci. Technol. Univ. 26(2), 61–65 (2012)
5. Kang, J.: The meaning and measurement of happiness. China Statistics 2006(9), 18–19 (2006)
6. Gong, C.-Z.: How to build the index system of GNH. J. Eastern Liaoning Univ. (Soc. Sci.) 8(6), 84–87 (2006)

7. Liao, H., Cao, H.: Social psychological mechanism produced by group events and its countermeasures. Innovation **2009**(1), 83–87 (2009)
8. Qiu, Z.: A social psychological foundation analysis on network public opinion in massive incidents. J. Gui Zhou Province Committee Party's School of C.P.C. 2011(3), 82–85 (2011)
9. Zhu, Q.: A comprehensive evaluation on index system of the harmonious development in economic society. Society of China Analysis and Forecast (2007)
10. Song, L., Appleton, S.: An empirical investigation into social discontent in urban China. China Econ. Q. **6**(4), 1339–1358 (2007)
11. Hu, L., Hu, A., Wang, L.: A empirical analysis on the changing situation in social unstable factors. Discovery **2007**(6), 105–114 (2007)
12. Yan, Y.: The measurement of the social stability and the construction of presentiment management system. Sociol. Stud. **2004**(3), 1–10 (2004)
13. Li, J., Li, X., Meng, T.: A universal and efficient language text mining. In: The 19th China Conference on Information Retrieval, vol. 7 (2013)

Energy Conservation Strategy for Big News Data on HDFS

Jiang Zhong[1,2], Hao Chen[2(✉)], and Lei Yang[2]

[1] Key Laboratory of Dependable Service Computing in Cyber Physical Society
(Chongqing University), Ministry of Education, Chongqing, China
zhongjiang@cqu.edu.cn
[2] College of Computer Science, Chongqing University, Chongqing 400030, China
{kenochen,yanglei}@cqu.edu.cn

Abstract. In this paper, an energy-conservation Hadoop Distributed File System
(HDFS) oriented to massive news data is proposed based on news access pattern,
in order to reduce energy consumption of big news data storage system. First,
divide all data nodes into real-time responding hot data nodes and standby cold
data nodes. To make a good balance between data access performance and energy-
conservation, this paper takes two strategies of priority allocation, named Active
State Node Priority (ASNP) and Lower Than Average Utilization Rate Node
Priority (LANP), to mostly guarantee the balance of data distribution in cluster
in order to obtain a good data access performance. It also confirms the opportunities to move data from hot data nodes to cold data nodes is based on the access
pattern of news data and develops a simulating experimental platform that can
evaluate energy consumption of any file accessing operation under any different
storage strategies and parameters. Simulation experiments shows that strategies
proposed in this paper saves 20 %–35 % energy than traditional HDFS and 99.9 %
responding time of reading files will not be affected, with an average of 0.008 %–
0.036 % time delay.

Keywords: File storage energy-conservation strategy · Data distribution
balance · Simulation platform

1 Overview

With the constant improvement and popularization of cloud computing technology,
distributed storage system, in the basic position, has great progress and development,
data centers of various sizes are developed rapidly. After the pursuit of performance,
capacity, fault tolerance and safety, green energy-conservation also become a new
standard in this field. According to the research of literature [1], IT industry's carbon
dioxide emissions takes 2 % of global emissions, which is expected to double by 2020.
In 2008, Internet equipment such as routers, switchers, severs and cooling equipment
consume a total of 868 billion KWH, which took 5.3 % of consumption of the world [2].
So how to reduce energy consumption of device nodes without affecting the performance
of service is a realistic question.

At present, some researchers have done some researches on how to reduce the energy
consumption of cluster. For example, Kaushik and Bhandarkar [3], Kaushik et al. [4]

© Springer Science+Business Media Singapore 2016
W. Chen et al. (Eds.): BDTA 2015, CCIS 590, pp. 59–73, 2016.
DOI: 10.1007/978-981-10-0457-5_7

divided cluster into cold and hot zone and reduce the energy of cluster through storing rarely accessed files in cold zone. Kaushik et al. [5] have studied how to forecast life cycle of files in the further research, reduce energy of cluster thorough migrating the files in hot zone to cold zone according to life cycle of them. In the research of Kaushik et al. [6], they proposed a method which reduces cooling energy consumption of cluster through scheduling tasks according to the temperature of nodes. Leverich and Kozyrakis [10], Lang and Patel [8] have proposed a energy saving strategy base on the utilization rate of node. Most of the research above just consider the problems of energy saving, but almost do not refer to that of the cluster load balancing.

This paper takes Hadoop distributed file system as the framework. Focus on the issues of energy-conservation and data distribution equilibrium, summers up the general pattern of news data access through studying news data access and proposed energy-conservation strategy oriented to big data news based on the pattern of news data access. In order to achieve a good balance between the performance and energy saving, we proposed Active State Node Priority (ASNP) strategy and Lower Than Average Utilization Rate Node Priority (LANP) strategy in this paper. Finally the experiment simulation platform is designed and implemented to verify the effectiveness of the proposed strategy in this article. The experiment result shows that the strategy, proposed in our paper, achieves a better balance between energy saving and data distribution equilibrium.

2 Access Pattern of News Data

The strategy, proposed in this paper, optimizes the HDFS stored procedure by using news data access pattern. So before expounding this strategy, we research on news data access pattern first. As we know, news has strong timeliness, so its access also has certain regularity. This section first explores news data access pattern through analyzing web site news access log with statistic method, finally gets the general pattern of news access.

We have obtained the page access log data set between January 2014 and June 2014 in each project page access statistics under Wikipedia [7]. Due to the large number of servers, Data obtained in this paper is only from one server access log, we hope that we can get news data access pattern through analysis of the access log data. So we choose 1000 news title as an observation set and count the superposition of daily visits about them in 15 days since they have been published. As shown in Fig. 1.

In order to quantitatively calculate the attenuation degree of visits, we use IBM SPSS to get the nonlinear fitting of the curve of Fig. 2, which shows the exponential model of daily total visits.

The value of R2 of the exponential model is 0.977, which means the model's fitting degree is very high. According to the fitting results of IBM SPSS, the constant of the model is 240493.456 and the value of b_1 is -0.264. The final obtained model can be expressed in Eq. 1

$$y = 240493.456 \times e^{-0.264x} \tag{1}$$

Equation 1 is for the mode of access quantity about 1000 news, the mode of access quantity about single news can be expressed in Eq. 2:

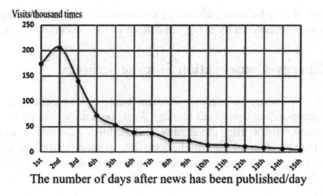

Fig. 1. Total daily visits of 1000 news

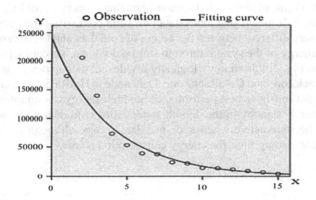

Fig. 2. The exponential model of total daily visits

$$y = 240.493 \times e^{-0.264x} \tag{2}$$

According to experience and statistics, we think that a single news will enter into the cold stage when its daily visits is below 5 times, which means the news will rarely be accessed and can be ignored. We can also know that, according to the results calculated by Eq. 2, visits of news will be below 5 times after 15 days form they have been published. We do not research the visits of news after 15 days.

In conclusion, the pattern of data access can be expressed as follows:

- Visits in 3 days takes more than 60 % of which in 15 days, while the visits within 7 days takes nearly 90 %.
- The visits of news is below 5 after 15 days form it has been published.

Based on the understanding of above the pattern above, we can propose corresponding energy-conservation optimization strategy. In the following parts we will adopt the Data Node Partition, Active State Node Priority (ASNP), Lower Than Average Utilization Rate Node Priority (LANP), File Migration and Node Standby strategies to achieve the goal of data balance and energy-conservation through the improvement of HDFS file storage mechanism by using the pattern of the news data access.

3   HDFS Energy-Conservation Storage Strategy

The current HDFS energy-saving management strategy is a generic model without proposing a corresponding optimized strategy focus on data access features. While the optimized strategy, considering the data access features, has better performance on energy saving than the general one. So the following parts proposed the energy-conservation storage strategy for big news data based on the news data access pattern in Sect. 2.

3.1   Data Nodes Partition Strategy

We can know that visits of news will decrease when time goes by according to the pattern obtained in Sect. 2. It means that news data will been cold with time goes by, and we know that it is very different between the access of cold data and the access of hot data. So it can save energy of the system through storing cold and hot data separately.

As shown in Fig. 3, this strategy logically divides all data nodes in the cluster into two parts HotRackZone and ColdRackZone. Data node in HotRackZone, called hot data node, is active all time, have best performance but high energy consumption. Data node in ColdRackZone is standby in the default state, called cold data node, which will be awakened and become active when node has data access, although it may affect the response time for reading files, the energy consumption is low.

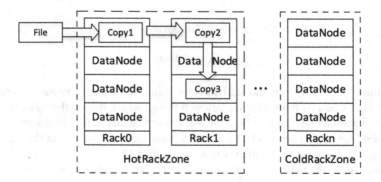

Fig. 3. Data node partition strategy

When a new file is created, file data will be written to node in HotRackZone to ensure good reading performance. The purpose of this strategy is making HotRackZone nodes guarantee low latency response of file access and reliability, while making ColdRack-Zone nodes standby as much as possible to reduce energy consumption.

Although writing data to cold data node may awake standby node first, but if considering the following two points, this situation will have least affect to the performance of the system when it happens.

- Adopt the strategy of Sect. 3.2, reduce the probability of waking up node while writing data to cluster.
- When writing data to ColdRackZone node, using methods in Sect. 3.3 which migrate data during low-access period.

For the convenience of description, this paper makes the following definitions:

Data nodes in HotRackZone and ColdRackZone are regarded as two sets, denoted by $HRZ = \{hdn_1, hdn_2, hdn_3, ..., hdn_m\}$ and $CRZ = \{cdn_1, cdn_2, cdn_3, ..., cdn_n\}$ hdn_i is one of the data nodes in HotRackZone, cdn_i is one of the data nodes in ColdRackZone, cdn_i has two kinds of states, one is standby, the other one is active after waked up, we remark the two states respectively as cdn_i^s and cdn_i^a, The state of cdn_i is remarked as $S(cdn_i)$.

For a data node set $DN = \{dn_1, dn_2, dn_3, ..., dn_m\}$, we define the operation functions of the set as follows:

- Active (DN) is expressed as a set whose elements are the active data nodes in DN.
- LRS (DN) is expressed as the nodes with the largest remaining space in the DN.
- URA (DN) is expressed as data nodes in DN whose utilization rate is lower than the average utilization rate of DN.

3.2 Two Strategies of Priority Allocation

3.2.1 Active State Node Priority (ASNP)

In our study, there are two cases in which we need to select target data node for writing the data. One is when a new file is created, data is written to hot data node, which are active, and have the best performance, that means we cannot reduce the energy consumption of cluster by selecting node, but can make data distribution equilibrium via directly select largest remaining space data node in HotRackZone as the target data node. The other one is in the process of file migration which will be introduced in Sect. 3.3. At this moment, data is written to cold data node in ColdRackZone, in which some nodes may have been awaken and become active if assigned tasks. For this reason, Active State Node Priority (ASNP) allocation strategy proposed in this paper reduces energy consumption of cluster by reducing the probability of waking up standby cold data node in ColdRackZone.

The key idea of ASNP:

- If system writes data to data node in HotRackZone, it will choose a target data node with the largest remaining space: LRS (HRZ).
- If system writes data to data node in ColdRackZone. First, it gets the active data nodes set in ColdRackZone, which is remarked as CRZ' = Active (CRZ).
  a. if CRZ' = Ø, it indicates that there are no active data node in ColdRackZone, So system will directly choose a target data node with the largest remaining space in CRZ: LRS (CRZ).

b. If CRZ' ≠ Ø, then system will choose a target data node with the largest remaining space in CRZ': LRS (CRZ').

It should be pointed out that if there are more than one node matched in the same time, system will choose one randomly.

The reasons of using this strategy are: First, selecting active nodes can avoid awakening standby nodes, thus reducing overall energy consumption of ColdRackZone. Second, in normal situation, larger remaining space of nodes means lighter load, writing data to this nodes can balance the load in the cluster.

3.2.2 Lower Than Average Utilization Rate Node Priority (LANP)

Although ASNP have good performance in energy-conservation, data distribution in ColdRackZone is imbalanced, which will have bad influence on system reading performance of files and increases the system response time. Therefore, we consider the utilization rate of nodes when choosing the target data node, system will obtain access performance through sacrificing some energy-saving effects in exchange for equilibrium of data distribution of cluster. Based on it, we proposes the Lower Than Average Utilization Rate Node Priority (LANP) strategy.

The key idea of LANP

- If system writes data to data node in HotRackZone. First, it chooses data node in the set HRZ whose utilization rate are less than the average, which are regard as set HRZ* = URA (HRZ). Then chooses a target data node with the largest remaining space in set HRZ*: LRS (HRZ*).
- If system writes data to cold data nodes in ColdRackZone. First, it choose data nodes whose space utilization rate is less than the average utilization rate of all nodes in ColdRackZone, these data node are regard as set CRZ* = URA (CRZ). Then gets the active data nodes set in set CRZ*, remarked as CRZ' = Active (CRZ*).
  a. If CRZ' = Ø, it indicates that there are no active data nodes in CRZ*, So system will directly choose a target data node with the largest remaining space in CRZ*, remarked as LRS (CRZ*).
  b. If CRZ' ≠ Ø, system will choose a target data node with the largest remaining space in CRZ': LRS (CRZ').

If system writes data to data node in ColdRackZone, system will choose data nodes whose space utilization rate is less than the average utilization rate of all nodes in ColdRackZone, then select the one which is active and with the largest remaining space among them as the target data node.

If there is no active node, system will select the one with the largest remaining space among them.

3.3   File Migration Strategy

Through the research on the pattern of news data access in Sect. 2, we can see that visits of each file decreases every day since it has been published. When time goes by, many news will be accessed very few, those news data will consume a large number of system energy if still stored in active nodes in HotRackZone. So we need to move files to cold data nodes if their residence time in hot data nodes is more than storage time threshold $T_{exsisted}$ and if their visits of previous day is lower than daily visits threshold $T_{accessed}$. We determine $T_{exsisted}$ according to the points with larger drops in a period of time on the access rule curve shown in Fig. 1 and $T_{accessed}$ based on experience, we suppose that a file belongs to cold files if its lowest daily visits is less than 5 times. In order to implement file migration strategy, system should get the files in hot data nodes whose storage time in HotRackZone is larger than $T_{exsisted}$ and the visit of previous day is less than $T_{accessed}$ first, and migrate them to cold data nodes.

The follows shows three points that need to be paid attention to when migration is implementing.

- Since data in hot data nodes need to be written to cold data nodes in the process of migration, system can select the target data nodes for migration of file data according to strategy in Sect. 3.2.
- Considering the news data access pattern, the probability of cold news become hot news and be accessed again is very small, so in this paper, the migration strategy is one-way, it means that data cannot migrate from cold data nodes to hot data nodes.
- It is certain that the file migration will affect the efficiency and performance of the whole system, so if system implements migration process during off-peak hours, it will minimize strategy impact on performance. We found that 2 PM to 4 PM is the slack period of site visits via statistical analysis of access log data. So file migration policies can be implemented in this period.

3.4   Nodes Standby Strategy

This strategy is only implemented on data nodes in ColdRackZone, with the purpose of reducing total energy consumption of system by making cold data nodes without reading and writing tasks to be standby. We think that it is wise to let a node to be standby if it has not been accessed for a long time, because this action almost will not cause any impact on system response to access, and can reduce energy consumption at the same time. So we set a threshold of node standby time in this paper known as T_{idle}, the last access time of cdn_i is represented as T_{la}^i and the current system time is represented as T_c, if $T_c - T_{la}^i > T_{idle}$, then $S(cdn_i) \leftarrow cdn_i^s$, which means system will make cdn_i enter standby. The process is shown in Fig. 4.

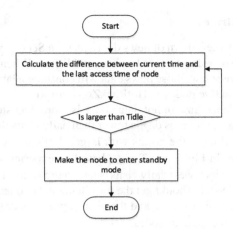

Fig. 4. Cold data node standby strategy process

4 Experiment Results and Analysis

In order to verify the strategy proposed in this paper, we develop a simulation experiment platform to verify its effectiveness. Figure 5 shows the functional structure chart of simulation platform we designed.

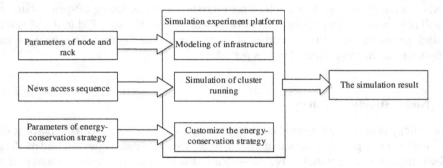

Fig. 5. The functional structure chart of simulation platform

In this simulation platform, we simulate a HDFS cluster with 120 data nodes and set the node power consumption according to the data in literature [4, 9], as listed in Table 1.

The data set used for experiments is from the access log of Wikipedia English news web site of one month. We preprocess the data set before experiment, assuming that the news contains pictures, videos and other multimedia information, so the size of files, in the log, is magnified 2000 times, eventually the visit record contains 283000 files, and the total data volume is 66 TB. Parameters used in experiments are listed in Table 2.

Table 1. Node power consumption

	Active power consumption (W)	Standby power consumption (W)	Wake up time consumption
CPU (Quad core)	80–150	12.0–20.0	30 us
DRAM DIMM	3.5–5	1.8–2.5	1 us
NIC	0.7	0.3	0 us
SATA HDD (1 TB)	11.16	9.29	10 s
PSU	50–60	25–35	300 us
DataNode (2 CPU, 8 DRAM, 4 × 1 TB HDD)	445.34	129.94	1000341 us

Table 2. Simulation parameters

Parameters	Value
Number of nodes	120
Number of hot data nodes	40
Number of cold data nodes	80
Node storage capacity	4 TB
Numbers of racks	15
$T_{existed}$	3, 7 days
$T_{accessed}$	5 times
T_{idle}	1 h
Upper limit of node storage space using	80 %
Time of file migration strategy implement	2 PM in the everyday

4.1 Analysis of Cluster Power Consumption

Energy consumption of traditional HDFS cluster or HotRackZone nodes, calculated by the following equation:

$$W(n) = \sum_{j=1}^{n} P_j \times T_j \tag{3}$$

n indicates the total number of data nodes in traditional HDFS cluster or HotRackZone, P_j indicates the power of the jth node (assume that the nodes are working in the rated power listed in Table 1), T_j represents the total working time of the jth node.

Energy consumption of ColdRackZone nodes is the sum of energy consumption in active mode and standby mode, which can be calculated by the following equation:

$$W_{cold} = \sum_{j=1}^{80} (P_j^{active} \times T_j^{active} + P_j^{idle} \times T_j^{idle}) \qquad (4)$$

W_{cold} indicates energy consumption of ColdRackZone nodes, P_j^{active} indicates the power of the jth node in the active mode, T_j^{active} indicates the working time of the jth node in the active mode. P_j^{idle} indicates the power of the jth node in standby mode, T_j^{idle} indicates the working time of the jth node in the standby mode.

So energy consumption of improved HDFS is calculated by the following equation:

$$W = W(40) + W_{cold} \qquad (5)$$

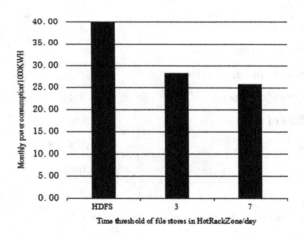

Fig. 6. Monthly power consumption of using ASNP

The result, shown in Figs. 6 and 7, is the comparison of monthly electricity consumption between improved and traditional HDFS. Figure 6 shows the cluster power consumption of using ASNP strategy within different time threshold of files stored in HotRackZone $T_{exsisted}$, we can see that power consumption of energy-conservation HDFS is less than 70 % of traditional HDFS. While Fig. 7 shows the cluster power consumption of using LANP strategy within different time threshold of file stored in HotRackZone $T_{exsisted}$, we can see that power consumption of using LANP is increased compared with which of using ASNP, but still saves 20 % energy than traditional HDFS. Whichever is used, the energy-conservation effect is better with the increase of $T_{exsisted}$.

Because the probability and time of nodes in standby mode will increase due to the visits of files which is moved to cold data node is lower as $T_{exsisted}$ increases.

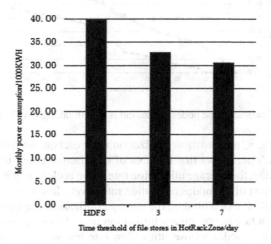

Fig. 7. Monthly power consumption of using LANP

4.2 Analysis of Cold Data Nodes Utilization Rate

We do not analyze the utilization rate of HotRackZone nodes, because they are active all time. Also because system always chooses one of the data nodes with the largest remaining space when data needs to be written in, which makes data distribution of HotRackZone has a good equilibrium. So we do not do the detailed analysis of them in this paper.

The result in Fig. 8 indicates the utilization rate of all ColdRackZone nodes when system takes ASNP strategy and set $T_{exsisted}$ as 7. And the result in Fig. 9 indicates the utilization rate of all ColdRackZone nodes when system takes LANP strategy and set $T_{exsisted}$ as 7. Curve with a triangle mark is the maximum one, curve with a dot mark is the minimum one, the unmarked curve indicates the average one.

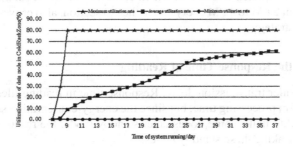

Fig. 8. ColdRackZone nodes utilization rate taking ASNP strategy

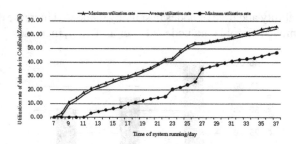

Fig. 9. ColdRackZone nodes utilization rate with taking LANP strategy

As shown in Fig. 8, nodes whose utilization rate reached the limit or zero always exists in ColdRackZone, shows that the data of distribution is not balanced. Between the 7th day to 21th day, the average utilization rate curve is close to the lowest utilization rate curve, means most of the nodes utilization rate is very low while only a few is very high. And in the 21th day to 37th day, the average utilization rate curve is closer to the maximum utilization rate curve, means utilization rate of most nodes is very high, but there still exists some nodes whose utilization rate are very low or zero. It further evidences that taking ASNP can lead data distribution imbalance.

As shown in Fig. 9, the average utilization rate curve almost overlaps with the maximum utilization rate curve. This suggests that taking LANP strategy can make data distribute more balanced. In the 7th day to 20th day, some nodes still with low utilization rate, this is because it will not write data to each node when the data quantity is not enough. After the 20th day, as more data moved in, the difference between the minimum and the maximum utilization rate will gradually decreases, and the distribution of the data will be more balanced.

4.3 Analysis of File Migration

Figure 10 shows the changes of quantities of migrated files in the everyday and data when $T_{exsisted} = 7$. The average is about 8000 files and 1.92 terabytes of data. Assume each cold data node have 4 hard disks (the volume of hard disk is 1 TB) and about 2 TB of data need to be moved to ColdRackZone every day, if bandwidth is 80 M/s, the operation can be completed within 2 h, which consists with the regularity that the low peak period is from 2 PM to 4 PM every day in the 3.3 section, also proves the feasibility of file migration strategy.

4.4 Impact on the Response Time of Reading

If we do not consider transmission delay, Response time in reading files of traditional HDFS is the system processing time, but since the energy-conservation storage strategy in this paper need to make some nodes enter standby mode, the response time should add the time of awaking these standby nodes.

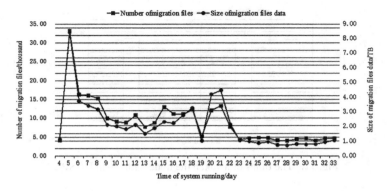

Fig. 10. Changes of migrated file number and migrated data size

As shown in Fig. 11, when system takes ASNP strategy and set $T_{exsisted}$ as 3, less than 0.1 % of the file access will awake data nodes and increase response latency, the percentage will decrease with $T_{exsisted}$ increasing.

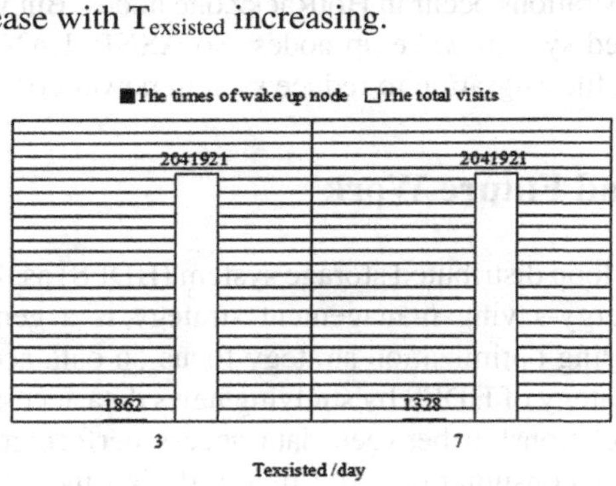

Fig. 11. The times of awaken the node with taken ASNP

As shown in Fig. 12, when system takes LANP strategy and set $T_{exsisted}$ as 3, about 0.13 % of the file access will awake data nodes and increase response latency, the percentage also decreases with $T_{exsisted}$ increasing.

We assume the reading bandwidth of a data node disk is 80 M/s and it does not exist queuing delay when reading file. There are 283000 files in the system, with a total data size of 66 TB. According to calculation, the average size of files is 246.6 M. As we know, time of reading files for active data nodes consumes in the process of reading disk, which is $T_{rh} = 3082500$ us, while for standby data nodes, it consumes in the process of waking up data nodes and reading disks because system needs to wake up standby nodes first. And the time of waking up data nodes is $T_{awake} = (30 + 1000000 + 1 + 310)$ us $= 1000341$ us, so the time of reading files is $T_{rh} + T_{awake}$.

Through analysis and calculation, ASNP, LANP's average delay increasing rates are 0.036 %, 0.008 % respectively.

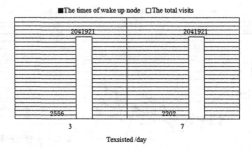

Fig. 12. The times of awaken the node with taken LANP

The result indicates that ASNP is slightly better than LANP in reading response time.

Whichever the system takes Energy-conservation strategies in this paper only have a very slight impact on reading response time.

They also will not affect the writing operations when system creates files, because at this moment these operations occur in HotRackZone nodes. But when migrating files, the operations may need system wake up nodes. So ASNP, LANP and file migration strategy can be used in file migration to reduce system power consumption.

5   Conclusions and Future Work

This paper chooses Hadoop distributed storage system (HDFS) as the based framework, the current HDFS energy-saving management strategy is a generic model without proposing a corresponding optimization strategy focus on data access features. So we improve the storage strategy of HDFS by studying news data access rules. In this paper, we also analyze the relationship between data access performance (balance of data distribution) and energy consumption. And finally design the simulation platform to support the research of energy consumption.

Because we have verified the strategy which is proposed in our paper through simulation platform and news access log data, there is a certain gap between the effect of actual experimental environment and effect of the simulation platform, So in the future work, we should obtain real big data of news and do the experiment in real cluster through implementing the strategy which is proposed in our paper, it can obtain more real experimental results.

References

1. Global Action Plan. An Inefficient Truth. Global Action Plan Report [EB/OL]. http://global-actionplan.org.uk, December 2007
2. Yun, D., Lee, J.: Research in green network for future internet. J. Korean Inst. Inf. Scientists Eng (KIISE) **28**(1), 41–51 (2010)
3. Kaushik, R.T., Bhandarkar, M.: GreenHDFS: towards an energy-conserving, storage-efficient, hybrid Hadoop compute cluster. In: Proceedings of the USENIX Annual Technical Conference, pp. 1–9. USENIX, Vancouver, BC, Canada (2010)

4. Kaushik, R.T., Bhandarkar, M., Nahrstedt, K.: Evaluation and analysis of greenhdfs: A selfadaptive, energy-conserving variant of the Hadoop distributed file system. In: 2010 IEEE Second International Conference on Cloud Computing Technology Science (CloudCom), pp. 274–287. IEEE (2010)
5. Kaushik, R.T, Abdelzaher, T., Egashira, R., et al.: Predictive data and energy management in Green- HDFS. In: 2011 International Green Computing Conference and Workshops (IGCC), pp. 1–9. IEEE (2011)
6. Kaushik, R.T., Nahrstedt, K.: Thermal-Aware GreenHDFS. University of Illinois, Urbana-Champaign (2011)
7. Wikipedia. Pagecount files for 2014 [EB/OL]. http://dumps.wikimedia.org/other/pagecounts-raw/2014/, June 2014
8. Lang, W., Patel, J.M.: Energy management for MapReduce clusters. In: Proceedings of the VLDB Endowment 2010, pp. 129–139. VLDB Endowment Inc., USA (2010)
9. Meisner, D., Gold, B.T., Wenisch, T.F.: PowerNap: eliminating server idle power. ACM SIGARCH Comput. Archit. News **37**(1), 205–216 (2009)
10. Leverich, J., Kozyrakis, C.: On the energy (In) efficiency of Hadoop cluster. In: ACM SIGOPS Operating Systems Review, pp. 61–65. ACM, New York, NY, USA (2010)

Research on Jukes-Cantor Model Parallel Algorithm Based on OpenMP

Kang Wan[1], Jun Lu[1,2(✉)], Jiaju Zheng[1], and Zihan Ren[1]

[1] College of Computer Science and Technology,
Heilongjiang University, Harbin, China
wankang1004@163.com,
lujun111_lily@sina.com,
{997238194,615726710}@qq.com
[2] Key Laboratory of Database and Parallel Computing
of Heilongjiang Province, Harbin, China

Abstract. The parallel algorithm of the evolutionary distance between different species is implemented by using OpenMP parallel technique in this paper. In order to get the best degree of algorithm parallelism, the method of making loop variable corresponding to the rower and column labels of distance matrix is adopted. It is to say that the double loop can be converted into single loop to improve the parallel efficiency. The serial algorithm and parallel algorithm are compared in this paper. The experiment result shows that the highest speedup is 14.1. It improves the running efficiency of the program. It is a great significance to dealing with massive bioinformatics data.

Keywords: OpenMP · Parallel · Evolutionary distance · Distance matrix

1 Introduction

DNA data play an important role on the research of molecular biology. One of the research achievements is evolutionary tree (phylogenetic tree) [1]. There are three steps to construct evolutionary tree from DNA data. The first one is sequence alignment. Then a nucleotide substitution model (distance model) is chosen to calculate distance matrix of different species. At last, distance method (Neighbor-Joining [2] or UPGMA) [3] is used to construct evolutionary tree from distance matrix [4]. The second step of constructing evolutionary tree (the calculation of distance matrix) is researched in this paper. The calculation of distance matrix refers to massive bioinformatics data and people have higher and higher requirement on the efficiency of data processing, so relevant serial algorithm cannot satisfy this requirement and research on parallel algorithm to this issue is very important. Jukes-Cantor model will be chosen in this paper and the method of making loop variable corresponding to the rower and column labels of distance matrix will be adopted to implement the parallel algorithm of the issue.

© Springer Science+Business Media Singapore 2016
W. Chen et al. (Eds.): BDTA 2015, CCIS 590, pp. 74–82, 2016.
DOI: 10.1007/978-981-10-0457-5_8

2 Jukes-Cantor Model

Jukes-Cantor model is the simplest nucleotide substitution model. This model assumes that the nucleotide substitution of every site takes place by the same probability. And the nucleotide of every site evolves to one of other three nucleotides by probability α every year (or other units of time). This can be shown in Table 1 [5, 6].

Table 1. Nucleotide substitution probability of Jukes-Cantor model

	A	T	G	C
A	–	α	α	α
T	α	–	α	α
G	α	α	–	α
C	α	α	α	–

Therefore, the probability of a nucleotide becoming one of other three nucleotides is $\gamma = 3\alpha$. γ equals the probability of nucleotide substitution on every site every year. It assumes that two nucleotide sequences X and Y were evolved from a common ancestor sequence before t years. q_t means the proportion of the same nucleotides between X and Y. p_t ($p_t = 1 - q_t$) means the proportion of different nucleotides. At the time $t + 1$(year), the proportion of the same nucleotides q_{t+1} can be gained by following method. Firstly, at the time t, the same site between X and Y will not change by the probability $(1 - \gamma)^2$ or its approximate value $(1 - 2\gamma)$ at the time $t + 1$. Secondly, at the time t, the site which has different nucleotides will become the same nucleotide by the probability $2\gamma/3$ at the time $t + 1$.

Hence, the following difference equation can be obtained,

$$q_{t+1} = (1 - 2\gamma)q_t + \frac{2}{3}\gamma(1 - q_t) \tag{1}$$

That is to say,

$$q_{t+1} - q_t = \frac{2}{3}\gamma - \frac{8}{3}\gamma q_t \tag{2}$$

If continuous model is applied and $q_{t+1} - q_t$ is replaced by d_q/d_t, remove the t (subscript of q_t), then the following differential equation can be gained.

$$\frac{d_q}{d_t} = \frac{2}{3}\gamma - \frac{8}{3}\gamma q \tag{3}$$

If the initial condition of this equation is $t = 0$ and $q = 1$, then

$$q = 1 - \frac{3}{4}(1 - e^{-\frac{8}{3}\gamma t}) \tag{4}$$

In this model, the expectation of nucleotide substitution on every site of the two sequences is $2\gamma t$. So d can be given by the following equation

$$d = -\frac{3}{4}\ln(1 - \frac{4}{3}\mathrm{p}) \tag{5}$$

Where $p = 1 - q$, p means the proportion of different nucleotides between sequence X and Y. d means the evolutionary distance (genetic distance) between X and Y [7].

3   Parallel Algorithm of Distance Model

3.1   The Design of Parallel Algorithm

3.1.1   OpenMP Parallel Technique

OpenMP(Open Multi-Processing) is a technique of shared memory programming model. The execution mode of OpenMP is the form of fork-join. "fork" creates new thread(s) or wakeup existing thread(s). "join" is the confluence of multiple threads [8].

The main scheduling mode of OpenMP is static and dynamic scheduling. Different scheduling modes have different influence on the execution of parallel program [9]. The following example implies the difference between the two scheduling mode. Assume that there are 10 tasks in a loop and a computer can run 4 threads at the same time. Then two tasks will remain finally. If static scheduling mode is used, then these two tasks will be scheduled to the first thread. If dynamic scheduling mode is used, then these two tasks will be scheduled to the first and the second thread averagely. Therefore, dynamic scheduling will improve the efficiency of the parallel program effectively. And static scheduling will aggravate the degree of load unbalance between different threads. So static scheduling will restrict the further promotion of program's execution efficiency. Hence, dynamic scheduling is adopted to dispatch tasks which will be paralleled in the experiments of this paper.

Amdahl's law is related to OpenMP. The law describes the relationship among speedup, the number of processor and serial factor (the proportion of serial part in the program) of the parallel program. Serial factor means the proportion of the serial part (It can't be parallelized) execution time occupied in the whole serial program's execution time. As shown in Fig. 1, when serial factor equals zero, the speedup of parallel program is ideal. If serial factor increases, then trend of speedup's increase will slow down. And it will tend to a limit value. While the number of processor is more than a specified value, increasing the number of processor has no positive influence on increasing speedup [10].

3.1.2   The Design of Parallel Algorithm

The design of parallel algorithm is based on analyzing the parallelism of serial algorithm, so serial algorithm must be analyzed before designing parallel algorithm. Serial program contains several parts. The first one is reading input file. The next one is

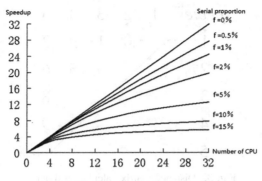

Fig. 1. Amdahl's law

preprocessing of data. The most important part is the calculation of distance matrix with double loop. Result is written into output file finally. It is possible that only the calculation of distance matrix can be parallelized. This part contains double loops. And the calculation of evolutionary distance between each two species is uncorrelated. Therefore, this part can be parallelized. According to above analysis, the parallelism of the program can be implemented by fork-join mode [11–13].

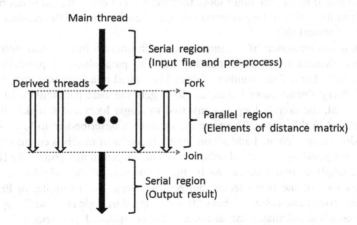

Fig. 2. Fork-join pattern of parallel implementation

As it is shown in Fig. 2, reading input file, preprocessing and output result of this program are serial, but the calculation of distance matrix can be parallel. In parallel part, parallel particle is the calculation of every element in distance matrix. Every element means the evolutionary distance between the two corresponding species.

According to the parallel model, the next is to analyze parallel part. At first, serial program contains double loop. Its calculation order is shown in Fig. 3.

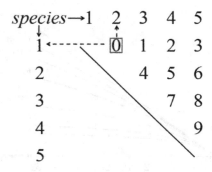

Fig. 3. Distance matrix calculation order

In Fig. 3, the rowers and column labels mean the ID of species. This example contains 5 species. Distance matrix is a symmetric matrix and elements in its diagonal line are zero, so only elements of upper triangle need to be calculated. Numbers in this matrix means the order of calculation. The task load of every execution of the outer loop is different in the double loop. For example in Fig. 3, the first execution in outer loop contains 4 calculations. They are 0, 1, 2 and 3 respectively. The second execution contains 3 calculations. They are 4, 5 and 6 respectively. If guidance commands of OpenMP are added before the outer loop, then the load of every thread is not balanced. It will result in the waste of thread resource. And the efficiency of the parallel program won't increase remarkably.

Under this circumstance, if a computer can run enough threads and every thread calculates an element of distance matrix, then the parallelism of program will be improved greatly. Even if the number of thread is limited in a computer, that will make the load of every thread more balanced. So it gets the best parallelism. In order to achieve the goal, the double loop is converted to single loop and it makes the single loop parallelized in this paper. The concrete method is described as follow. At first, a two-dimension array is created and according to the order of matrix's calculation, every element's corresponding rower and column label are stored in this array. The ID (rower and column label) of two species can be the parameters of the calculating function. Then the distance of the two species can be calculated. For example, in Fig. 3, the corresponding rower and column label of 0 in the upper triangle is 1 and 2 respectively. Then this execution calculates the distance between specie 1 and specie 2. The corresponding rower and column label of 1 in the upper triangle is 1 and 3 respectively. Then the execution calculates the distance between specie 1 and specie 3. And so on. So double loop becomes single loop and then the single loop is parallelized. Hence, the parallel program gets its best parallelism.

3.2   Parallel Algorithm Based on OpenMP

Input: DNA sequences of different species
Output: distance matrix

1. Preprocess: Making loop variable corresponding to rowers and column labels of distance matrix

```
Loop row=1 to specieNum
{
  Loop column=row+1 to specieNum
  {
    Record current rowers to index[indexCount][0];
    Record current column labels to idex[indexCount][1];
    indexCount= indexCount+1;
  }
}
```

2. The parallel algorithm

```
#pragma omp parallel num_threads(threadNum)
{
  #pragma omp for private(i) schedule(dynamic)
  //Parallel implementation
  Loop i=0 to specieNum*(specieNum - 1) / 2
  {
    boolean baddists = false;
    makev(index[i][0], index[i][1], &v,&baddists);
    v = fabs(v);      //compute the absolute value of v
    if baddists=true        //if a distance is invalid
    {
      v=-1;                 //distance equals -1
      baddists=false;       //reset the flag
    }
    //compute distance
    d[index[i][0]-1][index[i][1]-1] = v;
    //corresponding element in the lower triangle
    d[index[i][1]-1][index[i][0]-1] = v;
  }
}
```

Variable declaration:
    specieNum: the number of species
    row: current rowers
    column: current column labels
    index: array stored with rower and column label
    threadNum: the number of threads
    i: thread index
    baddists: a flag to describe that the distance is invalid or not
    v: the value of a distance between two different species
    d: distance matrix
    makev : a function that computes distance between specie index[i][0] and index[i][1] is calculated in this function

The hardware environment of these experiments is: CPU is Intel(R) Xeon(R) CPU E5-2670 v2 @2.50 GHz 2.50 GHz(2 processors); CPU op-mode(s) is 64-bit; the number of CPU cores is 20; the hard disk's space is 2.0T; operating system type is Linux 2.6.32-504.1.3.el.x86_64 GNOME 2.28.2.

4   Results of Experiments

The data of the experiments simulate different species' DNA sequences. These sequences are put into a file (input file). This dataset has 1200 species. The length of DNA of each species is 100000. And the file size is 114.5 M. The experiments test the execution time of serial program and parallel program. Then experiments are analyzed through comparing the results. Different numbers of thread are set in parallel program. They are 5, 10, 15, 20, 25, 30, 35 and 40 respectively. Every test with same numbers of thread is implemented for 5 times. Then the average value of the 5 times is chosen as the final results. The result is shown in Table 2.

Table 2. The execution time of serial and parallel programs (unit: min)

Times	Serial	10 threads	20 threads	30 threads	40 threads
1	58.040	8.085	4.614	4.334	4.113
2	57.892	8.189	4.729	4.287	4.060
3	57.498	8.053	4.743	4.334	4.076
4	57.342	8.075	4.627	4.288	4.095
5	57.494	8.050	4.745	4.250	4.095
Average	57.635	8.090	4.692	4.299	4.088

The serial/parallel running time and speedup of parallel with different thread number is shown in Fig. 4. The Figure implies that, with the increase of the number of thread, the program takes less and less time, and speedup becomes higher and higher.

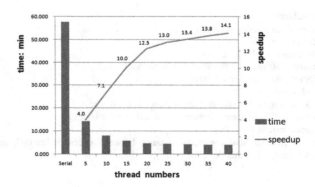

Fig. 4. The running time and speedup of programs with different threads

While the number of thread is 40, speedup is the highest one—14.1. But the increase speed of speedup becomes slower and slower with the increase of the number of thread. That fits Amdahl's law's ideas. At the same time, the serial factor of the program is approximately 1.36 %. According to Amdahl's law, when the number of CPU cores is 20, speedup is about 15. This parallel program's highest speedup is 14.1 while the number of CPU cores is 20. That result fits Amdahl's law.

5 Conclusion

1200 species' DNA sequences are simulated in this paper. And the evolutionary distances of each two species are calculated by using OpenMP parallel technique. The best degree of parallelism of the algorithm is obtained by converting double loop to single loop. Experiment result shows that the calculation efficiency of distance matrix is greatly improved by comparing parallel algorithm with serial algorithm. That has great significance on constructing evolutionary tree of different species quickly, analyzing the genetic relationship between different species and evaluating evolution of species.

Acknowledgment. This research is funded by a grant (No. F201138) from the Natural Science Foundation of Heilongjiang Province of China and a grant (No. 12521395) from the Scientific Research Fund of Heilongjiang Provincial Education Department of China.

References

1. Xu, Z.: Bioinformatics. Tsinghua University Press, Beijing (2008)
2. Saitou, N., Nei, M.: The neighbor-joining method: a new method for reconstructing phylogenetic trees. Mol. Biol. Evol. **4**(4), 406–425 (1987)
3. Zheng, X.: Distance Measures of Biological Molecular Data and Their Applications. Dalian University of Technology, Liaoning (2009)
4. Simonsen, M., Pedersen, C.N.S.: Rapid computation of distance estimators from nucleotide and amino acid alignments. In: Proceedings of the 2011 ACM Symposium on Applied Computing (2011)
5. Guo, Z.: Research on A Number of Problems of the Nucleotide Substitution Model. Northwest Agriculture & Forestry University, Shaanxi (2013)
6. Xie, X.: Research on Analysis and Evolution Model of Biological Sequences. Northwest Agriculture & Forestry University, Shaanxi (2012)
7. Nei, M., Kumar, S.: Molecular Evolution and Phylogenetics. Higher Education Press, Beijing (2002)
8. Luo, Q., Zhong, M., Liu, G., et al.: Compilation theory and implementation technique of OpenMP. Tsinghua University Press, Beijing (2012)
9. Li, Z.: Research and Analysis of Multi-thread Load Balance Scheduling Scheme Based on OpenMP. Xi'an Technological University, Shaanxi, China (2014)
10. Zhou, W.: Multi-core Computing and Programming. Huazhong University of Science and Technology Press, Wuhan (2010)
11. Gong, X., Zou, L., Hu, Y.: The research of multi-core system parallel program design methods based on OpenMP. J. Univ. South China (Sci. Technol.) **27**(1), 64–68 (2013)

12. Shi, Y., Lu, J., Shi, X., et al.: The key algorithms of promoter data parallel processing based on OpenMP. In: 2014 International Conference on Cyber-Enabled Distributed Computing and Knowledge Discovery (CyberC), pp. 154–157. IEEE Press, Shanghai (2014)
13. Shi, Y., Lu, J.: The parallel processing for promoter data base on OpenMP. In: Proceedings of the 2015 International Conference on Electronic Engineering and Information Science (ICEEIS 2015) (2015)

Rough Control Rule Mining Model Based on Decision Interval Concept Lattice and Its Application

Ailing Sun, Chunying Zhang$^{(\boxtimes)}$, Liya Wang, and Zhijiang Wang$^{(\boxtimes)}$

College of Science, North China University of Science and Technology,
Tangshan, Hebei, China
salangelbaby@163.com, {zchunying,wangzj}@ncst.edu.cn,
wang_liya@126.com

Abstract. Fusing the structure feature of interval concept lattice and the actual needs of rough control rules, we have constructed the decision interval concept lattice, further more, we also have built a rules mining model of rough control based on decision interval concept lattice, in order to achieve the optimality between rough control mining cost and control efficiency. Firstly, we have preprocessed the collected original data, so that we can transform it into Boolean formal context form, and then we have constructed the decision interval concept lattice in rough control; secondly, we have established the control rules mining algorithm based on decision interval concept lattice. By analyzing and judging redundant rules, we have formed the rough control association rule base in end. Analysis shows that under the premise of improving the reliability of rules, we have achieved the rough control optimization goal between cost and efficiency. Finally, the model of reservoir scheduling has verified its feasibility and efficiency.

Keywords: Rough control · Decision interval concept lattice · Attribute discretization · Decision rule mining

1 Introduction

Rough control is a new type method of intelligent rules control [1], rules extraction is the crucial link in intelligent control. The precision of rules directly affect the efficiency and accuracy.

Many researchers have adopted a lot of means to extract decision rules and a variety of algorithms have been developed. Dong et al. [2] presented the rough rule mining algorithm based on the theory of variable precision rough set. Huang [3] proposed the method of attribute reduction and rule extraction under the decision background. Rough control has achieved some success in industrial control applications [4, 5], however, when the variables in the control system is continuous, there are some problems, such as the establishment of a decision table, discretization of continuous variables, the consistency of rules and so on. Rough control always limited application, for its reason, there are some problems of rules in large number and low efficiency. In this paper, we put forward the decision interval concept lattice based on interval

© Springer Science+Business Media Singapore 2016
W. Chen et al. (Eds.): BDTA 2015, CCIS 590, pp. 83–92, 2016.
DOI: 10.1007/978-981-10-0457-5_9

concept lattice [6]. Based on the feature that concept lattice extension must meet a certain amount of the intension property, and its intension is determined by conditional attributes and decision attributes, so we adopt the theory of decision interval concept lattice to mine the decision rules, which become more decisive than traditional rules. The method we put forward not only reduce the mining cost, but also improve the control efficiency.

In this paper, we have designed the decision rule mining model of rough control based on decision interval concept lattice. Firstly we construct the decision interval concept lattice, then mine the decision interval rule based on the decision interval concept lattice. The model we constructed has achieved the optimum between rules mining cost, efficiency and reliability. The rationality of model analysis is presented, further more its feasibility and efficiency are proved by an example.

2 Decision Interval Concept Lattice

2.1 Basic Concepts

Definition 2.1. Let $(U, C \times D, R)$ be a decision context. $RL(U, C \times D, R)$ is the decision interval concept lattice constructed by $(U, C \times D, R)$, and (M, N, Y) is a decision interval concept based on RL. Where C is the set of conditional attributes, D is the set of decision attributes. In the interval $[\alpha, \beta](0 \leq \alpha \leq \beta \leq 1)$, α upper bound extension M^α and β lower bound extension M^β are defined respectively by Eqs. (1) and (2)

$$M^\alpha = \{x | x \in M, |f(x) \cap Y| / |Y| \geq \alpha, 0 \leq \alpha \leq 1\} \tag{1}$$

$$M^\beta = \{x | x \in M, |f(x) \cap Y| / |Y| \geq \beta, 0 \leq \alpha \leq \beta \leq 1\} \tag{2}$$

Where Y is the intension of concept, among them, $Y = C' \cup D'$, $C' \subseteq C$, $D' \subseteq D$. If $Y \neq \phi$, then $D' \neq \phi$, C' is the set of child conditional attributes of C, and D' is the set of child decision attributes of D. $|Y|$ is the number of elements in Y, namely cardinal number. M^α is the set of objects that may be covered by at least $\alpha \times |Y|$ attributes in Y; M^β is the set of objects that may be covered by at least $\beta \times |Y|$ attributes in Y.

Definition 2.2. Let $(U, C \times D, R)$ be a decision context. The ternary ordered pair (M^α, M^β, Y) is called decision interval concept, where Y is intension, it contains conditional intension and decision intension, namely decision concept description; M^α is α upper bound extension and M^β is β lower bound extension.

Definition 2.3. Let $L_\alpha^\beta(U, C \times D, R)$ be the set of decision interval concepts getting from $(U, C \times D, R)$ in $[\alpha, \beta]$. If $(M_1^\alpha, M_1^\beta, Y_1) \leq (M_2^\alpha, M_2^\beta, Y_2) \Leftrightarrow C_1 \supseteq C_2, D_1 \supseteq D_2$, then "$\leq$" is called partial order relation on $L_\alpha^\beta(U, C \times D, R)$.

Definition 2.4. Let $L_\alpha^\beta(U, C \times D, R)$ be the set of decision interval concepts getting from $(U, C \times D, R)$ in $[\alpha, \beta]$. If all of the concepts in $L_\alpha^\beta(U, C \times D, R)$ meet the partial order relation "$\leq$", then $L_\alpha^\beta(U, C \times D, R)$ is called decision interval concept lattice.

Definition 2.5. Let $G_1 = \left(M_1^\alpha, M_1^\beta, Y_1 \right)$ and $G_2 = \left(M_2^\alpha, M_2^\beta, Y_2 \right)$ be two nodes in decision interval concept lattice, and they have the relationship $G_1 \leq G_2 \Leftrightarrow C_1 \supseteq C_2$. If there is no G_3, which meet $G_1 \leq G_3 \leq G_2$, then G_2 is a parent node (immediate predecessor) of G_1 and G_1 is a child node (immediate successor) of G_2.

2.2 Decision Interval Rule

Definition 2.6. Let $(U, C \times D, R)$ be a decision context. D is the set of decision attributes, U is the set of rule objects, and $C \times D$ is the set of rule projects. R describes the relationship between U and $C \times D$. For $A \subseteq C$ and $B \subseteq D$, then $A_\alpha^\beta \Rightarrow B$ is a decision interval rule getting from $(U, C \times D, R)$ in $[\alpha, \beta]$.

Definition 2.7. For the decision interval rule $A_\alpha^\beta \Rightarrow B$, if RO_α is the set of objects in B that meets the degree of α, RO_β is the set of objects in B that meets the degree of β, then the set of objects in A that meets the degree of $[\alpha, \beta]$ also meets the possible degree in B, which is defined the roughness of interval rules by (3)

$$\gamma = \rho(RO_\beta / RO_\alpha) = |RO_\beta| / |RO_\alpha| \tag{3}$$

$0 \leq \gamma \leq 1$. For decision interval rule, the lower roughness, the more accurate.

Definition 2.8. For the decision interval rule $r : A_\alpha^\beta \Rightarrow B$, If $\forall p \in A$, the weight of p based on B is defined by (4)

$$\delta(p, r) = |g(p)| / |\cup g(y)|_\alpha, (y \in A) \tag{4}$$

$|\cup g(y)|_\alpha$ is the number of objects in A that meets the degree of α.

2.3 Construction Algorithm for Decision Interval Concept Lattice

Algorithm: *(DICLCA) Construction Algorithm for Decision Interval Concept Lattice*
INPUT: Decision Context $(U, C \times D, R)$
OUTPUT: Decision Interval Concept Lattice L_α^β
(1) Suppose α, β, determine the intension of decision interval concept, then get the initial node set G.
The intension of decision interval concept is determined by conditional attributes and decision attributes. If conditional attributes are $A = \{a_1, a_2, \ldots a_m\}$, $B = \{b_1, b_2, \ldots b_n\}$ and so on, and decision attributes is $D = \{d_1, d_2, \ldots d_l\}$, then for decision attribute d_2, the set of intension is $\{a_i b_j \ldots d_2\}$, $i = 0, 1 \ldots m, j = 0, 1 \ldots n$, $m, n \ldots$ can be 0 in a different time. $|\cdot|$ is the number of elements, then the number of intension is $|a_i b_j \ldots|$, $i = 0, 1 \ldots m$, $j = 0, 1 \ldots n$, $m, n \ldots$ can be 0 in a different time.

(2) Get the upper bound extension M_i^{α} and lower bound extension M_i^{β}.

(3) Construct the lattice. For the initial node set G, determine the layer and the parent-child relation according to the relationship between precursor and successor.

2.4 Mining Algorithm for Decision Interval Rule

Algorithm: (*DIRMA*) *Mining Algorithm for Decision Interval Rule*

INPUT: Decision Interval Concept Lattice $L_{\alpha}^{\beta}(U, C \times D, R)$, parameters α, β

OUTPUT: Decision Interval Rule

(1) For parameters α, β, use the breadth-first traversal method to get the set of decision concept node, namely *Dcset*. The object "x" of every concept node in *Dcset* must meet the requirement of conditional attributes A in $[\alpha, \beta]$.

Let $L_{\alpha}^{\beta}(U, C \times D, R)$ be the set of decision interval concept lattice, and $A \subseteq C$, if $\exists y$, meets $y \in f(x)$, $y \in A$, and $\alpha \le |y|/|A| \le \beta$ $(0 \le \alpha \le \beta \le 1)$, then "$x$" is called that it meets the conditional attributes A at the degree in $[\alpha, \beta]$.

(2) For each concept node in *Dcset*, mining rules $r : A_{\alpha}^{\beta} \Rightarrow B$, to form the set of decision Rules, namely *Diset*.

Let every intension of concept node contains conditional attributes and decision attributes, so one node can extract one rule, the former in rule is conditional attributes and the latter is decision attributes. Repeat the above steps, the set of decision Rules *Diset* be formed.

(3) For every rule in *Diset*, calculating its roughness and attribute contribution respectively, and then judging the accuracy and reliability based on its size of roughness and attribute contribution., so that we can extract the more optimal rules. That is, removing undesirable rules, the set of final decision rule be formed, namely *Disset*.

3 Mining Model of Decisions Interval Rule in Rough Control

3.1 Model Design

The simplest decision rules in rule base is used to realize the goal of rough control, which actually queries the same or similar condition attributes of decision rules in the rule base, to be applied to rough control. Using the mining algorithm of decision interval concept lattice can get a set of decision interval rules in rough control, the mining model is shown in Fig. 1.

The most prominent advantages of rough control rules mining model based on decision interval concept lattice is that it guarantees the reliability of the rules, at the same time, realizes the optimality between the scale of rule base and the mining cost. The optimality depends on the interval parameter Settings, because of interval parameter determines the structure of constructed decision interval concept lattice, and then affects the number of interval association rules and its accuracy. So adopting the decision rule extraction method to extract rules based on decision interval concept

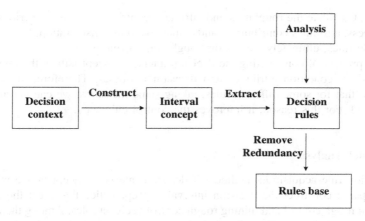

Fig. 1. Decision rule acquisition process

lattice, we can realize the optimization between mining cost and efficiency. Specific steps are as follows:

Step 1, Collect and preprocess the original data so as to get a decision table.

(1) According to the practical background in rough control, the observation and control quantity be considered as condition attributes and decision attributes respectively. Record the control strategy adopted by dispatch staff on representational state in rough control so as to form the original decision table;

(2) In most cases the data in industrial control is continuous. Using the method of decision interval concept lattice to mining control rules, we can discretize the continuous variables. With the aid of the background knowledge in industrial processes, we discrete continuous data, and mark each discrete interval with numbers;

(3) Using the method of rough set theory to reduce and combine data, after that we get the processed original decision table.

Step 2, Build the Boolean form of decision context table.

(1) According to the attributes condition showed by original decision table, which had got in above steps, we mark attributes and construct the set of attributes. The marking numbers are subscript of set elements. For example, an attribute is marked by A, the appeared marking number in the table are $2, 3, 4$, then the set of attribute is $A = \{a_2, a_3, a_4\}$;

(2) According to the corresponding attribute set, we build the Boolean form of decision context table.

Step 3, Suppose α, β, use the algorithm of Sect. 2.3 to construct the decision interval concept lattice, which is matched by practical background in rough control.

Step 4, Use the algorithm of Sect. 2.4 to extract decision interval rules in rough control.

Step 5, Calculate the roughness and attributes contribution. Taking various factors on roughness, attributes contribution, and actual cost into consideration. Removing the undesirable rules, ultimately we get the rough control rule base.

In the process of constructing the decision interval concept lattice, the intension is determined by condition attributes and decision attributes. Therefore, the extracted rules mean that for some different rules, taking same measures, we can get all results we expected. For this reason, it improves the control efficiency.

3.2 Model Analysis

The decision rule acquisition methods of decision interval concept lattice is divided into two parts: construct the decision interval concept lattice that meet the practical meaning in rough control and mining rough control decision rules. Among them, let the original data table be transformed into the Boolean context table is the key step. So we should discrete those continuous attributes. Such as a condition attribute marked by A, discrete it, it will become $a_1, a_2, \ldots, a_k$. Each object has attributes at most one. Therefore, the $DICLCA$ is different from the reference [7], and it can form $|a_i b_j \ldots|$ decision interval concepts.

The decision interval concept extension contains a certain amount or proportion of objects set in intension, so the extracted rules become more targeted. Removing undesirable rules, consequently the reliability of the rules and control efficiency have been improved. In the process of traditional rough control rules extraction, in some respects, the establishment of decision table, the discretization of continuous, and the consistency of decision rules, its time complexity becomes more larger, and mining cost highly. The model guarantees the reliability of the rules, at the same time, improves the mining precision of rough control rule and control efficiency.

4 Case Study

Reservoir is a complex system. In this example, under the premise of the function is mainly electricity, the first hydropower station is a large hydropower station that power generation and flood play important roles simultaneously. An upstream hydropower station, that is, the second hydropower station have great influence on the first hydropower station. Mining decision rules of reservoir dispatching, specific steps are as follows:

(1) Collect and preprocess the original data

Analysis specific conditions of the first hydropower station. Considering observation about the main factors influencing the hydropower station scheduling as condition attribute set, the control quantity as decision attribute set. Discrete those continuous attribute. The specific process is as follows:

Condition attributes: a– water discharge by the second hydropower station
b– the natural runoff condition
Decision attributes: d– daily average electricity by hydropower station

(a) The range of discretized water discharge by the second hydropower station (m^3/s)

$$1 - [100 - 200) \quad 2 - [200 - 300) \quad 3 - [300 - 400) \quad 4 - [400 - 500)$$

(b) The range of discretized natural runoff condition

$$1-\text{dry} \quad 2-\text{moderate} \quad 3-\text{flood}$$

(d) The range of discretized daily average electricity (kWh/d)

$$1 - [250 - 300) \quad 2 - [300 - 350) \quad 3 - [350 - 400) \quad 4 - [400 - 450)$$

Record the measure that is taken by scheduling persons from January to June, merge the same decision, finally the original data table is formed. As is shown in Table 1.

Table 1. Original data table

U	a	b	d
1	3	1	2
2	2	1	2
3	2	2	2
4	3	2	3
5	2	3	3
6	1	3	3

(2) Construct the decision formal context table of reservoir scheduling. According to the data shown in Table 1, mark discretized attribute with numbers and build attribute set. It can get $A = \{a_2, a_3, a_4\}$ $B = \{b_1, b_2, b_3\}$ $D = \{d_3, d_3, d_4\}$. The decision form table is shown in Table 2.

Table 2. Decision form table

U	a_1	a_2	a_3	b_1	b_2	b_3	d_2	d_3
1	0	0	1	1	0	0	1	0
2	0	1	0	1	0	0	1	0
3	0	1	0	0	1	0	1	0
4	0	0	1	0	1	0	0	1
5	0	1	0	0	0	1	0	1
6	1	0	0	0	0	1	0	1

(3) Construct the decision interval concept lattice in reservoir scheduling.

For the decision context in Table 2, supposing α, β, and then get the concept intension, the upper bound M_i^{α} and lower bound extension M_i^{β}. It is concluded that

Table 3. Decision interval concept

Concept	M^α	M^β	Intension	Concept	M^α	M^β	Intension
F_1	ϕ	ϕ	ϕ	F_{17}	$\{456\}$	$\{6\}$	$a_1 d_3$
F_2	$\{1236\}$	ϕ	$a_1 d_2$	F_{18}	$\{23456\}$	$\{5\}$	$a_2 d_3$
F_3	$\{1235\}$	$\{2\}$	$a_2 d_2$	F_{19}	$\{1456\}$	$\{4\}$	$a_3 d_3$
F_4	$\{6789\}$	$\{1\}$	$a_3 d_2$	F_{20}	$\{12456\}$	ϕ	$b_1 d_3$
F_5	$\{123\}$	$\{12\}$	$b_1 d_2$	F_{21}	$\{3456\}$	$\{4\}$	$b_2 d_3$
$\vdots$	$\vdots$	$\vdots$	$\vdots$	$\vdots$	$\vdots$	$\vdots$	$\vdots$
F_{13}	$\{235\}$	ϕ	$a_2 b_3 d_2$	F_{29}	$\{14\}$	ϕ	$a_3 b_1 d_3$
F_{14}	$\{12\}$	$\{1\}$	$a_3 b_1 d_2$	F_{30}	$\{4\}$	$\{4\}$	$a_3 b_2 d_3$
F_{15}	$\{134\}$	ϕ	$a_3 b_2 d_2$	F_{31}	$\{456\}$	ϕ	$a_3 b_3 d_3$
F_{16}	$\{1\}$	ϕ	$a_3 b_3 d_2$	F_0	ϕ	ϕ	Ω

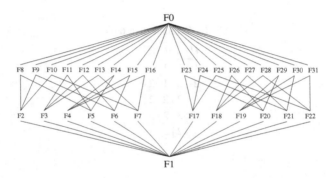

Fig. 2. Decision interval concept lattice based on reservoir scheduling

decision interval concept is shown in Table 3. The decision interval concept lattice is shown in Fig. 2

(4) Extract control rules in reservoir scheduling.

For parameters α, β, use the breadth-first traversal method to get the set of decision concept node. Because of the feature that cardinal number of condition attribute must meet $\alpha \leq |y|/|A| \leq \beta$, so we can get the set of decision concept node $Dcset = \{F_3 F_4 F_5 F_6 F_{11} F_{12} F_{14} F_{17} F_{18} F_{19} F_{21} F_{22} F_{25} F_{28} F_{30}\}$. Calculate its roughness and attribute contribution, and all the association rules of the roughness and attribute contribution as is shown in Table 4.

(5) The construction and optimization of rule base in reservoir scheduling

According to the roughness and contribution, we remove those undesirable rules. For example, comparing the rules $a_2 \Rightarrow d_3$ with $a_3 \Rightarrow d_3$, under the premise that they have the same attribute contribution, but the former roughness is lower than the latter, thus, the latter is a desirable rule. Repeat steps, the final set of rules $Disset$ can be

Table 4. Decision interval rules

Rules	Roughness	Attribute contribution	Rules	Roughness	Attribute contribution
$a_2 \Rightarrow d_2$	25 %	$(a_2, \delta) = 100\%$	$a_2 \Rightarrow d_3$	20 %	$(a_2, \delta) = 100\%$
$a_2 b_2 \Rightarrow d_2$	25 %	$(a_3, \delta) = 100\%$	$a_3 \Rightarrow d_3$	25 %	$(a_3, \delta) = 100\%$
$b_1 \Rightarrow d_2$	67 %	$(b_1, \delta) = 100\%$	$b_2 \Rightarrow d_3$	25 %	$(b_2, \delta) = 100\%$
$b_2 \Rightarrow d_2$	67 %	$(b_2, \delta) = 100\%$	$b_3 \Rightarrow d_3$	34 %	$(b_3, \delta) = 100\%$
$a_2 b_1 \Rightarrow d_2$	34 %	$(a_2, \delta) = 75\%$	$a_1 b_3 \Rightarrow d_3$	50 %	$(a_1, \delta) = 50\%$
		$(b_1, \delta) = 50\%$			$(b_3, \delta) = 100\%$
$a_2 b_2 \Rightarrow d_2$	50 %	$(a_2, \delta) = 75\%$	$a_2 b_2 \Rightarrow d_2$	50 %	$(a_2, \delta) = 75\%$
		$(b_2, \delta) = 50\%$			$(b_2, \delta) = 50\%$
$a_3 b_1 \Rightarrow d_2$	50 %	$(a_3, \delta) = 67\%$	$a_3 b_2 \Rightarrow d_3$	100 %	$(a_3, \delta) = 67\%$
		$(b_1, \delta) = 67\%$			$(b_2, \delta) = 67\%$
$a_1 \Rightarrow d_3$	34 %	$(a_1, \delta) = 100\%$			

obtained. Under the practical cost, remove redundant rules secondly. The final reservoir scheduling rule base in rough control is

$$\{a_2 \Rightarrow d_2, b_1 \Rightarrow d_2, a_2 b_1 \Rightarrow d_2, a_3 b_1 \Rightarrow d_2, a_1 \Rightarrow d_3, b_2 \Rightarrow d_3, a_1 b_3 \Rightarrow d_3, a_2 b_3 \Rightarrow d_3, a_3 b_2 \Rightarrow d_3\}$$

5 Conclusions

In this paper, we put forward the rule mining model based on the theory of the decision interval concept lattice in rough control. From the initial data preprocessing to the decision interval concept lattice construction in rough control, and the decision control rules extraction, the model has carried on the detailed design. Model analysis summarizes the two main parts of the model, constructing decision interval concept lattice and decision rule mining. Compared with the traditional methods in rough control, model highlight the reliability and effectiveness. It is verified by a case that the model improves its feasibility.

Acknowledgements. This work is partially supported by the National Natural Science Foundation of China (Grant No. 61370168, 61472340), Conditional Construction Project of Hebei Province Technology Hall (Grant No. 14960112D). The authors also gratefully acknowledge the helpful comments and suggestions of the reviewers, which have improved the presentation.

References

1. Aixian, P., Yun, G.: Problems in rough control. J. Ocean Univ. China **39**(6), 1315–1320 (2009)
2. Dong, W., Wang, J., Gu, S.: The rule acquisition algorithm based on theory of variable precision rough set. Comput. Eng. **14**(1), 73–75 (2007)

3. Huang, J.: Attribute reduction and rule acquisition based on the rough concept lattice. Software **32**(10), 16–23 (2011)
4. Peters, J.F., Skowron, A., Suraj, Z.: An application of rough set methods in control design. Fundamenta Informaticae **43**(1), 269–290 (2011)
5. Wang, H., Rong, Y., Wang, T.: Rough control for hot rolled laminar cooling. **In:** International Conference on Industrial Mechatronics and Automation (2010)
6. Liu, B., Zhang, C.: New concept lattice structure—interval concept lattice. Comput. Sci. **39**(8), 273–277 (2012)
7. Zhang, C., Wang, L.: The incremental generation algorithm of interval concept lattice based on attribute power set. Appl. Res. Comput. **31**(3), 731–734 (2014)

Research on Association Analysis Technology of Network User Accounts

Yunpeng Guo, Yan Liu$^{(\boxtimes)}$, and Junyong Luo

Zhengzhou Science and Technology Institute, Zhengzhou, China
wudigyp@126.com, ms_liuyan@aliyun.com, Luojunyong@china.com.cn

Abstract. As large numbers of user information on various network services are becoming valuable resources for social computing, how to associate user information scattered in a variety of services turns to one of the key issues of user identity mining. First this paper outlines the difficulties of dealing with account association, then describes the basis of account association method. After that, this paper sums up account association methods mainly from five aspects which are user naming conventions, user profiles, user writing styles, user online behavior, and user community relationship. Finally, this paper points out the development direction and prospects of account association.

Keywords: User identity · Mining association analysis · User account

1 Introduction

Currently, with billions of internet users, the data generated is growing exponentially, and the consequent massive data brings both opportunities and challenges to user identity mining [1]. Because of no association between various network services, the user information is dispersed as fragmentation and unable to be shared in different communities, leading to a great challenge to fetch the intact user information. Jing et al. [2] described that, as to an individual user, with the information fragmentally scattered in various communities, the information provided by a single community is incomplete. Therefore, how to solve the problem of accounts association becomes one of the key issues of user identity mining.

Meanwhile, account association is attracting more and more peoples interest. First, due to the limitations of user information provided by a single community, the commercial organizations are more willing to associate and aggregate the users activity information among multiple community to get full user data. Secondly, the ISP also has interests to find and identify multiple accounts created by the same user in the same services. According to the terms of the network service, the same user in the same network services should only register an account, however, most users tend to register multiple accounts. Therefore, account association has become one research object of the multiple institutions.

W. Chen et al. (Eds.): BDTA 2015, CCIS 590, pp. 93–101, 2016.
DOI: 10.1007/978-981-10-0457-5_10

However, for network anonymity, most ISPs allow users to choose a name instead of their real identities freely. And, as a result of using single sign-on between different network services, the same user can use a single username to login to different network service, different usernames to login to the same network services, or different usernames to login to different network services. It means that the same username may involve more than one user and the same user may also have more than one username as well. Therefore, the correlation analysis of user account is increasingly difficult.

According to the principle of account correlation analysis, This article summarized five methods of account association analysis: (1) the analysis method based on user naming conventions; (2) the analysis method based on user profiles; (3) the analysis method based on user writing styles; (4) the analysis method based on user online behavior; (5) the analysis method based on user community relationship.

In this paper, the username refers to the account name presented in the regulations for the internet user account name released by the National Internet Information Office. In this provision, internet user account names refers to the account name registered or used by the organizations or individuals in the blog, microblog, instant communication tools, bbs, post bars, post comments and other internet information services.

2   Basis of Association Analysis on Network User Accounts

2.1   User Naming Conventions

Personality is different from others, manifested in different circumstances, relatively stable, as the sum of the psychological characteristics affecting the person's explicit and implicit behavior patterns, while habits are formed gradually and uneasily changed behaviors. People tend to use the same or similar name because they can reduce their memory load on the one hand, and on the other hand can allow other users to easily identify themselves. When researchers analyzed the properties of the user, the username is found to be more special, because the user usually registered for certain network services with personal habits, such as adding prefixes, suffixes, abbreviating some or all, inserting letters, numbers, characters, etc. on the basis of the original username to generate new usernames. These personalized habits make the usernames owned by the same user between in different network services be of some similarities.

2.2   User Profiles

Dr. Liu, came from Harbin Institute of Technology, noted that users have average 3.92 communities and 2.44 usernames by analyzing About.me data. At the same time, Irani et al. [3] also pointed out that each user exposes 4.3 user attribute fields averagely in a community, and this average will be increased to 8.25 for those registering eight or more communities.

2.3   User Writing Styles

In a strict sense of word, the writing style refers to the writers unique artistic personality gradually formed and reflected on the overall creation, while writing style in this article refers to the feature set, which is extracted from text information (usually including e-mail and "friends group", "mood" published in social networking, etc.) left by the same network user in his network activity, can be used to represent users personal writing characteristics.

2.4   User Online Behavior

Each network user is a natural person, a natural person's life trajectory exhibits a certain regularity, which influences and determines the trajectory of his Internet surfing also presents a certain regularity primarily reflected in his location information.

Location information can be divided into geographical location information and network location information. With the improvement on the wireless network and the popularity of intelligent terminas, the mobile phone replaces PC as personal information centers, meanwhile, Mobile Internet Operators, Online Community Services, APP applications have added the personal location information and produced a large amount of location data. These data played a certain role in researchers to percept and identify the user behavior, and people's geographical location information generally exhibits some regularities in the time sequence. By analyzing the location information regularity of different services and different applications, it can help researchers associate the user accounts utilized in different services and applications.

Secondly, the personal data acquisition nowadays greatly surpasses the past either in terms of scales, variety and accurate degree [4]. Researchers also found some regular stuff in the collection, sorting, analysis of use network activity. These laws may allow researchers to better understand the user, to better service for the user. Entitle with "The Personal Analytics of My Life" blog [5] and published by Stephen Wolfra, an article shows his results of personal data collection and disposition in the past 20 years.

2.5   User Community Relationship

In addition to behavioral data, the user's social network can also provide a wealth of information. Since real-life human social relationships get extended, developed and maintained in the virtual world [6,7], this allows that the same users relationship diagram in multiple social networks have a high degree of similarity. Facebook, Twitter, Sina Microblog, Renren, etc. are typical examples to extension of users real-world social relations.

3   Method of Association Analysis on Network User Accounts

3.1   The Analysis Method Based on User Naming Conventions

Since there's a certain naming habits of users naming a new account name, the researchers beware of the point have studied the account association from the perspective of the username similarity measure.

Zafarani et al. [8] first put forward the issue of associating username among several communities into the same ones, but the accuracy is only 66 %, due to the initial solution of building a model of search based on Web. Then, Zafarani et al. [9] further proposes a new model - MOBIUS (Modeling Behavior for Identifying Users across Sites), used for discovering and identifying the same user identity mapping issue between different sites. From the habits of the user to select a username, it finds the redundant information between the usernames, makes the feature model of redundant information through analyzing selected ten featured characteristics, and perform efficient user identification with machine learning method adopted. At the end of the essay, the author gave two conclusions: First, numbers [0–9] are on average ranked higher than English alphabet letters [a–z], showing that numbers in usernames help better identify individuals; Secondly, non-English alphabet letters or special characters are among the features across sites have higher odds-rations on average. In terms of accuracy, the value was enhanced from the original 66 % [8] to 91.38 % [9].

Different from username modeling [9], Perito et al. [10] implemented mathematical method of Markov-Chains to estimate the probability of the username for the uniqueness analysis of username, and adopted Markov-Chains and TF-IDF respectively to link different username string. Finally the author points out that the probability of two usernames linking to the same user has a high dependence on the username string matching entropy.

However, the analysis method based on the user naming habits highly relies on the user's personalized custom, its accuracy will greatly reduce in case of dealing an irregular set of username.

3.2   The Analysis Method Based on User Profiles

In all network services, user profile can be seen as a N-dimensional vector, of which each property, such as nicknames, avatars, gender, address, telephone, preferences, is seen as a one dimension [11]. Use the information field, a user attribute vector is generated to solve the matching problem of two accounts through similarity measure. In general, this method relies on the preset threshold, but the choice of the threshold is very difficult [12].

Vosecky et al. [11] putted forward a vector comparison based algorithm (Vector-Based Comparison Algorithm), in which with multiple attributes of User Profile was denoted as vector, and assigning weights according to the level of the attributes depicting the user, whether the two accounts belonging to the same user are determined by computing the similarity between two vectors containing weights.

Iofciu et al. [13] also gave a similarity linear combination method of the username edit distance and user attributes distribution to infer whether two accounts belong to the same user. It is dependent on the two basic assumptions: the username used by the same user in different communities are same or similar; the attribute value given by the same user in two network service system is also similar.

From similarity measure of the attributes as well, Malhotra et al. [14] performed similarity calculation of the features extracted from user profiles using Jaro Winkler similarity and other methods, and then solve the account association problem with machine learning.

As the analysis method based on user profiles, however, is restricted by the attribute value submitted by the user, the method will not be applicable in case of a user without habit of consummating the user profile, or different user profiles submitted by the same user between different network services.

3.3   The Analysis Method Based on User Writing Styles

In the staff engaging into the analysis and research of users writing style, Iqbal's team has made outstanding contributions. Through analyzing each text message content of the users, Iqbal et al. [15,16] extracted Lexical Features, Syntactic Features, Structural Features, Domain-Specific Features and other characteristic information, then generated feature vectors. The banks of feature vectors are banks of transactional data for frequent pattern mining, while a number of transaction data obtained by analyzing the same users multiple mail messages forms a frequent pattern mining transaction set. For the same users transaction set, the frequent writing features, namely writing style, is eventually acquired using the basic frequent pattern mining algorithms, such as Apriori, FP-Growth. Writing style eventually associates the same user of multiple mail service systems.

At present, due to the parallel computing developed by leaps and bounds, the parallel algorithm for mining frequent patterns [17,18], can be utilized to improve the method of Iqbal et al. for enhanced efficiency of analysis.

Although the user's writing style analysis can only be applied to the network service account association with text information released, it is one indispensable method for the network user accounts association analysis.

3.4   The Analysis Method Based on User Online Behavior

Compared to static user attributes, the user behavior is dynamic, in formation of a timeline sequence.

3.4.1   The Analysis Method Based on Location Information

Chi et al. [19] enunciated three questions: What the location-based social perception and its computing framework is; What the relation of human behavior and the positional sociality is; What methods for identifying and data minding are generally applied in the practical analysis and position system applications especially confronted with the large data analysis of location. Liu et al. [20] also analyzed regularities of groups from the perspective of location information, with certain reference significance for the correlation analysis of accounts.

Based on the trajectory weighted network diagrams, Mao et al. [21] proposed a new algorithm for the analysis and recognition of the user behavior, and this algorithm looks for user behavior regularities through the analysis of the users location data produced in the process of mobility. In order to extract characteristic behavior and key locations included in people's daily activities, they first modeled the trajectories for position based on weighting network diagrams, and then use graph algorithms to diving the positional track and acquired several location centers, which means key positions. On this basis, they performed feature subspace analysis of the key locations to identify behavior.

With mobility of the geographic location, the network location of the internet users in network services will have corresponded change, thus the accounts association can be done by matching the statistical results of the network position sequence of two accounts.

3.4.2   The Analysis Method Based on Historical Traces

Yuan et al. [22] put forward a computational framework LifeSpec to acquire data and information published by users through grabbing the network information. With the similarity analysis of user profile and historical traces together, they found that the users all have more than two usernames through the analysis of more than 50 million pieces of behavioral data from captured more than 1.4 million users. Goga et al. [23] also made use of content published by the users for correlation analysis.

Researchers [24, 25] also pointed out that the statistically rare attribute values can help us to better confirm the correlation between the two accounts, such as a less popular movie viewing records, a similar "friends loop", both with great probability to associate the accounts in two databases.

3.5   The Analysis Method Based on User Community Relationship

In many research institutions and industries, social graph analysis has already successfully applied to antifraud, influence analysis, sentiment monitoring, market segmentation, engagement optimization, experience optimization, and other applications where complex behavioral patterns must be rapidly identified. And researchers Backstrom et al. [26] from Cornell University for the first time clearly proposed the social network topology based re-identifying issue, that despite node attributes removed, the attacker can still infer to the identity of the node through the network topology and node-related characteristics.

Narayanan et al. propose a statistical methods [24] capable of being applied to anonymize correlation analysis across data sets, and later, they again put forward an anonymous method [27] based on social relations graphs among multiple social network firstly. Labitzke et al. [28] also found that the use of social network structures diagram information is capable of successfully identifying the anonymous network. The reason is that the user's social network relationship is an extension and development of his real-world social relationship, thus the structure diagrams of the same user in different online social network are similar.

Jing et al. [4] mentioned that, the user matching relations between the two networks can be understood by matching two social network graph structures. And Jing et al. [29] in the anonymity of the chart data match hosted by the international conference on ACM web search and data mining acquired better results and achievements by performing anonymization through method based on subgraph isomorphism and the similarity of the node.

4   Future Direction

As network user accounts association technology is the basis of the user information aggregation across the network service, however, the diversity and heterogeneity of user information about different network services has brought about great challenge to account association [2]. Existing correlation analysis method usually just considers a problem of a single aspect, but each method has limitations. In the future, fusion of multiple analysis methods will be the tendency to improve the accuracy of the accounts association. By merging two methods of user's online behavior and writing style, Goga [12] shows how to match the accounts through choosing three types of the features of building activity fingerprints which are presented on many social networks: the location, timing, and language characteristics. In addition, former researches took more the accuracy of correlation results into account. However, it's often inefficient. The arrival of the era of big data brings a landmark change to correlation analysis. Faced with huge amounts of data, how to optimize the algorithm, how to improve the parallel degree of algorithms, will be the development direction of improving associated account efficiency.

Meanwhile, correlation analysis of account also opens up new research directions, such as, understanding the user migration patterns among social networks [30]. And we analyses the correlation between account from the perspective of multidimensional space and integrate into information resources from different network services. It helps to collect user's interests and personal habits help, also contributes to improve the quality of the recommendation system. In addition, due to convenient and rapid dissemination of the network information, the network has become an important means to criminals engaged in criminal activities, and related criminals are increasing day by day. Since many internet users personal and property safety is seriously threatened, results of the correlation analysis can apply to cyber-crime investigation and evidence collection process.

Acknowledgments. This work was supported by the National Natural Science Foundation of China (61309007).

References

1. Xiaobin, H.: Network Information Mining. Publishing House of Electronics Industry, Beijing (2005)
2. Jing, L., Ting, L.: A survey of linking users across multiple online communities. Intell. Comput. Appl. **4**(4), 39–42 (2014)
3. Irani, D., Webb, S., Li, K., et al.: Large online social footprints-an emerging threat. In: International Conference on Computational Science and Engineering, vol. 3, pp. 271–276. IEEE Press, Vancouver (2009)
4. Jing, Y., Xing, X.: Users understand based on the large-scale behavior data. Commun. China Comput. Fed. **5**(4), 14–18 (2014)
5. Wolfram, S.: The personal analytics of my life. Stephen Wolfram Blog (2012)
6. Pengzhi, F.: From chaos to symbiosis - think about the relationship of the nature between the virtual world and the real world. Stud. Dialectics Nat. **18**(7), 44–47 (2002)
7. McLuhan, M.: Understanding Media: The Extensions of Man. The Commercial Press, Beijing (2000)
8. Zafarani, R., Liu, H.: Connecting Corresponding Identities Across Communities. ICWSM, San Jose (2009)
9. Zafarani, R., Liu, H.: Connecting users across social media sites: a behavioral-modeling ap-proach. In: Proceedings of the 19th ACM SIGKDD International Conference on Knowledge Discovery and Data Mining, pp. 41–49. ACM, Chicago (2013)
10. Perito, D., Castelluccia, C., Kaafar, M.A., Manils, P.: How unique and traceable are usernames? In: Fischer-Hübner, S., Hopper, N. (eds.) PETS 2011. LNCS, vol. 6794, pp. 1–17. Springer, Heidelberg (2011)
11. Vosecky, J., Hong, D., Shen, V.Y.: User identification across multiple social networks. In: International Conference on Networked Digital Technologies, pp. 360–365, Ostrava (2009)
12. Goga O.: Matching User Accounts Across Online Social Networks: Methods and Applications, Paris (2014)
13. Iofciu, T., Fankhauser, P., Abel, F., et al.: Identifying Users Across Social Tagging Systems. ICWSM, Barcelona (2011)
14. Malhotra, A., Totti, L., Meira Jr., W., et al.: Studying user footprints in different online social networks. In: International Conference on Advances in Social Networks Analysis and Mining, pp. 1065–1070. IEEE Computer Society, Istanbul (2012)
15. Iqbal, F., Binsalleeh, H., Fung, B., et al.: Mining writeprints from anonymous e-mails for forensic investigation. Digit. Invest. **7**(1), 56–64 (2010)
16. Iqbal, F., Binsalleeh, H., Fung, B., et al.: A unified data mining solution for authorship analysis in anonymous textual communications. Inform. Sci. **231**, 98–112 (2013)
17. Leung, C.K.-S., Hayduk, Y.: Mining frequent patterns from uncertain data with MapReduce for big data analytics. In: Meng, W., Feng, L., Bressan, S., Winiwarter, W., Song, W. (eds.) DASFAA 2013, Part I. LNCS, vol. 7825, pp. 440–455. Springer, Heidelberg (2013)
18. Lin, M.Y., Lee, P.Y., Hsueh, S.C.: Apriori-based frequent itemset mining algorithms on MapReduce. In: International Conference on Ubiquitous Information Management and Communication, p. 76. ACM, Kuala Lumpur (2012)

19. Chi, G., Yuan, F., Jingnan, L., et al.: Social awareness computation methods for location based services. J. Comput. Res. Devel. **50**(12), 2531–2542 (2013)

20. Fen, L., Hongbo, T., Guodong, G.: Group discovery method in mobile communication network based on location information. Appl. Res. Comput. **30**(5), 1471–1474 (2013)

21. Mao, X., Qi, G., Li, S., et al.: Location trace based weighted network analysis algorithm for user behavior analysis and recognition. In: The 6th Joint Conference on Harmonlous Human Machine Environment, The 5th Pervasive Computing Conference (2010)

22. Yuan, N.J., Zhang, F., Lian, D., et al.: We know how you live: exploring the spectrum of urban lifestyles. In: Conference on Online Social Networks, pp. 3–14. ACM, Boston (2013)

23. Goga, O., Lei, H., Parthasarathi, S.H.K., et al.: Exploiting innocuous activity for correlating users across sites. In: International conference on World Wide Web, pp. 447–458, Rio de Janeiro, Brazil (2013)

24. Narayanan, A., Shmatikov, V.: Robust de-anonymization of large sparse datasets. In: IEEE Symposium on Security and Privacy, pp. 111–125. IEEE, Oakland (2008)

25. Frankowski, D., Cosley, D., Sen, S., et al.: You are what you say: privacy risks of public mentions. In: International ACM SIGIR Conference on Research and Development in Information Retrieval, pp. 565–572. ACM, Seattle (2006)

26. Backstrom, L., Dwork, C., Kleinberg, J.: Wherefore art thou r3579x?: anonymized social networks, hidden patterns, and structural steganography. In: International Conference on World Wide Web, pp. 181–190. ACM, Banff (2007)

27. Narayanan, A., Shmatikov, V.: De-anonymizing social networks. In: IEEE Symposium on Security and Privacy, pp. 173–187. IEEE Press, Oakland (2009)

28. Labitzke, S., Taranu, I., Hartenstein, H.: What your friends tell others about you: low cost linkability of social network profiles. In: International ACM Workshop on Social Network Mining and Analysis. AAAI, San Diego (2011)

29. ACM WSDM 2013 data challenge (de-anonymization track), Rome, Italy (2013)

30. Kumar, S., Zafarani, R., Liu, H.: Understanding User Migration Patterns in Social Media. AAAI, San Francisco (2011)

A Distributed Query Method for RDF Data on Spark

Minru Guo and Jingbin Wang[✉]

College of Mathematics and Computer Science,
Fuzhou University, Fuzhou 350108, China
`guo_mr@foxmail.com`,
`wjbcc@263.net`

Abstract. With the upcoming data deluge of semantic data, the fast growth of RDF data has brought significant challenges in query. A new distributed RDF query algorithm RQCCP (RDF data Query combined with Classes Correlations with Property) on Spark platform is proposed to solve the problem of low efficiency for RDF data query. It splits and stores RDF data by the class of Subject, Predicate and the class of Object, simultaneously building index file of classes correlations with property; the index is applied to narrow the scope of input for query, filtering out irrelevant triples in advance and intermediate results of query cached in memory as resilient distributed dataset to reduce disk and network I/O. The results of experiments conducted on large-scale RDF datasets show that RQCCP has high query performance.

Keywords: Distributed · Spark · RDF · Index · Query

1 Introduction

RDF (Resource Description Framework) is a framework proposed by WWW to describe the information of World Wide Web, which provides information description specification for various applications on the Web [1]. With the rapid development of semantic web, the volume of RDF data grows fast. How to manage large-scale RDF data efficiently has become a crucial problem. In order to process large-scale RDF data, many RDF systems have been proposed, i.e., Hexastore [2], SW-store [3], RDF-3X [4, 5], etc. They store RDF data in relational tables and process queries using relational operators. The main problem of relation-based RDF stores is that they need too many join operations for evaluating queries. To address this problem, many techniques have been brought forward, i.e., the clustered property table, vertical partitioning, multiple indexing. But these centralized systems have poor scalability.

With the development of cloud computing technology, there are some researches of RDF data storage and management based on Hadoop platform, such as HadoopRDF [6], H2RDF [7], Liu L [8], IMSQ [9], etc. Most of them convert queries into MapReduce jobs whose intermediate results are persisted to disk, thus result in lots of disk and network I/O.

© Springer Science+Business Media Singapore 2016
W. Chen et al. (Eds.): BDTA 2015, CCIS 590, pp. 102–115, 2016.
DOI: 10.1007/978-981-10-0457-5_11

A new distributed RDF query algorithm RQCCP on Spark platform is proposed in this paper. It splits and stores RDF data into HDFS by the class of Subject, Predicate and the class of Object, simultaneously buildings index file of classes correlations with property and Instance-Class mapping file; the index is applied to narrow the scope of input for query, filtering out irrelevant triples in advance, intermediate results of query cached in memory to reduce disk and network I/O and greedy policy is adopt to join the intermediate results. The results of experiments conducted on large-scale RDF datasets show that our method has high query performance.

The rest of the paper is organized as follows. In Sect. 2, we introduce RDF, Spark and some necessary definition. In Sect. 3, we present the proposed RQCCP in detail, including data splitting, storage, indexing, query processing, etc. The experimental analysis is shown in Sect. 4 and our conclusions in Sect. 5.

2 Preliminaries

2.1 RDF and SPARQL

RDF is based on the idea of identifying resources using URIs (Uniform Resource Identifiers), and describing resources in terms of simple properties and property values. RDF data can be represented as triples of the form (subject, predicate, object). The interpretation of a triple is that <subject> has a property <predicate> whose value is <object>. The <predicate> is a property of the resource, and the <object> is the value of the property for the resource.

Definition 1 (RDF Triple). Assume the existence of two pairwise disjoint sets: a set of URIs U and a set of literals L. A triple $t \in U \times U \times (U \cup L)$ is called an RDF triple.

Definition 2 (Triple Pattern). Assume the existence of three pairwise disjoint sets: a set of URIs U, a set of literals L and a set of variables VAR. A triple $tp \in (U \cup VAR) \times U \times (U \cup L \cup VAR)$ is called a triple pattern. By the definition of the RDF triple, we know RDF triple is special triple pattern, i.e. RDF triple $\subseteq$ Triple pattern.

Definition 3 (Instance). We define an instance as the subject or object which is a non-variable in triple pattern, i.e. $\forall (S_i, P_j, O_k)(1 \leq i, j, k \leq n) \in$ Triple pattern, where n is the number of triple pattern and $S_i, O_k \notin VAR$. Then $\forall v \in \{S_i, O_k\}, v \in$ Instance.

Definition 4 (Class). Given a $v \in$ Instance, the class of v is denoted by $v.Class$. If exist triple pattern whose predicate is rdf:type, then the class of subject in the triple pattern is object in the triple pattern, i.e., $(S, P, O) \in$ Triple pattern $(P \in \{rdf : type\})$, by the definition of Instance, we know $S, O \in$ Instance, so $S.Class = O$.

SPARQL [10] was recommended by W3C as the standard query language for RDF data. A SPARQL query is a set of triple patterns. The evaluation of a SPARQL query consists of finding all possible variable bindings that satisfy the given query patterns.

2.2 Spark and RDD

Apache Spark is a fast and general engine for big data processing [11]. Unlike Hadoop, Spark holds intermediate results in memory rather than writing them to disk. It also provides concise APIs for writing programs fast.

The main abstraction Spark provides is a RDD (Resilient Distributed Dataset), which is a collection of elements partitioned across the nodes of the cluster that can be operated on in parallel. Any data is represented as RDDs in Spark. RDDs achieve fault tolerance through a notion of lineage: if a partition of an RDD is lost, the RDD has enough information about how it was derived from other RDDs to be able to rebuild just that partition. RDDs support two types of operations: transformations, which create a new dataset from an existing one, and actions, which return a value to the driver program after running a computation on the dataset.

2.3 MemSQL

MemSQL is the real-time database for transactions and analytics with an in-memory, distributed, relational architecture [12]. MemSQL provides both in-memory row-based and on-disk column-based stores. It is wire-compliant with MySQL but faster than MySQL.

MemSQL 4 has integration with big data processing frameworks like Spark, HDFS, and Amazon S3, so any organization can get real-time insights from its data. MemSQL Spark Connector can be used to operationalize Spark and perform real-time insights on

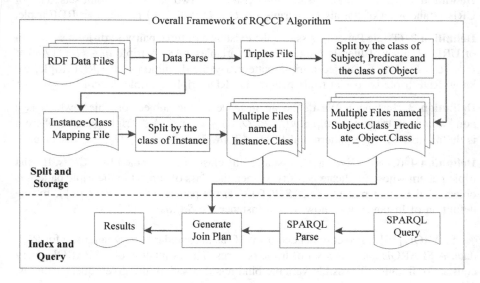

Fig. 1. Overall framework of RQCCP algorithm

the data stored in HDFS or S3 distributed data stores. By integrating Spark and MemSQL technologies, users can load data in parallel between MemSQL and Spark clusters, deploy models using an operational data store, and serve results of analytics via a SQL interface.

3   RQCCP Algorithm

In order to improve the efficiency of RDF data query, we propose RQCCP algorithm which includes two parts: data splitting and storage; data indexing and query. The overall framework of RQCCP algorithm is shown in Fig. 1.

3.1   Split and Storage

The RDF data used in this paper is Lehigh University Benchmark (LUBM) [13]. The RDF data files are in XML format. In order to convert it into triple format, we use Jena [14] API to parse the data file. During parsing phase, the class of subject S.Class, predicate P and the class of object O.Class in RDF triples are obtained and written to data file. In the meantime, write each instance and its corresponding class into Instance-Class mapping file.

Then split two files on Spark platform and upload files after splitting to HDFS. First of all, the generateActualKey() function and generateFileNameForKeyValue() function in class MultipleTextOutputFormat should be overwritten. Multiple data files named S.Class_P_O.Class and Instance-Class mapping files named Instance.Class would be generated finally.

Algorithm 1 shows the overall process of RDF data Splitting and Storage. Line 6 obtains all subjects. The classes of subject are obtained in Lines 8–10. Line 11 writes subject and its corresponding class into Instance-Class mapping file. Then get predicate and object in RDF triple (Lines 14–15). Write < S.Class_P_Literal,S,O > into data file if object is Literal, else get the class of object then write < S.Class_P_O.Class,S,O > into Instance-Class mapping file (Lines 16–25). Line 32 loads lines from data file. We use map() function to split each line on comma, the first string after splitting is used as key, the rest is used as value, then save value to file in HDFS according to key (Lines 32–36). The splitting process of Instance-Class mapping file is the same as data file (Lines 37–41).

Algorithm 1. RDF data splitting and storage

Input: RDF Data Files
Output: Multiple RDF data files and instance-class mapping files

```
1:   files ← RDF Data Files
2:     for each file ∈ files do
3:        model ← ModelFactory.createDefaultModel()
4:        in ← FileManager.get().open(file.getPath)
5:        model.read(in, "")
6:        subs ← model.listSubjects()
7:        for each S ∈ subs do
8:           subPros ← S.listProperties(RDF.type)
9:             for each subPro ∈ subPros do
10:               S.Class ← subPro.getObject()
11:               Write <S,S.Class > to classfile
12:               stmtIter ← model.listStatements(S, null, null)
13:               for each stmt ∈ stmtIter do
14:                  P ← stmt.getPredicate()
15:                  O ← stmt.getObject()
16:                  isLiteral ← O.isLiteral
17:                  if isLiteral is false then
18:                     objPros ← O.listProperties(RDF.type)
19:                     for each objPro ∈ objPros do
20:                        O. Class← objPro.getObject()
21:                        Write <S.Class_P_O.Class,S,O> to datafile
22:                     end for
23:                  else
24:                     Write <S.Class_P_Literal,S,O> to datafile
25:                  end if
26:               end for
27:            end for
28:         end for
29:      end for
30:  conf ← new SparkConf()
31:  sc ← new SparkContext(conf)
32:  lines ← sc.textFile(datafile)
33:  splits ← lines.map(_.split(","))
34:  rdd.key ← splits[0]
35:  rdd.value ← splits[1]+","+splits[2]
36:  rdd.saveAsHadoopFile()
37:  lines ← sc.textFile(classfile).distinct
38:  splits ← lines.map(_.split(","))
39:  rdd.key ← splits[0]
40:  rdd.value ← splits[0]+","+splits[1]
41:  rdd.saveAsHadoopFile()
```

Multiple data files and Instance-Class mapping files would be generated when data splitting and storage finished.

The purpose of the data splitting and storage is to reduce the data input of the query task by filtering out irrelevant data using index.

3.2   Index and Query

RQCCP algorithm improves the RDF data query performance from three aspects: (1) reducing the data input; (2) reducing the disk I/O; (3) reducing the network overhead.

Determine the Input Files. There is not always exist correlation between two classes in RDF data. From file names of data files after splitting and storage, we can find class is only associated with a few classes through specific property. Therefore, the data file names can be used as an index for indexing query input file. The index is called the CCP (Classes Correlations with Property) index.

The process of determining the input files according to index is shown in Algorithm 2.

Algorithm 2. Determine the input files based on index

Input: SPARQL query clause and index file
Output: input file for query

Step1. If there is a query clause with single variable, goto Step2. Otherwise, goto Step3.

Step2. Obtain the subclasses of class to which variable belongs and predicate of query clause. Then obtain the scope of class to which non-variable instance belongs combined with CCP index. If only one class in the class scope, goto Step5. Otherwise, goto Step4.

Step3. Obtain the subclasses of class to which subject variable belongs, predicate and the subclasses of class to which object variable belongs, then combine with CCP index to determine the input files.

Step4. Union MemSQL tables whose name corresponding to the class scope of non-variable instance and query the class of non-variable instance.

Step5. File is determined by the subclasses of class to which variable belongs, predicate and the subclasses of class to which non-variable instance belongs.

Example 1. Take LUBM Q1 as an example.

```
SELECT ?X
WHERE
{ ?X rdf:type ub:GraduateStudent .                              //1
   ?X ub:takesCourse
       http://www.Department0.University0.edu/GraduateCourse0 }//2
```

?X is known to belong to GraduateStudent class from query clause 1 and has takesCourse property from query clause 2. Since GraduateStudent class has no subclasses, we can obtain GraduateStudent class is only associated with GraduateCourse class through takesCourse property by using pattern matching (GraduateStudent_-takesCourse_*) combined with CCP index. So non-variable instance (http://www. Department0.University0.edu/GraduateCourse0) must belong to GraduateCourse class. Thus the input of query clause 2 is GraduateStudent_takesCourse_GraduateCourse file.

We import Instance-Class mapping files from HDFS to MemSQL tables in order to get the class of non-variable instance in query clause quickly in the case that there is more than one class obtained by using pattern matching combined with CCP index. The advantages are: (1) Don't need to read the file from disk; (2) Data can be indexded quickly. First of all, jar file of MemSQL Spark connector should be added to classpath. Then import com.memsql.spark.connector._ package to program and import data to MemSQL tables using saveToMemsql() function. The structure of the table is shown as below:

```
create table class_i (
  instance varchar(128) primary key,
  class varchar(128)
);
```

Various class names in RDF data are used as table name. Every table has two fields: instance which is used as primary key and class. Thus index is created automatically in instance field. Then if there exists instance in query clause, we can obtain the class to which instance belongs quickly by union tables corresponding to class scope. Using MemSQL database to store Instance-Class mapping files is equivalent to building a distributed shared-memory mapping table.

Example 2. Take LUBM Q3 as an example.

> SELECT ?X
> WHERE
> { ?X rdf:type ub:Publication .   //1
>   ?X ub:publicationAuthor
>   http://www.Department0.University0.edu/AssistantProfessor0}//2

?X is known to belong to Publication class from query clause 1 and has publicationAuthor property from query clause 2. Since Publication class has no subclasses, we can obtain Publication class is associated with 7 classes, i.e., AssistantProfessor, AssociateProfessor, FullProfessor, GraduateStudent, Lecturer, ResearchAssistant and TeachingAssistant through publicationAuthor property by using pattern matching (Publication_publicationAuthor_*) combined with CCP index. So non-variable instance (http://www.Department0.University0.edu/AssistantProfessor0) must belong to one of these 7 classes. Union tables corresponding to these 7 classes, then obtain the class to which instance belongs quickly by using where clause to select rows where the instance is http://www.Department0.University0.edu/AssistantProfessor0. Thus the class of ?X is determined, the input file is Publication_publicationAuthor_AssistantProfessor. Since the input data is reduced greatly, on the one hand can reduce disk I/O, on the other hand can relieve memory pressure, the efficiency of RDF data query is improved.

Query Planning and Execution. In order to reduce the execution time of the query, the strategies adopted by RQCCP algorithm are as follows:

    I. Query clause with single variable whose predicate is not rdf:type has the highest priority.

   II. Query clause with multiple variables whose predicate is not rdf:type and which contains the variable of query clause in I has second priority.

 III. Query clause with multiple variables whose predicate is not rdf:type and which does not contain the variable of query clause in I has the lowest priority.

The process of RDF data query is shown in Algorithm 3.

Algorithm 3. RDF data query

Input: SPARQL query statement, data files
Output: SPARQL query results

Step1. Analyze the SPARQL query statement. If there is a query clause with single variable whose predicate is not rdf:type, goto Step2. Otherwise, goto Step3.

Step2. Call Algorithm 2 to obtain the input files, then goto Step4.

Step3. If there is only one query clause in query statement, goto Step9. Otherwise, goto Step10.

Step4. Load the input file into memory as RDD, then filter data which does not satisfy the non-variable instance to obtain the variable (denoted as *var*) result collection (denoted as *varSet*).

Step5. If there is a query clause with multiple variables whose predicate is not rdf:type, goto Step12. Otherwise, goto Step6.

Step6. In order to use map-side join, the *varSet* is distributed to each node of cluster.

Step7. Get query clauses containing *var*, call Algorithm 2 to obtain the input files and then fitler data.

Step8. If there is a query clause without *var* whose predicate is not rdf:type in the remaining query clauses, goto Step10. Otherwise, calculate the number of query clauses with *var*. If there is only one query clause, goto Step12. Otherwise, goto Step11.

Step9. Obtain the subclasses of class to which object variable belongs, then select instance field from MemSQL tables corresponding to the classes, goto Step12.

Step10. Get remaining query clauses whose predicate is not rdf:type, then call Algorithm 2 to obtain the input files and load the input files into memory.

Step11. Call Algorithm 4, then output the query results.

Step12. Output the query results.

Example 3. Take LUBM Q8 as an example.

> SELECT ?X, ?Y, ?Z
> WHERE
> { ?X rdf:type ub:Student .   //1
>   ?Y rdf:type ub:Department .   //2
>   ?X ub:memberOf ?Y .   //3
>   ?Y ub:subOrganizationOf <http://www.University0.edu> . //4
>   ?X ub:emailAddress ?Z}   //5

First of all, analyze query clause 4 which is with single variable and whose predicate is not rdf:type. ?Y belongs to Department class and has subOrganizationOf property. Next, we can obtain http://www.University0.edu belongs to University class combined with CCP index. Thus the input of query clause 4 is Department_subOrganizationOf_University file. All data files have two columns, one is subject, another is object. So we can obtain the intermediate result of ?Y (denoted as Y_set) by select data whose object is http://www.University0.edu. Then Y_set is distributed to each node of cluster. And then, analyze query clause 3 which is with variable ?Y and whose predicate is not rdf:type. Since Student class has two subclasses, i.e., GraduateStudent class and UndergraduateStudent class, the input files are GraduateStudent_memberOf_Department and UndergraduateStudent_memberOf_Department. Then obtain intermediate result of ?X and ?Y (denoted as X_Y) by filtering data whose object is not in Y_set. After that, analyze query clause 5. We can obtain GraduateStudent class and UndergraduateStudent are both associated with Literal class through emailAddress property combined with CCP index. So ?Z belongs to Literal class. Thus the input files of query clause 5 are GraduateStudent_emailAddress_Literal and UndergraduateStudent_emailAddress_Literal. Load the input files into memory to get intermediate result of ?X and ?Z (denoted as X_Z). Finally, join X_Y and X_Z on X. Because the join operator on Spark requires shuffling, in order to reduce the network overhead for shuffling, we repartition the two intermediate result sets by the join key before joining. In this way, data with same key will be in the same partition. Then the final results can be obtained by join the repartitioned intermediate result sets quickly.

If the number of intermediate result sets is greater than two such as Q2 and Q4, then greedy policy is adopt to join the intermediate results which sorts the intermediate results by their size and join two smallest intermediate results each time.

The process of joining intermediate results is shown in Algorithm 4.

Algorithm 4. Join intermediate results

Input: intermediate results
Output: final results

1:   midResultes ← intermediate results
2:   list ← new mutable.MutableList()
3:   **for each** midResulte ∈ midResultes **do**
4:      list ← list+midResulte.count
5:   **end for**
6:   array ← list.zipWithIndex
7:   sortedArray ← array.sortByKey()
8:   i ← 1
9:   sum ← midResultes.size-1
10: rdd ← midResultes.get(array(0)._2)
11: **while**(sum>0)
12:    tmp ← midResultes.get(array(i)._2).repartitionByKey()
13:    res ← rdd.repartitionByKey().join(tmp)
14:    i ← i+1
15:    sum ← sum-1
16:    rdd ←res
17: **end while**
18: finalResult ← rdd

---

4 Experiment and Evaluation

4.1 Cluster Configuration

All the experiments were conducted on a cluster with four machines. Each node of cluster has four processors at 3.2 GHz, 8 GB RAM and a 1T hard disk. The cluster runs Hadoop version 1.0.4, Spark version 1.4.0 and MemSQL version 4.0.34.

4.2 Datasets

The LUBM is a very frequently used synthetic benchmark for RDF query. It represents a network of university systems, in which all the entities in the university, such as student, professor and course are described in a format of triples. It provides data generator and a set of 14 SPARQL queries which are intensively used to test SPARQL query engine performance.

We use LUBM data generator to create datasets with academic domain information, enabling a variable number of triples by controlling the number of university entities. By varying this parameter between 100 to 1000, we create datasets vary from 15.24

Table 1. Datasets summary

Name	# of triples	Size
LUBM100	15.24 million	1.1 GB
LUBM200	30.34 million	2.13 GB
LUBM500	75.86 million	5.35 GB
LUBM1000	151.86 million	10.7 GB

million to 151.86 million RDF triples with sizes ranging from 1.1 GB to 10.7 GB. Table 1 shows a numerical summary of the datasets.

The space occupied by LUBM1000 after parsing and splitting using RQCCP, Liu L [8] and IMSQ [9] is shown in Fig. 2.

RQCCP algorithm needs less storage space than Liu L and IMSQ, as can be seen from Fig. 2. Because the predicate of RDF triple is contained in the file name, only subject and object are needed to store in file. IMSQ splits data by different subject, predicate and object which has redundant storage. So IMSQ consumes more storage space than Liu L and RQCCP.

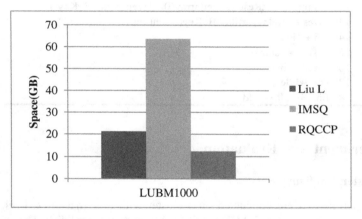

Fig. 2. Space occupied by LUBM1000 after parsing and splitting

4.3 Experimental Results

We compare the query evaluation performance of RQCCP against Liu L and IMSQ on different datasets. The input statistics of three algorithms when executing queries Q1–Q8 on LUBM1000 are listed in Table 2.

Since Liu L does not split data and build index, all triples are input no matter which query is executed. Compared with Liu L algorithm, IMSQ consumes lots of storage space, but it can reduce the input data of query by indexing, so the efficiency of IMSQ is better than Liu L. Because there exists redundant storage in IMSQ algorithm, the number of data files split by RQCCP algorithm is far less than by IMSQ algorithm. And except for Q6, the input data of IMSQ is more than RQCCP, as can be seen from Table 1.

Table 2. Number of input triples (million)

	Q1	Q2	Q3	Q4	Q5	Q6	Q7	Q8
Liu L	151.86	151.86	151.86	151.86	151.86	151.86	151.86	151.86
IMSQ	10.52	7.39	16.06	8.6	19.74	10.45	44.61	34.47
RQCCP	5.05	5.07	1.43	1.99	1.14	10.45	29.54	20.91

Thus the index of RQCCP has good filtering effects. Execution time for querying Q1–Q8 of three algorithms on different datasets is shown in Figs. 3, 4, 5 and 6.

As can be seen from Figs. 3, 4, 5 and 6, execution time of Liu L and IMSQ are longer than RQCCP for every query, especially complex queries such as Q2, Q4 and Q8. RQCCP firstly filters out irrelevant triples in advance according to index file. And intermediate results of query are cached in memory to reduce disk and network I/O. Then map-side join is used in the case of query with instance and greedy policy is

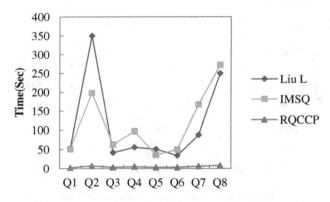

Fig. 3. Query execution time on LUBM100

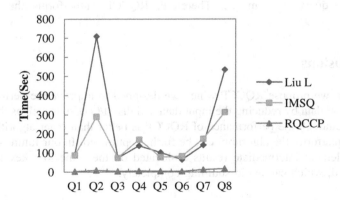

Fig. 4. Query execution time on LUBM200

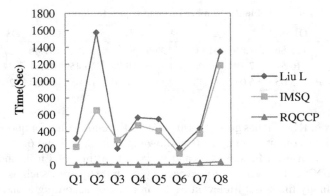

Fig. 5. Query execution time on LUBM500

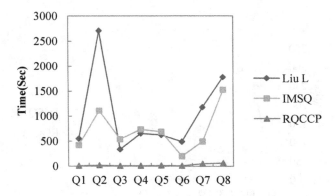

Fig. 6. Query execution time on LUBM1000

adopted to join the intermediate results. Finally repartition by join key can further improve the query performance. Therefore, RQCCP outperforms the other two algorithms.

5 Conclusions

In this paper, we propose RQCCP which we designed to improve the performance of query for RDF data by reducing the input data and disk I/O cost as well as the network overhead. Although the performance of RQCCP is better than other algorithms based on Hadoop platform, the algorithm can be further optimized. In our future studies, we plan to implement intermediate results co-located on the same join key rather than co-partitioned, which can avoid shuffling completely.

References

1. Du, F., Chen, Y., Du, X.: Survey of RDF query processing techniques. J. Softw. **24**(6), 1222–1242 (2013)
2. Weiss, C., Karras, P., Bernstein, A.: Hexastore: sextuple indexing for semantic web data management. Proc. VLDB Endowment **1**(1), 1008–1019 (2008)
3. Abadi, D.J., Marcus, A., Madden, S.R., et al.: SW-Store: a vertically partitioned DBMS for Semantic Web data management. VLDB J.—Int. J. Very Large Data Bases **18**(2), 385–406 (2009)
4. Neumann, T., Weikum, G.: Scalable join processing on very large RDF graphs. In: Proceedings of the 2009 ACM SIGMOD International Conference on Management of Data, pp. 627–640. ACM (2009)
5. Neumann, T., Weikum, G.: The RDF-3X engine for scalable management of RDF data. VLDB J. **19**(1), 91–113 (2010)
6. Husain, M.F., McGlothlin, J., Masud, M.M., et al.: Heuristics-based query processing for large RDF graphs using cloud computing. IEEE Trans. Knowl. Data Eng. **23**(9), 1312–1327 (2011)
7. Papailiou, N., Konstantinou, I., Tsoumakos, D., et al.: H2RDF: adaptive query processing on RDF data in the cloud. In: Proceedings of the 21st International Conference Companion on World Wide Web, pp. 397–400. ACM (2012)
8. Liu, L., Yin, J., Gao, L.: Efficient social network data query processing on MapReduce. In: Proceedings of the 5th ACM Workshop on HotPlanet, pp. 27–32. ACM (2013)
9. Wang, J., Fang, Z.: Distributed optimized query algorithm based on index. Comput. Sci. **41**(11), 233–238 (2014)
10. Prud'Hommeaux, E., Seaborne, A.: SPARQL query language for RDF. W3C Recommendation, 15 January 2008
11. Spark. http://spark.apache.org/
12. MemSQL. http://www.memsql.com
13. Guo, Y., Pan, Z., Heflin, J.: LUBM: a benchmark for OWL knowledge base systems. Web Seman. Sci. Serv. Agents World Wide Web **3**(2), 158–182 (2005)
14. Carroll, J.J., Dickinson, I., Dollin, C., et al.: Jena: implementing the semantic web recommendations. In: Proceedings of the 13th International World Wide Web Conference on Alternate Track Papers & Posters, pp. 74–83. ACM (2004)

Real-Time Monitoring and Forecast of Active Population Density Using Mobile Phone Data

Qi Li$^{(\boxtimes)}$, Bin Xu, Yukun Ma, and Tonglee Chung

Department of Computer Science and Technology,
Tsinghua University, Beijing, China
zhongguoliqi@163.com, xubin@tsinghua.edu.cn, yukunma@aliyun.com,
tongleechung86@gmail.com

Abstract. Real-time monitoring and forecast of large scale active population density is of great significance as it can warn and prevent possible public safety accident caused by abnormal population aggregation. Active population is defined as the number of people with their mobile phone powered on. Recently, an unfortunate deadly stampede occurred in Shanghai on December 31th 2014 causing the death of 39 people. We hope that our research can help avoid similar unfortunate accident from happening. In this paper we propose a method for active population density real-time monitoring and forecasting based on data from mobile network operators. Our method is based solely on mobile network operators existing infrastructure and barely requires extra investment, and mobile devices play a very limited role in the process of population locating. Four series forecasting methods, namely Simple Exponential Smoothing (SES), Double exponential smoothing (DES), Triple exponential smoothing (TES) and Autoregressive integrated moving average (ARIMA) are used in our experiments. Our experimental results suggest that we can achieve good forecast result for 135 min in future.

Keywords: Real-time forecast · Population density · Public safety · Mobile phone data

1 Introduction

Active population density is vital in determining the possibility of accidents in public safety and there are many areas that has attracted a lot of attention, such as railway stations, coach stations, scenic spots and squares. However, abnormal population aggregation in areas that are least paid attention to poses a greater threat of causing safety problem in public. An effective way to monitor population at large scale in real-time and forecast the distribution of active population have not been proposed to our knowledge. In our research, we propose a method that makes use of mobile network operators existing facilities to achieve monitoring and real-time forecast of population density with high accuracy at large scale.

© Springer Science+Business Media Singapore 2016
W. Chen et al. (Eds.): BDTA 2015, CCIS 590, pp. 116–129, 2016.
DOI: 10.1007/978-981-10-0457-5_12

Mobile phones now have an extremely high penetration rate across the globe. The global mobile user penetration rate has reached 96 % in 2014 [1]. In developed countries, the number of mobile phone has surpassed the total population. In developing countries, it is as high as 90 % and continuing to rise. Because of the high penetration rate of mobile phone usage, the numerous types of data generated by mobile phone have become a promising source of data for researchers.

Network-based and handset-based localization technologies are two main categories of wireless location. Handset-based technology, which is based on sensors on the mobile phone such as GPS receiver, have a relatively higher location accuracy. However, high accuracy location collection sometimes violates privacy rights of mobile users, and users are not willing to share their location data unconditionally. Large scale participation of users in location recording is not applicable. In this paper, we propose a real-time active population density monitor and forecast framework which is based on network-based data and technology. Cellular network services are provided by base stations and each base station covers an area with a fixed radius. To provide high quality service, network operator usually build more base stations in densely populated areas and less on lower populated areas. Moreover, network operator collect data from all base stations, which include name, id, latitude, longitude, communication technology type etc. When mobile users use their cell phones to make a call, send a message or access the Internet, a cell phone will choose the nearest and most suitable service provided by surrounding base station. When a user moves to a new area, the cell phone will re-choose an appropriate service automatically according to signal strength and network standard. Network operator then store the records, which includes user identification, base station identification, time stamp etc., of communication between cell phone and base station to database or other storage devices.

A kind of Received Signal Strength Indication (RSSI) data was used in our experiments, also known as measurement report (MR) data. MR data is collected by digital cell mobile communications network sent by users' equipment. MR data is generated by each mobile devices every 480 ms. Because of this high frequency in data, trajectory of each user has a strong continuity in terms of time and space. So the MR data has very high research value for researchers.

In this paper, we propose a method for real-time monitoring and forecasting of active population density. We use Thiessen polygon to divide the coverage of service provided by each base station and calculate the area of the service coverage. We find a way to display population density by combining Thiessen polygon and population data provided by base stations. Four series forecasting methods,namely Simple Exponential Smoothing (SES), Double exponential smoothing (DES), Triple exponential smoothing (TES) and Autoregressive integrated moving average (ARIMA) are used in our experiments. We find that simple methods like SES and DES have a performance of validity of over 0.8 when time span is less than an hour with threshold set to 0.4. We can also see the ARIMA is better at forecasting active population density with longer time span.

The rest of the paper is organized as follows. In Sect. 2, we describe related work on forecast of population density. In Sect. 3, we give an overview of the experimental data. Section 4 describes the four methods we use in the experiments. Section 5 shows experimental results, and gives detail analysis of the result. Finally we conclude this paper in Sect. 6.

2   Related Work

Until now, research and information on the area of population distribution and population density remains scarce [2]. However, both of these are very important to government policy making, urban management [3], academic research [4–6] and other fields. The development of techniques, such as simple areal weighting, dasymetric model [7–10], improves accuracy [7,10–15] and reduces workload. Despite these improvement, previous methods are constrained by population data from censuses reports [7,8,10–15]. In fact, because of the mobility of people, the number of people within a region is impossible to remain unchanged, and this change may occur very fast, both regularly or non-regularly. The population density of the area also changes rapidly correspondingly. So we may acquire some data at a certain time, but such data are out-dated already that they cannot meet our need. Recently, technologies based on cellular network [16] which make use of Call Detail Record (CDR) collected by the network operators and location information of base stations, are used to estimate the population and can produce a higher precision. CDR are data records produced by user when making a call. The estimation of population based on CDR cannot reflect the actual population and it also cannot be used to forecast future population.

Our research is also based on cellular network, but the data we used in our experiments is MR data which is collected by mobile network operators from mobile phone every 480 ms. The active population can be reflected precisely without extra methods or simulations. The MR data is very suitable for active population forecast.

3   Data Description

In this research, two different datasets are used in our experiments. Measurement report (MR) data, which are a kind of RSSI data sent by user equipment, are collected by digital cell mobile communications network. MR data generated by a cellphone usually contains the signal strength data measured from surrounding base stations and MR data are sent by mobile equipment to the base station which provides the communication service. Network operator can gather all the MR data from all the base stations. MR data are generated every 480 ms, so the high frequency in data, trajectory of each user has a strong continuity in terms of time and space. Our dataset contains 90720 records collected from two base stations from February 2015 to March 2015. Each record has been preprocessed and only contains useful features that can be expressed as a tuple in the form $[phoneID, stationID, timestamp]$. The interval between each record in our dataset is 1 min and time span of each record is 480 ms.

Another dataset contains 9,136 records of locations of distinct base stations in Hainan Province. Each record can be expressed using a tuple in the form [*stationID, latitude, longitude*].

In our method, it is unnecessary to get precise position information of each person and the population in service coverage of each station is our concern.

4   Real-Time Active Population Monitoring and Forecasting

The main research goal of this paper is population density forecast. Population density is generalized as:

$$\rho = \frac{N}{S} \tag{1}$$

where, N is the population within service coverage and S is the area of service coverage. So population statistics and area of service coverage calculation are two important parts of our work.

4.1   Area of Service Coverage for a Base Station

Because of the uneven distribution of population, more base stations are built in areas with higher population density than in that with lower population density. Thus, the distribution of service coverage of base stations is not a standard honeycomb shape. Figure 1 shows a demonstration of a standard honeycomb shaped cellular network. Theoretical base station locations does not help us calculate the actual base service coverage area. An alternative way to get service coverage areas is by Thiessen polygon [17,18]. Figure 2 shows how service coverage can be divided by Thiessen polygon where each polygon represents the service coverage area of a base station.

The essence of Thiessen polygon is simply polygon, thus the process of getting the area of a Thiessen polygon is as follows. First, we divide the Thiessen polygon into some triangles by connecting one vertex to all the other vertices of polygon. Second, we convert the latitude and longitude coordinates into geometric distance. We assume the vertices coordinates are represented as (*latitude1, longtitude1*) and (*latitude2, longtitude2*). The geometric distance d of two coordinates is calculated using the formula:

$$d = 2 \times R \times \sqrt{\sin^2(\frac{a}{2}) + \cos(radLat1) \times \cos(radLat2) \times \sin^2 \frac{b}{2}} \tag{2}$$

where, R is earth radius, and

$$radLat1 = \frac{latitude1 \times \pi}{180.0} \tag{3}$$

$$radLat2 = \frac{latitude2 \times \pi}{180.0} \tag{4}$$

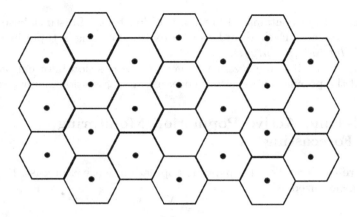

Fig. 1. Standard Cellular Network.

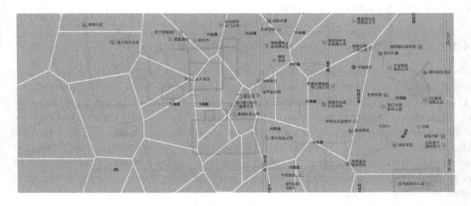

Fig. 2. Actual service coverage divided by Thiessen polygon in cellular network.

$$radLng1 = \frac{longtitude1 \times \pi}{180.0} \tag{5}$$

$$radLng2 = \frac{longtitude2 \times \pi}{180.0} \tag{6}$$

$$a = radLat1 - radLat2 \tag{7}$$

$$b = radLng1 - radLng2 \tag{8}$$

Based on the assumption that the earth is a standard sphere, we can obtain the actual geometric distance between two geometric coordinates using the above conversion. Finally, after calculating all the sides of the triangle, the Heron's formula is applied to get the area of every divided triangles. Where the Herons formula is as follows:

$$s = \sqrt{p \times (p - a) \times (p - b) \times (p - c)} \tag{9}$$

$$p = \frac{1}{2}(a + b + c) \tag{10}$$

where a, b and c represents the three sides of one triangles respectively. Ultimately, the area of Thiessen polygon can be obtained by a summation of all the divided triangles.

$$area = s_1 + s_2 + \cdots + s_n \tag{11}$$

where $s_i(i = 1, 2, \ldots, n)$ represents the area of the ith triangles in Thiessen polygon.

4.2   Active Users Within Service Coverage of a Base Station

Measurement report (MR) is sent to a base station from a cell phone every 480 ms, so a base station can receive measurement reports from all cell phones within its service area in 480 ms. If mobile phones are turned on in the service area, they must send measurement reports back to the base station following service provider's requirement. So the number of mobile users calculated by parsing MR data is equal to the real number of mobile users whose mobile phone is powered on.

The Number of Active Mobile Users. We define mobile phone users with their phone turned on as active mobile users. People can either be in resting state or non-resting state in their daily lives. People with non-resting state have a larger range of activities so they are the main objectives of our research. The power on and power off state of a mobile phone have a strong connection with its user's state. When a mobile user is ready to have a rest or take a break and do not want be disturbed, such as sleeping at night, he or she will turn off the mobile phone. In this situation, the number of mobile users collected should be smaller than the actual number and this is normal, also this part of mobile users are active mobile users. It is usually that at a non-resting time or in a non-resting venue, mobile phones are usually turned on, so the number of active mobile users are almost the same as the actual number of mobile users. Therefore the number of active mobile users is our main research point.

Two Data Sets of Active Population. We obtain two datasets as our experimental data. Figure 3 shows our first dataset *MRDS1*, which is the coverage of a base station which is most probably a residential areas. February 18th 2015 is the eve of Chinese New Year. All family members will gather to celebrate the arrival of the New Year at home. At 20:00, the number of people reaches its peak. And at the beginning of February 19, there are still many people in active state. This makes sense if we understand the activity schedule of a typical Chinese family

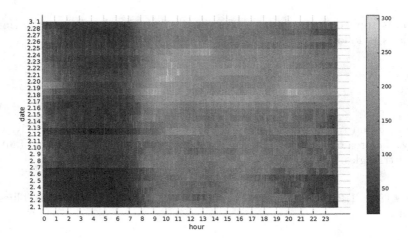

Fig. 3. *MRDS1* is a dataset of the population over a period of 29 *days* × 24 *hours*. The y axis refers to the date and x axis refers to the time in hours. A darker data point refers to a lower population number and vice versa.

on Chinese New Year. A strong activity presence is displayed between 9:00 and 22:00 from February 19 to February 25, this period is the official Chinese New Year statutory holiday.

Figure 4 shows our second dataset *MRDS2*, that is collected from a base station near shopping malls. On February 18th, activity presence reached a minimum, and in the following 7 days Chinese New Year statutory holidays, activity existence is at a very low state. In the other periods, it has a high degree of activity presence. Overall, the two graphs shows two obvious opposite trends.

The two figures accurately describe the activity pattern of Chinese people during the Chinese New Year holiday in two different areas.

4.3   Real-Time Monitoring and Forecast of Active Population Density

The active population data can be expressed in time series as shown in Fig. 5. $t_{current}$ is the current time and t_{latest} is the time of the latest data that can be observed while data before t_{latest} can also be observed. The situation after time $t_{current}$ is what we want to estimate and are interested in, which can be forecast according to history data. As we know, collecting and processing of data requires time, so the data that we can observed is history data and the delay time $T_{delay} = t_{current} - t_{latest}$. If we want to forecast the population density at time t, the time span is $T_{span} = t - t_{latest}$. Then we define that the actual value at time t as V_t.

Based on the above, for a base station, we can obtain the area of service coverage and the active population. The active population density can be calculated accurate. The real-time monitoring of active population density can

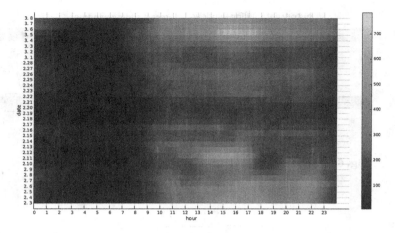

Fig. 4. *MRDS2* is a dataset of the population over a period of $35\,days \times 24\,h$. The y axis refers to the date and x axis refers to the time in hours. A darker data point refers to a lower population number and vice versa.

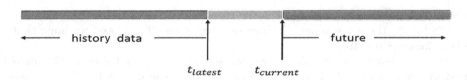

Fig. 5. Description of a time series in our experiment.

be done easily at a large scale. Obviously, the latest data we can observed at time t_{latest}.

The definition of real-time population density forecast is as follows:

If a method $F(t_k)$ uses the data before t_{latest} to forecast data at the future time $t_k(t_k \geq t_{current})$. The error of the forecast value and actual value $err_{T_{span},k}$ is define as the actual absolute value of the difference between the forecast value $F(t_k)$ and the actual value V_{t_k} divide by the actual value V_{t_k}. We use an indicator function 1_k that output 1 if $err_{(T_{span},k)}$ is less than or equal to a threshold ϵ and 0 if greater than threshold ϵ. Finally, we calculate a validity rate p of our forecast by taking a mean of the sum of all indicator function $1(k)$, $k = 1, \ldots, n$.

A formal definition can be expressed in following formulas:

$$err_{(T_{span},k)} = \frac{|F(t_k) - V_{t_k}|}{V_{t_k}} \tag{12}$$

$$1(k) = \begin{cases} 1, err_{(T_{span},k)} \leq \epsilon \\ 0, err_{(T_{span},k)} > \epsilon \end{cases} \tag{13}$$

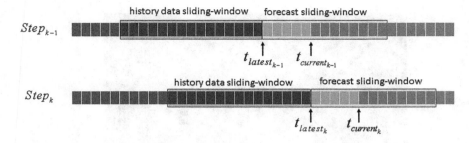

Fig. 6. Demostration of sliding window on a time series data.

$$p = \frac{\sum_{k=1}^{n} 1(k)}{n} \tag{14}$$

conditioned by,

$$t_{current} < t_1 < t_2 < \cdots < t_n \tag{15}$$

$$t_{current_k} \geq t_{latest_k} \tag{16}$$

where, $T_{span} = t_k - t_{current}$ and ϵ is the threshold.

Obviously, there are some relations between p and T_{span}, and we hope that p is as large as possible.

Figure 6 shows more detail. As we can see, we use a fixed sliding window is used in our experiment. The width of history data sliding-window W_h is fixed and time of each step is set to T_{delay}. Multiple time points after $t_{current}$ will be detected according to definition of real-time forecasting.

4.4   Forecast Methods

Four series forecasting methods,namely Simple Exponential Smoothing (SES), Double exponential smoothing (DES), Triple exponential smoothing (TES) and Autoregressive integrated moving average (ARIMA) are used in our experiments. Next, we will give a brief introduction to each these methods.

Exponential Smoothing. Exponential smoothing is usually applied to smooth data, which is used by many window functions in signal processing, acting as low-pass filters to remove high frequency noise.

1. Simple Exponential Smoothing (SES) [19, 20]. The raw data sequence is represented by $\{x_t\}$ beginning at time $t = 0$, and the result generated by the algorithm is written as $\{s_t\}$. When the sequence begins at time $t = 0$,

$$s_0 = x_0 \tag{17}$$

$$s_t = \alpha x_t + (1 - \alpha)s_{t-1}, t > 0 \tag{18}$$

where α is the smoothing factor, and $0 < \alpha < 1$.

2. Double Exponential Smoothing (DES) [21]. The raw data sequence is represented by $\{x_t\}$, beginning at time $t = 0$. $\{s_t\}$ is the smoothed value for time t , and $\{b_t\}$ is the best estimate of the trend at time t. F_{t+m} is an estimate of the value of x at time $t + m(m > 0)$ based on the raw data up to time t. Double exponential smoothing is defined to be

$$s_1 = x_1 \tag{19}$$

$$b_1 = x_1 - x_0 \tag{20}$$

And for $t > 1$ by

$$s_t = \alpha x_t + (1 - \alpha)(s_{t-1} + b_{t-1}) \tag{21}$$

$$b_t = \beta((s_t - s_{t-1}) + (1 - \beta)b_{t-1}) \tag{22}$$

where α is the data smoothing factor, $0 < \alpha < 1$, and β is the trend smoothing factor, $0 < \beta < 1$.

To forecast beyond x_t

$$F_{t+m} = s_t + mb_t \tag{23}$$

the initial value b_0 can be set according to needs. F_0 is undefined and $F_1 = s_0 + b_0$.

3. Triple Exponential Smoothing (TES) [22]. A sequence of observations is $\{x_t\}$, beginning at time $t = 0$ with a cycle of seasonal change of length L. $\{s_t\}$ is the smoothed value of the constant part for time t. $\{b_t\}$ is the sequence of best estimates of the linear trend that are superimposed on the seasonal changes. $\{c_t\}$ is the sequence of seasonal correction factors. $\{c_t\}$ is the expected proportion of the forecast trend at any time $t \bmod L$ in the cycle that the observations take on. The output of the algorithm is F_{t+m}, an estimate of the value of x at time $t + m$, $m > 0$ based on the raw data up to time t.

$$s_0 = x_0 \tag{24}$$

$$s_t = \alpha \frac{x_t}{c_{t-L}} + (1 - \alpha)(s_{t-1} + b_{t-1}) \tag{25}$$

$$b_t = \beta(s_t - s_{t-1} + (a - \beta)b_{t-1}) \tag{26}$$

$$b_t = \beta(s_t - s_{t-1}) + (1 - \beta)b_{t-1} \tag{27}$$

$$c_t = \gamma \frac{x_t}{s_t} + (1 - \gamma)c_{t-L} \tag{28}$$

where α is the data smoothing factor, $0 < \alpha < 1$, β is the trend smoothing factor, $0 < \beta < 1$, and γ is the seasonal change smoothing factor, $0 < \gamma < 1$.

Autoregressive Integrated Moving Average (ARIMA) [23]. ARIMA is used to time series data either to better understand the data or to forecast future points in the series. It can be defined as following,

$$Y_t = (1 - L)^d X_t \tag{29}$$

$$(1 - \sum_{i=1}^{p} \phi_i L^i)Y_t = (1 + \sum_{i=1}^{p} \theta_i L^i)\varepsilon_t \tag{30}$$

where X_t is a raw data sequence, t is an integer index, L is the lag operator, the θ_i are the parameters of the moving average part, The error terms ε_t are independent, identically distributed variables sampled from a normal distribution with zero mean. parameters p, d, and q are non-negative integers, p is the order of the Autoregressive model, d is the degree of differencing, and q is the order of the Moving-average model. ARIMA models are generally denoted $ARIMA(p, d, q)$.

5   Experiments

In this section, we describe our experiment in detail which includes our environment setting, our evaluation setting and our findings and results.

5.1   Environment Setup

We use two datasets for our experiments. One dataset consist of a time span of 34 days and the the other 29 days. The actual data collection frequency in application is one data per minutes. Altogether, we gather 90720 data points for our experiment. We set our history data sliding window W_h to 3000. The time of each step is set to T_{delay} to 15. The time for each step is set according to the data collection time cycle of the actual situation.

We test four different methods for forecasting, namely, simple exponential smoothing (SES), double exponential smoothing (DES), triple exponential smoothing (TES) and autoregressive integrated moving average (ARIMA).

5.2   Evaluation Method

We evaluate the methods using the error function that we defined in Sect. 4.3. We set four different threshold values $\epsilon = 0.05, 0.1, 0.2, 0.4$. Ten different forecast time is used, we choose $t_{span} = 15, 45, 75, 105, 135, 165, 195, 225, 255, 285$ respectively. We present our findings in the next section.

5.3   Results

Figure 7 shows there results for our first dataset MRDS1. We can that forecasting result for a time span of 15 min is satisfying. We can see that when we set ϵ to 0.4, validity rate for SES and DES is over 0.9. Even if we set ϵ to 0.05, we can see a validity rate of almost 0.5 for SES and DES. That means if we set the threshold to a strict condition, we can still get good results with some of the forecasting methods. The same results can be observed when time span is set 45 min. But we can see that the longer the time span, the harder it is to get good results. With time span over 135 min, the results are below expectation. We can conclude that our methods have a fair performance within 1 h. In other words,

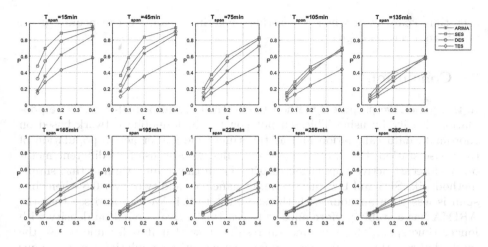

Fig. 7. Experimental results for MRDS1.

our forecasting methods can satisfy our requirement for real-time population density forecasting.

Figure 8 shows our result for the second dataset MRDS2. From our observation, the results are better than that of the MRDS1. Even with time span of 105 min, we can still obtain a satisfying result. But if we compare the two results, we find a similar trend, that is the longer the time span, the least accurate our forecast results. Moreover, we find that the simple methods like SES and DES have better performance when time span is small. Some other observations we find is that the more complex method ARIMA has a more stable performance

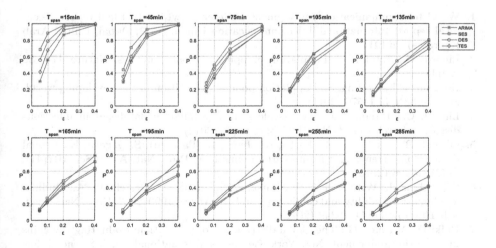

Fig. 8. Experimental results for MRDS2.

over time span, even when forecasting long time span, we can acquire good results to a certain degree.

6   Conclusion

In this paper, we propose a method to effectively monitor real-time active population density by using Thiessen polygon to divide mobile network based on coordinate data calculation and at the same time gather active population of each base station. Based on the above results, we also use four different methods to forecast active population density. From the results, we find that simple methods like SES and DES have a performance of validity of over 0.8 when time span is less than an hour with threshold set to 0.4. We can also see that the ARIMA has a stable performance at forecasting active population density with longer time span. The experimental results show that it is feasible to use the method of real-time monitoring and forecasting active population density in our research to avoid some possible public safety accidents.

Our work can be a foundation for more future work that uses complex models to better forecast active population density and for developing high level applications.

Acknowledgement. This work is supported by China National Science Foundation under grant No. 61170212, China National High-Tech Project (863) under grant No. 2013AA01A607, Ministry of Education-China Mobile Research Fund under grant No. MCM20130381, and Tsinghua University Initiative Scientific Research Program (No. 20131089190). Beijing Key Lab of Networked Multimedia also supports our research work.

References

1. World telecommunication development conference (wtdc-14): Final report. Technical report, International Telecommunication Union, (ITU, Dubai, United Arab Emirates) (2014)
2. Linard, C., Tatem, A.: Population mapping of poor countries. Nature **474**, 36 (2011)
3. Bongaarts, J., Sinding, S.: Population policy in transition in the developing world. Science **333**, 574–576 (2011)
4. Linard, C., Tatem, A.J.: Large-scale spatial population databases in infectious disease research. Int. J. Health Geogr. **11**(7) (2012)
5. O'Neill, B.C., Dalton, M., Fuchs, R., Jiang, L., Pachauri, S., Zigova, K.: Global demographic trends and future carbon emissions. Proc. Nat. Acad. Sci. **107**(41), 17521–17526 (2010)
6. OLoughlin, J., Witmer, F.D.W., Linke, A.M., Laing, A., Gettelman, A., Dudhia, J.: Climate variability and conflict risk in East Africa, 1990–2009. Proc. Nat. Acad. Sci. **109**(45), 18344–18349 (2012)
7. Balk, D., Yetman, G.: The global distribution of population: evaluating the gains in resolution refinement. New York: Center for International Earth Science Information Network (CIESIN), Columbia University (2004)

8. Deichmann, U., Balk, D., Yetman, G.: Transforming Population Data for Inter-disciplinary Usages: From Census to Grid. Center for International Earth Science Information Network, Washington DC (2001)

9. Stevens, F.R., Gaughan, A.E., Linard, C., Tatem, A.J.: Disaggregating census data for population mapping using random forests with remotely-sensed and ancillary data. PloS one **10**(2), e0107042 (2015)

10. Gaughan, A.E., Stevens, F.R., Linard, C., Jia, P., Tatem, A.J.: High resolution population distribution maps for Southeast Asia in 2010 and 2015. PloS one **8**(2), e55882 (2013)

11. Tobler, W., Deichmann, U., Gottsegen, J., Maloy, K.: World population in a grid of spherical quadrilaterals. Int. J. Popul. Geogr. **3**(3), 203–225 (1997)

12. Dobson, J.E., Bright, E.A., Coleman, P.R., Durfee, R.C., Worley, B.A.: Landscan: a global population database for estimating populations at risk. Photogram. Eng. Remote Sens. **66**(7), 849–857 (2000)

13. Balk, D.L., Deichmann, U., Yetman, G., Pozzi, F., Hay, S.I., Nelson, A.: Determining global population distribution: methods, applications and data. Adv. parasitol. **62**, 119–156 (2006)

14. Linard, C., Gilbert, M., Snow, R.W., Noor, A.M., Tatem, A.J.: Population distribution, settlement patterns and accessibility across Africa in 2010. PloS one **7**(2), e31743 (2012)

15. Azar, D., Engstrom, R., Graesser, J., Comenetz, J.: Generation of fine-scale population layers using multi-resolution satellite imagery and geospatial data. Remote Sens. Environ. **130**, 219–232 (2013)

16. Deville, P., Linard, C., Martin, S., Gilbert, M., Stevens, F.R., Gaughan, A.E., Blondel, V.D., Tatem, A.J.: Dynamic population mapping using mobile phone data. Proc. Nat. Acad. Sci. **111**(45), 15888–15893 (2014)

17. Aurenhammer, F.: Voronoi diagrams survey of a fundamental geometric data structure. ACM Comput. Surv. **23**(3), 345–405 (1991)

18. Okabe, A., Boots, B., Sugihara, K.: Spatial Tessellations. Concepts and Applications of Voronoi Diagrams. John Wiley & Sons, New York (1992)

19. Holt, C.C.: Forecasting seasonals and trends by exponentially weighted moving averages. Int. J. Forecast. **20**(1), 5–10 (2004)

20. Brown, R.G.: Smoothing, Forecasting and Prediction of Discrete Time Series. Courier Corporation, New York (2004)

21. Kalekar, P.S.: Time series forecasting using holt-winters exponential smoothing. Kanwal Rekhi Sch. Inf. Technol. **4329008**, 1–13 (2004)

22. Winters, P.R.: Forecasting sales by exponentially weighted moving averages. Manage. Sci. **6**, 324–342 (1960)

23. Makridakis, S., Hibon, M.: Arma models and the boxcjenkins methodology. J. Forecast. **16**(3), 147–163 (1997)

Ranking-Based Recommendation System
with Text Modeling

Chuchu Huang[✉] and Guang Chen

School of Information and Communication Engineering,
Beijing University of Posts Telecommunications,
Beijing 100876, China
chuchu@bupt.edu.cn

Abstract. In the study of recommendation systems, many methods based on predicting ratings have been put forward. However, the rating-predicting methods have some shortages. It pays too much attention to predicting, instead of the nature of recommendation, which is predicting the order of ratings. Thus, we use a pairwise-based learning algorithm to learn our model and take the zero-sampling method to improve our model. In addition, we propose a text modeling method making the recommendations more explicable. It is proved that our system performs better than other state-of-art methods.

Keywords: Ranking based model · Text modeling · Learning to rank · Pairwise

1 Introduction

In the web 2.0 generation, plenty of researches have been done on recommendation system. Because of the development of the Internet, it takes more time to find the exact thing that we need among large scale of information. Since recommendation system can recommend items before searching which save users a lot of time, it has become indispensable.

Most current recommendation systems use the hybrid recommendation method which fusion the collaborative filtering method and content-based method [1–5]. The basic assumption of collaborative filtering is that, if two users have the similar behavior on a serious of items, they will probably behave similar on other items. Generally, collaborative filtering systems only consider users' ratings or click information. GroupLens system adopts predict-based collaborative filtering method to recommend news and movies [6]. Xing et al. combine collaborative filtering and timing analysis method to build their system [7]. Content-based recommendation system is the earliest recommendation system, which is based on the analysis of users and items [8–11]. Hybrid recommendation, however, combines the advantages of both, and takes both the rating information and the context information.

Recently, either collaborative filtering model or content-based model takes the method of matrix factorization to build the recommendation system. In the early years, the traditional SVD model was put up which later found not suitable for sparse matrixes [12]. In the year of 2006, Simon Funk proposed the F-SVD model to solve the

© Springer Science+Business Media Singapore 2016
W. Chen et al. (Eds.): BDTA 2015, CCIS 590, pp. 130–143, 2016.
DOI: 10.1007/978-981-10-0457-5_13

problem [13]. Later, many fellows put forward different methods to improve the traditional SVD [14] and [15]. Besides, some recommendation systems factorize the rating matrix into two matrixes, user matrix and item matrix [16] and [17].

On this basis, Yehuda Koren added the bias of users and items which improves the performance of system by 32 % [18]. On general, matrix factorization based method maps the high-dimensional rating matrix into two lower dimensional matrixes. This saved memory spaces and decreases the complexity of model, but still has some problems. One of the problems is that the result is unexplainable. It is difficult to tell why we recommend this item not another one. Though every dimension of the matrix probably has some sort of meaning, it is difficult to determine the certain meaning of a certain dimension.

From another perspective, the existing recommendation systems mostly minimum the root mean square error between real ratings and predicted ratings [19–21]. This reduces the complexity of algorithm; however, in the task of recommendation, it is more important to predict the ranking of the ratings than the ratings. It is the relative value that tells the preference of the user. Thus, the primary task is to learn the ranking of items in a user's opinion. LTR (Learning to Rank) is the most suitable model for this task. There are three kinds of LTR model (point wise, pairwise and list wise). Point wise model is based on single point, which is quite similar with the traditional rating predicting model. List wise model is based on all samples. Pairwise is based on every pair of data and it is the most popular model in recommendation systems [22–24]. Compare these three methods, point wise model and list wise model do not fit much with the recommendation task. On the contrary, pairwise models do have good performance. However, in some pairwise based models, samples that are not rated will be ignored or take zero instead. Neither of them is a reasonable way of dealing with the unrated items; the former ignores the reason why the user does not choose the item and the latter will over-fit.

In addition, the information that the recommendation systems take will usually include ratings, items' properties, users' properties, users' behaviors [25] and reviews. It is obvious that reviews are very import, for these texts may explain the choosing behavior of the users. However, current recommendation systems have very limited means when modeling review texts. The most seen method is LDA, which is also used by Julian M in HFT. Nevertheless, LDA has the same problem with matrix factorizing, that is hard to explain every dimension of the vector.

For the existing of the three issues mentioned above, our work aims to improve the performance of the recommendation system as well as the interpretability of the final recommendation. Thus, we put forward three solutions. First, use word vector and define dimensions artificially to endow every dimension with a certain meaning. Second, aim at predicting rankings instead of ratings to build the system and take zero sampling to improve the performance of the model. Third, train the words' vectors to model the review texts.

2 Related Work

2.1 Rating Predicted Model

Collaborative filtering systems that only consider users' ratings usually use matrix factorization method, which factorize the rating matrix R into user matrix and item matrix. In some other systems, offset α, β_u and β_i are added to become (1). $\vec{\gamma_u}$ is the user vector and $\vec{\gamma_i}$ is the item vector.

$$rec(u, i) = \alpha + \beta_u + \beta_i + \vec{\gamma_u} \cdot \vec{\gamma_i} \tag{1}$$

The lost function is as (2). $\alpha, \beta_u, \beta_i, \vec{\gamma_u}, \vec{\gamma_i}$ are the parameters.

$$R(\alpha, \beta_u, \beta_i, \vec{\gamma_u}, \vec{\gamma_i}) = \sum_{u,i} \left(rec(u, i) - r_{u,i}\right)^2 \tag{2}$$

2.2 HFT Model

The HFT model combines the information of ratings and reviews.

As to the ratings, HFT factorizes the rating into the dot product of user's vector γ_u and the item's vector γ_i. γ_u and γ_i are both k-dimensional vector; each dimension is abstract and can be seen as an aspect. On the k-th dimension, γ_{uk} represents how much user u cares about this aspect and γ_{ik} represents how outstanding item i is on this aspect. Therefore, if an item performs well on the aspects that a user cares, it will probably get a high rating from this user.

In HFT model, it uses LDA to model the review texts. It regards the review from user u to item i as a document which denotes $d_{u,i}$; puts all review of user u together into a document which denotes d_u; puts all users' reviews towards item i together into a document and denotes d_i. HFT learns the item i's topic distribution θ_i from d_i and the user u's topic distribution θ_u from d_u.

It fusions the ratings and reviews through connecting θ_i and γ_i. Instead of training θ_i and γ_i separately, HFT combines them through formula (3). The topic number K and the rating factorize dimension K are set the same.

$$\theta_{i,k} = \frac{\exp(\kappa \gamma_{i,k})}{\sum_{k'} \exp(\kappa \gamma_{i,k'})} \tag{3}$$

The objective function is as shown in formula (4), the former is the error sum of squares and the latter is the likelihood function of the topic model.

$$R\left(\alpha, \beta_u, \beta_i, \vec{\gamma_u}, \vec{\gamma_i}\right) = \sum \left(rec(u, i) - r_{u,i}\right)^2 - \mu \cdot l(T|\theta, \varphi, z) \tag{4}$$

The notations used in this paper are explained in Table 1.

Table 1. Notation explanation

Notation	Explanation
α	The overall score offset
β_u	The score offset of user u
β_i	The score offset of item i
$\vec{\gamma}_u$	User vector
$\vec{\gamma}_i$	Item vector
$rec(u, i)$	The predicted rating of user u to item i
$r_{u,i}$	The real rating of user u to item i
γ_{uk}	The kth dimension of $\vec{\gamma}_u$
γ_{ik}	The kth dimension of $\vec{\gamma}_i$
$d_{u,i}$	Reviews of user u to item i
d_u	Reviews of user u to all items
d_i	Reviews of all users to item i
θ_i	The topic distribution of d_i.
θ_u	The topic distribution of d_u
$\theta_{i,k}$	The kth dimension of θ_i
L_j	The length of jth review
q	The quality of the review
ρ	The normalized coefficient
w	The weight of the review

3 Model

Our system is a hybrid system, which combines the ratings and reviews of users to items. As framework, we take the method of matrix factorization. However, there are two main differences between other model and other models. One is the optimization target. Our Model aims at learning the real ranking of items, meanwhile other systems mostly aim at predicting the ratings. The other is the way of defining the dimensions. Unlike other models, our model gives each dimension a certain meaning and makes the recommendation more interpretable.

Our model can be divided into two sub models that respectively modeling for ratings and reviews. On one hand, the rating model uses the ranking based matrix factorization to build the users' model and items' model. On the other hand, the review model combines the word vector and artificial definition of dimensions to model the review texts. Finally, we combine the whole model by relating the items' model and the texts' model.

3.1   Ranking-Based Matrix Factorization Model

In the rating predict model, the objective is to minimize the RMSE. While in this work, we define the object to be learning the real ranking of items. In other words to minimize the error rate of predicting the size relationship of every pair, as shown in formula (5)

$$\sum_u \sum_{r_{u,a} > r_{u,b}} [[rec(u, i_a) \leq rec(u, i_b)]] \tag{5}$$

$[[A]]$ means when condition A is true, it equals one, otherwise it equals zero.

Nevertheless, this objective function is not continuous, so we take formula (6) instead to train the model. Reason for that is the existence of formula (7).

$$\sum_u \sum_{r_{u,a} > r_{u,b}} exp(rec(u, i_b) - rec(u, i_a)) \tag{6}$$

$$[[x \leq 0]] \leq e^{-x} \tag{7}$$

u represents a user, $r_{u,a}$ is the real rating that user u gives to item a, $rec(u, i_a)$ is the value that user u gives to item a predicted by our model, which is predicted by the following formula.

$$rec(u, i) = \alpha + \beta_u + \beta_i + \vec{\gamma_u} \cdot \vec{\gamma_i} \tag{8}$$

We take the method of L-BFGS to learn the model. While training, we need to set the dimension number K; we randomly choose some unrated items to be the negative samples and contain them in the training process.

3.2   Text Modeling

While considering the un-interpretability of LDA model, we come up with a new model that combines the word vector and artificial definitions of dimensions. This method not only makes each dimension explainable, but also filters some noise in the texts.

Our text modeling methods include two steps, building the attribute word sets and modeling review texts.

Firstly, building the attribute word sets, the process is shown in Fig. 1. Take all the reviews as training data to train the words' vectors; at the meantime, POS tag all the words and choose the nouns and the adjectives of high frequency. After that, classify these words manually, define their types and choose the core attribute words to be the seed words. By calculating the cosine distances, we choose the words near every type of seed words to be part of the attribute word sets. Finally, we take the method of KNN to classify all the words in the data set to determine the rest words' attribution. One word can only belongs to one attribute set.

Modeling the reviews is based on the work above. To extract the attribute words is to extract the features. Every type of attribute is a feature of the reviews. In this way,

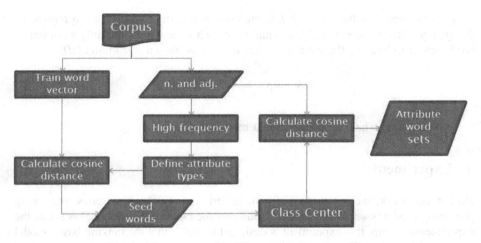

Fig. 1. The process of building the attribute word sets

we achieve the goal to explain the result of recommendation. During the modeling of a single piece of text, we calculate the proportion of words in each category and regard it as each dimension of the vector. Finally, normalize the vectors to get the final texts' model.

3.3 Model Merging

The last part of the model is to combine the ratings and reviews to build the recommendation model.

In the first part of our model, we get the user vector $\vec{\gamma}_u$, item vector $\vec{\gamma}_i$ and three biases α, β_u, β_i by factorizing. To combine the ratings and the reviews, we try to replace $\vec{\gamma}_u$ or $\vec{\gamma}_i$ by the vectors calculated by the text model. When replacing user vector $\vec{\gamma}_u$, we need to learn the other parameters $\vec{\gamma}_i, \alpha, \beta_u, \beta_i$; when replacing item vector $\vec{\gamma}_i$, we need to learn the other parameters $\vec{\gamma}_u, \alpha, \beta_u, \beta_i$. And we still take formula (1) as objective function while training.

When building the text model $\vec{\gamma}_u$ of user u, we take all the reviews from user u and using the method in 2.1.2 to model all the reviews by user u. Later, a linear weighted method is taken to combine all the review vectors of user u. The weight depends on the quality of the reviews, and it is believed a sentence with a suitable length is more likely to have high quality. Neither the short sentences nor the long sentences perform well enough in the evaluation of review's quality. Formula (9) is put forward to evaluate the quality of the reviews.

$$q = \frac{L_j}{L_j + |L_j - \bar{L}|} \tag{9}$$

L_j is the length of the review j; $\bar{L}$ is the average length of all reviews; q represents the quality of the review. Finally, the quality of each review must multiply a normalize coefficient ρ to become the weight of each review, as shown in formula (10).

$$w = \rho \cdot \frac{L_j}{L_j + |L_j - \bar{L}|} \tag{10}$$

It is the same when we build the text model of item i.

4   Experiment

During our work, we set three series of experiment. In the first series, the rating predicting model is set as the control group and the ranking based model is set as the experimental group; this experiment is designed to verify that the ranking based model performs better than the rating predicting model on ranking. In the second series, we aim to test the effect of our text modeling method. We do that by comparing our model with LDA. Besides, we compare the ones with and without text modeling to see whether our text model is effective. In the third series, we compare our model to HFT model to analyze the strength and weakness.

The datasets we use is from Amazon [29], which contains both ratings and reviews.

This paper mainly measures the ranking performance of the system, so we use NDCG and MRE to evaluate the system performance. NDCG is one of the indexes for the ranking performance in information retrieval field; the specific formula is showed below as (11) and (12). Under the situation of recommendation, $r(j)$ is the j-th biggest rating in the predicted values, MAX is the DCG when the ranking is totally the same as real values.

$$DCG = \sum_{j=1}^{n} \frac{2^{r(j)} - 1}{\log(1+j)} \tag{11}$$

$$NDCG = DCG/MAX \tag{12}$$

MRE is from [7], the formula is show in (13), to add all the pairs when real rating $r_{u,i_a} > r_{u,i_b}$ but the predicted values are on the opposite.

$$MRE = \frac{1}{n} \sum_{x} [[rec(u, i_a) \leq rec(u, i_b)]] \tag{13}$$

4.1   Rating-Based Model and Ranking-Based Model

Model in this paper is based on two main points. The first is that predicting based model did not achieve the best performance on ranking, which will affect the subsequent recommendations. To verify this, in the first experiment, we compare the ranking performance of rating-based model and ranking-based model. In this experiment, three groups are set, one is the rating-based model, one is the ranking based model without zero sampling, the other is the ranking based model with zero sampling.

The results are shown in Table 2, the first item is the reverse rate, the second is NDCG, "*" represents for the best effect. In the three groups of experiments, generally speaking, ranking based model with zero sampling achieves the best results, which has a lower MRE and a higher NDCG than the rating-based model. On the contrary, ranking based model without zero sampling performs badly in the test. This may be due to ignoring the unrated items, which will probably cause the loss of users' preference information. For example, some users choose item A instead of B which is quite similar with A. When this happens, it may tell us users prefer A than B for some reason and we would lose this information if we choose totally ignoring the unrated items. However, sampling the unrated samples randomly help us restore this part of the information.

Table 2. Comparison of rating-based model and ranking-based model

	Rating based	Ranking based	Ranking based (with zero samping)
Art	0.1566/0.8876	0.1380/0.8601	*0.1349/*0.9238
Automotive	0.1417/*0.9181	0.1564/0.8863	*0.1143/0.9035
Baby	*0.2367/0.7688	0.2645/0.7554	0.2493/*0.8579
Beauty	0.2351/*0.7947	0.2350/0.7763	*0.2238/0.7812
Books	0.1603/0.8794	*0.1498/*0.8900	0.1598/0.8665
Electronics	0.1612/0.8220	0.1749/0.8331	*0.1447/*0.8868
Gourmet_Foods	0.2544/*0.7994	0.2433/0.7438	*0.2130/0.7577
Health	0.1459/0.8767	0.1669/0.7767	*0.1432/*0.8782
Jewelry	*0.1834/0.8510	0.2049/0.8437	0.1877/*0.8601
Office_Products	*0.1012/0.8744	0.1107/0.8712	0.1132/*0.8834
Pet_Supplies	0.2468/0.7561	0.2611/0.7467	*0.2390/*0.7759
Shoes	*0.2951/0.7313	0.3021/0.7061	0.2956/*0.7786
Video_Games	0.1907/0.7891	0.2091/0.7522	*0.1851/*0.7940

4.2 Text Modeling

This experiment contains two parts. The first part verifies the effectiveness of the text modeling method we put up by comparing it with LDA. The second part judges whether the combination of word vector and manually defined dimensions is effective by comparing models with and without text modeling.

Using method in Sect. 3.2, we classify nouns and adjectives into six classes of attributes, partly shown in Table 3. Using the LDA model, we could get the words under each topic, Table 4 shows part of the words when topic number K is set 6. By compare words in the two tables, we can clearly see that, our model performs quite well. Using the method put forward in this article, we can get attribute words quite relevant to each attribute and each class of attribute words has an obvious semantic meaning. By contrast, when using the topic model, words under each topic do not have a very clear subject.

Table 3. Attribute words extracted by word vector

Attribute	Words
Quality	lack products **excellent** product **better performance** texture **quality** precision **good poor consistent** guidelines **design** specs errors **resilient consideration strong brilliant** slow experience **sharpening**
Practicability	**particular** certain type example **purposes** types **easier used** standard need **device common** techniques practice **available useful** addition **uses** possible **necessary specific** form purpose **application limited consider**
Price	**bargain value seller expensive lot refund deal supply money cheap sale cost investment issue lots card supplies spending worth store stock affordable buy premium business bonus dollar price cheaper pocket**
Size	**approx less size** areas **smaller** splitting **underside** front **shape sizes large larger weight giant blocks thickness length** tail spread **height bigger heavier number inches distance far small** average **older shape**
Material	**layers material** shoes **packaging** stains **plastic scrapbooking** artwork **fabric mixing paper glue piece matte** crack **items process adhesive materials** brittle **durable matter** embellishments **notebooks waste nylon**
Looking	**looks spot green shade color purple cute** years **interesting features** alter **shape performance width black shadows white bar** made **layer look** shown **looking colors shows monster show cutest** little quite

Table 4. Topic words extracted by LDA

Attribute	Words
1	machine sewing thread brother bobbin sew serger needle stitches foot machines stitch singer tension manual threading needles features beginner case basic problems reviews easy threader bobbins quilting
2	glue pencils gun pencil prismacolor sticks glitter pastels guns trigger pastel colored blend drip lead crayons prisma heats 72 blending tin glued heat premier crayola elmers temp hot charcoal drawings
3	punch pens pen punches ink hole puncher noodlers fountain tape punching nib punched belts nibs holes calligraphy lamy books 3hole spine punchers converter swingline writing bleed safari waterproof leather flows
4	epson stamps printer stamp stickers beads prints pad print photo matte t5846 picturemate ink album paper sticker magnets photos t5845 m printers glossy printing premium roll inkjet scrapbook luster
5	scissors blades cutting blade fiskars cutter shears sharp brushes rotary pair mat scissor handed titanium cutters mats lefthanded rulers markers Westcott ruler pinking xacto fiskar cuts earrings comfortable
6	dye yarn chair cricut loom wool scarf compressor airbrush dyes dyed acid putty skein roving necklace faded silver spray air mold jacquard blanket iron stain necklaces wax it soaked baby

After joining the reviews, the performance have improved on most of the data sets compared to the model without text information, MRE has improved over eight data sets, NDCG has improved on 10 data sets. The MRE performances on 13 data sets of

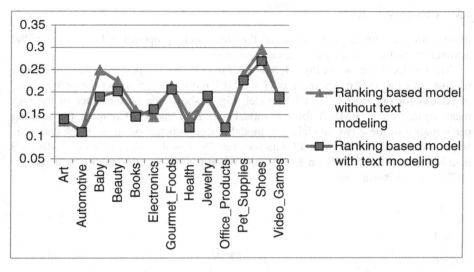

Fig. 2. MRE performance of two models (Color figure online).

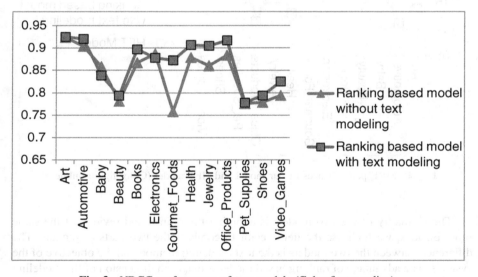

Fig. 3. NDCG performance of two models (Color figure online).

two models are shown in Fig. 2 and The NDCG performances are shown in Fig. 3. This illustrates the text modeling method we proposed is effective, and by fusion with the rating part it can effectively use the rating information and the text information.

4.3   Compare to HFT

In this part, we compare our model and the HFT model proposed in [1] to analyze the advantages and disadvantages of these two models.

The results are shown in Figs. 4, 5 and 6, Fig. 4 shows the performance of the mean reverse error rate (MRE) for our model and HFT model, Fig. 5 shows the performance of NDCG for our model and HFT model, Fig. 6 shows the performance of the root mean square error (RMSE) for our model and HFT model. Compare the ranking performance of these two models, our model performs better on most of the data sets. In total, we do experiments on 13 data sets; ranking based model has lower MRE on 9 data sets and better NDCG on 8 data sets. When referring to the predicting accuracy, HFT model performs better.

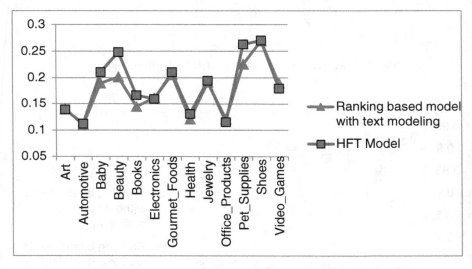

Fig. 4. MRE performances of our model and HFT model (Color figure online).

The similarity of the two models is adopting the ratings and reviews at the same time. Besides, we both use the item vector to combine the two parts of models. The difference between the two models is the text modeling method and the objective of the systems. The advantage of ranking based model is that each dimension of the modeling result has a certain meaning and this improves the interpretability of recommendation. On the other hand, though HFT model performs worse on the ranking and do not have a strong interpretability, it has a better predicting accuracy.

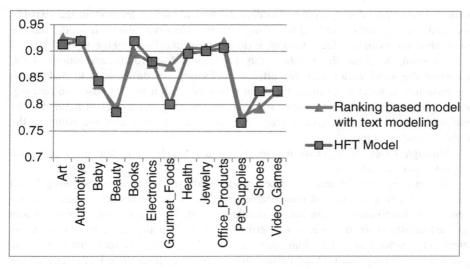

Fig. 5. NDCG performances of our model and HFT model (Color figure online).

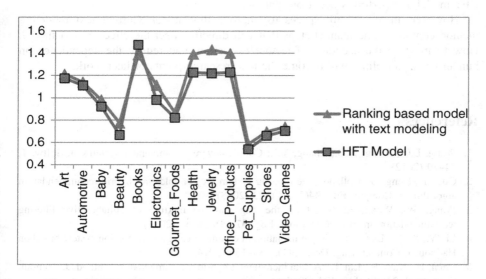

Fig. 6. RMSE performances of our model and HFT model (Color figure online).

5   Conclusions and Future Work

In our work, we study the state of art recommendation systems deeply and find three common problems in these models. Firstly, by decomposing the rating matrix to build the system, the results of recommendation will not have very good interpretability. Secondly, the recommendation systems based on predicting ratings do not fit directly with the recommending purpose. Thirdly, the text modeling method in the existing recommender systems is limited to the topic model. However, the topic model has the same shortage as matrix factorization, it do not have excellent interpretability.

According to these three points, this paper puts forward three solutions. First, combine the word vector with the artificial definitions of dimensions to make each dimension interpretable. Second, build the recommendation system based on ranking. Aim at learning the correct ranking of items and making the model fit better for the ranking. Third, train the word vectors and model the review text according to the trained words' vectors.

Through series of experiments, we verify the feasibility of the three proposed improvement means.

We can see that the ranking based model performs better than the rating based model on ranking in the first comparable experiment. In the second experiment, we verify the effectiveness of out text modeling method. The combining of word vectors and artificially defined dimensions performs better than the LDA model on interpretability. In addition, after adding our text modeling method, system improves on the predicting accuracy and ranking performance, which suggests the effectiveness of our text modeling method. In the last set of experiments, by comparing the model with HFT model, we found that the proposed model have better interpretability compare to HFT model and performs better on ranking.

However, there are still spaces to improve this model. Current text modeling method contains some manual steps that will directly affect the effect of text model. How to improve the accuracy of feature extraction, and reduce the dependence on human when modeling, will become the next research points of later work.

References

1. Wang, L.C., Meng, X.G., Zhang, Y.J.: Context-aware recommender systems. J. Softw. **1**, 24–29 (2012)
2. Guo, Y., Deng, G.: Collaborative filtering system project cold start hybrid recommendation algorithm. J. Comput. Eng. **34**(23), 11–13 (2008)
3. Wang, W., Wang, J.: Based on the Tag and the mixture of collaborative filtering recommendation method. J. Comput. Eng. **37**(14), 34–35 (2011)
4. Li, W., Xu, S.: Design and implementation of e-commerce recommendation system based on Hadoop. J. Comput. Eng. Des. **35**(1), 130–136 (2014)
5. Chen, H., Xiao, R., Lin, L.: A data incremental hybrid recommendation method. J. Comput. Syst. Appl. **23**(10), 119–124 (2014)
6. Resnick, P., Iacovou, N., Suchak, M.: GroupLens: an open architecture for collaborative filtering of Netnews. In: CSCW 1994, pp. 175–186, New York (1994)
7. Xing, C.X., Gao, F., Zhan, S.: To adapt to changes in user interest in the collaborative filtering recommendation algorithm. J. Comput. Res. Develop. **44**(2), 296–301 (2015)
8. Pazzani, M.J., Billsus, D.: Content-based recommendation systems. In: Brusilovsky, P., Kobsa, A., Nejdl, W. (eds.) Adaptive Web 2007. LNCS, vol. 4321, pp. 325–341. Springer, Heidelberg (2007)

9. Pazzani, M.J.: A framework for collaborative, content-based and demographic filtering. Artif. Intell. Rev. **13**(5), 393–408 (1999)
10. Chun, Z., Xing, C.X., Zhou, L.Z.: The personalized search algorithm based on content filter. J. Softw. **14**(5) (2003)
11. Basu, C., Hirsh, H., Cohen, W.: Recommendation as classification: using social and content-based information in recommendation. In: AAAI, pp. 714–720 (1998)
12. Vozalis, M.G., Margaritis, K.G.: Applying SVD on generalized item-based filtering. J. IJCSA **3**(3), 27–51 (2006)
13. Funk, S.: F-SVD. http://sifter.org/~simon/journal/20061211.html
14. Kurucz, M., Benczúr, A.A, Csalogány, K.: Methods for large scale SVD with missing values. In: KDD, vol. 12, pp. 31–38 (2007)
15. Jannach, D., Gedikli, F., Karakaya, Z.: Recommending hotels based on multi-dimensional customer ratings. In: ENTER, pp. 320–331 (2012)
16. Liu, N.N, Yang, Q.: Eigenrank: a ranking-oriented approach to collaborative filtering. In: SIGIR Conference on Research and Development in Information Retrieval, pp. 83–90. ACM (2008)
17. Jean-Francois, P., Tuong-Vinh, T., Nicolas, U., Massih-Reza, A., Patrick, G.: Learning to Rank for Collaborative Filtering (2007)
18. Koren, Y., Bell, R., Volinsky, C.: Matrix factorization techniques for recommender systems. J. Computer **8**, 30–37 (2009)
19. Bobadilla, J.S., Ortega, F., Hernando, A.: A collaborative filtering approach to mitigate the new user cold start problem. J. Knowl.-Based Syst. **26**, 225–238 (2012)
20. Zhang, Y., Cao, B., Yeung, D,Y.: Multi-domain collaborative filtering (2012). arXiv preprint 1203.3535
21. Sarwar, B., Karypis, G., Konstan, J.: Item-based collaborative filtering recommendation algorithms. In: World Wide Web, pp. 285–295. ACM (2001)
22. Alexandros, K., Linas, B., Yue, S.: Learning to rank for recommender systems. In: RecSys 2013, Hong Kong (2013)
23. Steffen, R., Christoph, F.: Improving pairwise learning for item recommendation from implicit feedback. In: WSDM 2014, New York (2014)
24. Julian, M., Jure, L.: Hidden factors and hidden topics understanding rating dimensions with review text. In: RecSys 2013, Hong Kong (2013)
25. Fang, Y., Si, L.: A latent pairwise preference learning approach for recommendation from implicit feedback. In: CIKM 2012, Shanghai (2012)
26. Shi, Y., Martha, L., Alan, H.: Collaborative filtering beyond the user-item matrix a survey of the state of the art and future challenges. In: 2014 ACM, Chicago (2014)
27. Zhang, W., Chen, T., Wang, J., Yu, Y.: Optimizing Top-N collaborative filtering via dynamic negative item sampling. In: SIGIR 2013, Dublin (2013)
28. Shi, Y., Martha, L., Alan, H.: List-wise Learning to rank with matrix factorization for collaborative filtering. In: RecSys 2010, Barcelona (2010)
29. Amazon Review Datasets. http://snap.stanford.edu/data/web-Amazon.html

Design and Implementation of a Project Management System Based on Product Data Management on the Baidu Cloud Computing Platform

Shenghai Qiu$^{(\boxtimes)}$, Yunxia Wang, Wenwu Jin, and Jiannan Liu

Department of Mechanical Engineering, Nanjing Institute of Technology,
Nanjing 211167, Jiangsu, China
{qiush2000, wang-yunxia}@njit.edu.cn,
{649861584, 1242684580}@qq.com

Abstract. Aiming at enterprises without commercial project management systems (PMS) in a product data management (PDM) environment, and using a cloud computing platform, this research analyses the business process and function of complex product project management in PDM, and proposes a PMS-based organizational structure for such a project. This model consists of a task view, user view, role view, and product view. In addition, it designs the function structure, E-R model and logical model of a PMS database, and also presents an architecture based on the Baidu cloud platform, describes the functions of the Baidu App Engine (BAE), establishes the overall PMS software architecture. Finally, it realizes a revised product design project by using EasyUI, J2EE and other related technologies. Practice shows that the PMS designed for PDM has availability, scalability, reliability and security with the help of the Baidu cloud computing platform. It can provide a reference for small- and medium-sized enterprises seeking to implement information systems with high efficiency and at low cost in the age of big data.

Keywords: PDM · PMS · Cloud computing · Big data · BAE

1 Introduction

Advanced Management Information Systems (MIS) can help enterprises to carry out effective scientific management, reduce risk, and improve market competitiveness. However it requires significant funding for enterprises to develop MIS, and improve safety management and maintenance. How to reduce software maintenance and management costs has been a pressing problem. Recently, cloud computing, as a hot technology, has become the business computing model which distributes computing tasks in a resource pool (the cloud) consisting of a large number of computers, so as to enable users to obtain computing power, storage space and information services [1]. The cloud is generally a large server cluster consisting of resources such as calculating servers, storage servers and broadband. As special software is used to realize automatic management, cloud computing can greatly reduce the maintenance costs of a software

© Springer Science+Business Media Singapore 2016
W. Chen et al. (Eds.): BDTA 2015, CCIS 590, pp. 144–154, 2016.
DOI: 10.1007/978-981-10-0457-5_14

system. To maintain consistency, accuracy, and the sharing and security of products, PDM is used to organize and manage relevant data and processes at each stage of the product life cycle [2]. Project management is an important function of a PDM system which operates on the basis of workflow management. Several foreign project management software, such as Primavera's P3, NIKU's Open WorkBench, Welcom's OpenPlan etc., these software are suitable for large and complex project management, and Sciforma's ProjectScheduler, the Primavera's SureTrak, Microsoft's Project and IMSI's TurboProject are suitable for small and medium-sized project management. There are some project management software, such as iMIS-PM, i6P, and TGPMS and so on in china. All of these software can decompose complex tasks, calculate the critical path, provide resource usage report and provide easy to learn and easy to use client interface. However, existing PMS systems is expensive, not suitable for PDM system and the internet environment. With the development of cloud computing technology, it is becoming more common to establish the PMS on the internet platform. It is of practical significance, when developing a PMS, to realize effective integration of the PDM system and Enterprise Resource Planning (ERP) system, progress control, and comprehensive monitoring of product manufacturing projects. The main contribution of this study is the analysis, design and implementation of PMS in PDM environment on cloud computing platform.

2 Project Management in the PDM

2.1 The Basic Concept of Project Management

PDM integrates related information (such as Bills of Materials (BOM), process data, processing data, and engineering data) with a related process (*e.g.* design, examination, and approval) of a product. Its functions include document management, workflow and process management, product structure and configuration management, parts classification management, project change management, project management, and other functions [3]. Project management of PDM refers to project managers coordinating activities for all tasks in a project involving design and manufacture of a product. It includes management and configuration of project planning, organization, personnel and related data, monitoring and tracking the running status of the project to realize the goals of the project using a system viewpoint, and methods and theories within limited resources constraints. A PDM system deals with complicated product design projects and the complicated management processes thereof, multiple levels of product, more participants, multiple tasks and multiple departments in modern machinery manufacturing industries. This produces huge data (design drawing, design specifications, *etc.*) and consumes more computing and storage resources. So it is suitable, when building the PMS of a PDM, to use a cloud computing service platform. PMS projects include process design and approval of new products, or revised products in process design, quasi-archive design processes, and processes for quasi-archive conversion to a full archive design, product design changes, *etc.* [4]. The quasi-archive design process is shown in Fig. 1. Each project decomposes into tasks in accordance with the project contents including: design, audit, approval, standardized examination, archiving, *etc.*

Large tasks are apt to be divided into sub-tasks. Each sub-task is designed to be an independent work package, so it is easy to allocate tasks and calculate workload: each task has a sequence according to the management of the workflow [5].

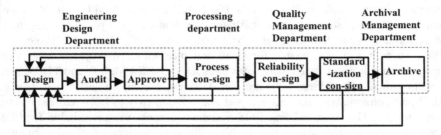

Fig. 1. Project management process: a quasi-archive product design

2.2   Organization Structure Model of Project Management

Project management mainly includes two parts: organization management and control management in the PDM platform. The former functions as follows: defining the project reasonably, decomposition of project tasks, division of project groups, definition of project roles, assignation work to staff, and establishing the relationship between groups, users, roles, and tasks according to their characteristics and the project plan. The latter mainly executes functions, such as tracing and controlling progress on a project, allocation and adjustment of resources, outlay and cost control, *etc*. All those functions are applied to complete the project within the planned time and resource constraints.

Figure 2 shows the organization structure model for project management. The model consists of four views (product view, task view, role view, and user view) and their inter-relationships. The product view is a generalized BOM, including both the hierarchy structure of the product tree, and all kinds of product documents in the design and manufacturing process. All of these objects are called controlled objects, such as design documents for parts, design documents for a process, engineering variation notices, *etc*. Product view can be represented by 4-tuple pv, such as pv ::= <PRO,ASS, PAR,DOC>, wherein PRO indicates product, ASS is assembly, PAR denotes parts, and DOC indicates all types of related documents. Role view mainly refers to all kinds of roles participating in the project, such as: the chief designer, deputy chief designer for structural design, team leader for structural design, structural designer, *etc*. It can be represented by 3-tuple rv, such as rv ::= <P,R,PR>, wherein P is a project, R indicates the participation role of P, and PR is the parent role of R. User view describes how a project manager assigns staff to different working groups according to duties or tasks. A project has multiple working groups: a working group consists of multiple people. User view can be presented by 3-tuple uv, such as uv ::= <P,G,U>, wherein P is a working group, G indicates a working group of P, and U is all users of G. The task view of a project describes the work breakdown structure (WBS) of the project by task. Tasks need to be decomposed in a more complete and independent way so as to confirm responsibilities, and calculate workloads and costs easily. Tasks usually exist in the form of a work package. This view can be represented by 3-tuple tv, such as tv ::= <P,T,PT>, wherein P indicates project, T is the specific task or sub-task of P, and PT indicates the parent task of T. The relationships between product view, project view, view, and user view are complex: they jointly comprise the project network diagram with multiple tasks according to completion order, as shown in the intermediate region of Fig. 2. U21, as a member, may authorize multiple different roles (such as R111, R211) and perform different project tasks (*e.g.* T13, T16) in different working groups (*e.g.* G1 N, G21, G2 N). Those tasks could be design, approval, or archived according to different controlled objects of the product (PAR1, DOC1), wherein task T13 can be realized by the workflow management system. This relationship can be represented by 4-tuple pos, such as pos ::= <pv,rv,uv,tv>.

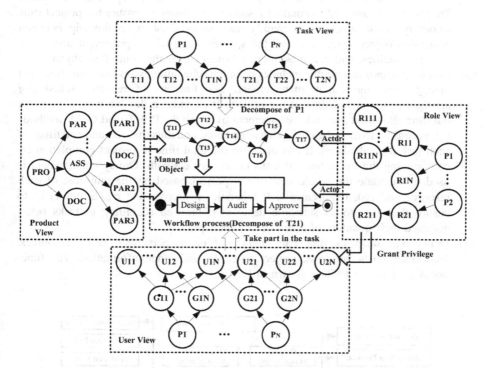

Fig. 2. Project organizational structure model of PDM

2.3   Function Analysis of PMS

According to the definition and content of project management, the main functions of project management are:

(1) *Project management:* this includes definition, modification, supervision, and task management of a project. The definition is to define the name, nature, start time, end time, main content, status and header of a project based on the work to be done. The supervision is to monitor project progress and resource allocation;

while task management aims to divide the contents of the project into several independent task lists which establishes the task number, name, content, start time and end time, precursor tasks, subsequent tasks, leader, task status, *etc.*

(2) *Organisation management:* this consists of workgroup definition, role definition, user management, assigning roles to users, task allocation, permission set-up, *etc.* According to the needs of the project, its organisation is divided into several working groups, such as product structure design, process design, quality control, other standardised groups, *etc.*, and according to the type of job, defines roles such as chief designer of the structure, process designer, and standardised engineer [6]. Finally, the project manager assigned each staff member to a group to act in their corresponding role. Over the project life cycle, the same user may play different roles and can participate in many different working groups simultaneously. Therefore it is essential to conduct resource scheduling to ensure the project runs smoothly. Task allocation is mainly used to establish a relationship between controlled object, role, user, and the specific tasks of the project: it also adds, deletes, modifies, and executes query privileges over the controlled object.

(3) *Resource management:* this includes resource scheduling, workload statistics, cost management, and outlay management. The function of resource scheduling arranges progress and time of the project. It provides a flexible way of carrying out time allocation and task management by using the PERT, and CPM, methods [7], according to the relationship between tasks, and thus the start and end times of each task are calculated. It can optimise the total time to project completion by compressing the finish time of tasks on the critical path. Outlay management is used to calculate the whole cost of the project based on the unit cost of each resource and task requirements. It schedules resources and reallocates tasks to ensure that the project schedule delay is minimised according to tasks being assigned their quantum of resource and their priority.

(4) *Query and statistics:* this can query and count personnel, tasks, resources, workloads, costs, progress, project schedule, and other information. The functional structure of PMS is shown in Fig. 3.

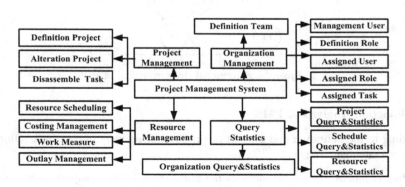

Fig. 3. Functional structure of the PMS

2.4   Data Analysis in the PMS

According to the DFD description of the PMS, this research determined the following entities and relationships among them: project, task, workgroup, roles, users, user organisation, task assignment, controlled objects, operation permissions, user log-in information, task scheduling, project cost, *etc.* The relationships between these entities are mostly one-to-many relationships, and an E-R database concept model is designed as shown in Fig. 4.

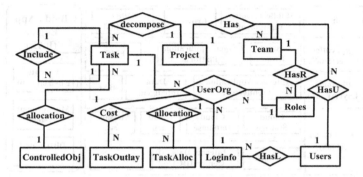

Fig. 4. The entity-relationship model of the PMS database

Figure 4 omits all the attributes of each entity set and relationship set. The corresponding database logical data model is outlined thus: (the underlined attributes are defined as primary keys, wavy lines denote foreign keys):

Project(projectNO, pname, pmanager, pstime, petime, pcontents, pftime, pstatus), Team(teamNO, projectNO, teamname, workdesc), Users(userNO, teamNO, uname, address, gender, e-mail, tele, qq), Loginfo(loginNO, userNO, logpwd, isadmin, ischarger), Roles(roleNO, teamNO, role name, roledesc), Controlled Obj(object NO, objname, objType, objlvl, isvalid), Task(taskNO, projectNO, ptaskNO, taskname, startDate, endDate, contents, ldays, tcontents, tstatus, taskType), UserOrg(roleNO, taskNO, loginNO, permit, cost), Taskalloc(allocNO, roleNO, taskNO, loginNO, ttaskstate, performance, ftime, isvalid), TaskOutlay(costNO,roleNO,taskNO, loginNO, costType, realCost, makedate).

This database logical model is the basis for the physical database design and is realized through a particular database, such as Oracle, MySQL, *etc.*, which can be used to create tables, views, procedures, triggers, *etc.*

3   System Architecture: Baidu Cloud Computing Platform

Cloud computing is a distributed computing model, including a hardware platform (the cloud infrastructure layer), the cloud platform layer, and a cloud service application layer. Cloud computing provides a hardware and software service mode for an enterprise according to need and it charges according to use [8]. Cloud computing takes infrastructure as a service (IaaS), platform as a service (PaaS), and software as a service (SaaS) according to service type. The Baidu cloud can handle software applications, photos, documents, music, and contacts can be used on all types of equipment. This public cloud architecture is described in Fig. 5: the architecture uses the distributed computing model and is divided into an infrastructure layer, a platform layer, an application layer, and client layer from bottom to top. To let users develop and deploy applications, the cloud provides an open application platform.

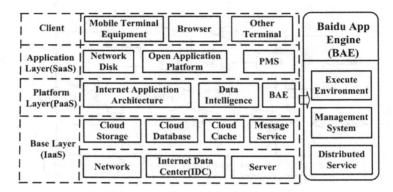

Fig. 5. Baidu public cloud architecture

BAE provides a multilingual, elastic distributed server running platform, a cloud database, cloud storage, cloud messages, a cloud pipeline, a cloud trigger, and other services, as well as support for PHP, Java, and other high-level languages. This helps developers to rapidly develop and deploy applications in the PaaS level of the Baidu cloud. The BAE architecture is divided into an execution environment, a management system, and multiple distributed services. Among which, the execution environment includes the access layer, code execution layer, and data layer. The management system includes: a user management platform, system management platform, scheduling service, monitoring service, resource audit services, *etc*. Distributed services include many basic services and business components, such as databases, caches, cloud storage, FetchUrl, and other services. The latest BAE technology adopts light-weight virtual machine technology in the underlying software to solve resource isolation problems in older versions. It is not limited to a specific running environment and programming language, Its local development environment is consistent with the running environment of the cloud for developers to reduce the costs of learning, development, and transfer: this also enhances developer productivity [9, 10].

4 Software Architecture Design of a PMS System Based on BAE

The architecture of PMS systems needs to be designed by selecting a J2EE technology match according to the Baidu cloud open platform, as shown in Fig. 6. This framework uses five layers: the user browser layer, user presentation layer, control layer, data access layer, and database layer. The functions of each layer are described below:

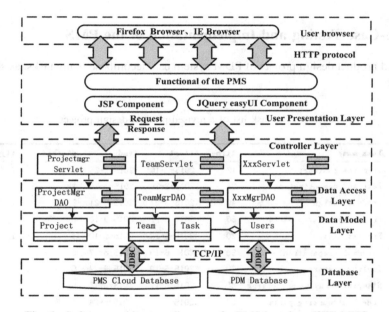

Fig. 6. Software architecture diagram of a PMS based on J2EE MVC

The user browser layer is for a client with only a browser installed, yet who can connect to the internet.

The user presentation layer is the user-interface of the PMS system, and is realised by JSP technology and JQuery easyUI component technology.

The controller layer is realised by Servlet technology on the J2EE platform, and is mainly responsible for the parameters of the user interface and user requests: it executes corresponding commands according to requests, then returns operation results to the client in a certain data format. It operates the database by accessing the DAO layer therein, so this layer acts as a bridge between the presentation layer and data access layer.

The data access layer mainly realises CRUD operations on the database using JDBC technology, it describes how to access the database, avoids the controller layer directly operating the database, reduces the coupling of management systems, makes the program more logical, provides data format in JSONArray type for the upper layer, and allows this to then be accepted and shown by a JQuery easyUI component.

The database layer mainly provides data storage services and query and statistical functions. The PMS cloud database uses the MySQL database, which currently fails to support trigger and stored procedures, so the MySQL database has limited functionality. The PMS accesses only the database through SQL statements. The user must log-in to the Baidu cloud open platform, manually create a cloud database, and complete the work (e.g. by creating tables, indices, and views). Baidu Inc. and users are responsible for cloud database security, thus reducing the cost of user management and operation. User data safety on a public cloud is low and has been a hot research topic.

5 Use-Case Testing and Implementation of the PMS

A revised product design project of PDM system can decompose many tasks, as shown in Table 1.

Table 1. The task list of a revised product design project

tno	Task name	ttime	tstart	tend	ptaskid	ttype
1	Sign the contract	12	2015-2-10	2015-2-25		Y
2	Test and demonstration	21	2015-2-26	2015-3-26	1	Y
3	Product design	40	2015-3-27	2015-5-21	2	Y
4	Procure purchased parts	80	2015-5-22	2015-9-10	3	
5	Process design	20	2015-5-22	2015-6-18	3	Y
6	Material preparation	50	2015-6-19	2015-8-27	5	
7	Tooling prepare	70	2015-6-19	2015-9-24	5	Y
8	Stamping abrasive processing	40	2015-6-19	2015-8-13	5	
9	Order outsourcing parts	24	2015-5-22	2015-6-24	3	
10	Outsourcing processing	46	2015-6-25	2015-8-27	9	
11	Parts processing	45	2015-9-25	2015-11-26	6, 7	Y
12	Parts Heat treatment	45	2015-11-27	2016-1-28	11	Y
13	Stamping parts processing	28	2015-9-25	2015-11-3	8, 6, 7	
14	Test equipment processing	63	2015-5-22	2015-8-18	3	
15	Product assembly	45	2016-1-29	2016-3-31	12, 10, 13, 4	Y
16	Inspection and testing	38	2016-4-1	2016-5-24	15, 14	Y
17	Product acceptance	9	2016-5-25	2016-6-6	16	Y

According to Table 1, it can calculate earliest start time, earliest finish time, latest start time and latest finish time. It can determine the critical path, calculate the total time limit of the project by using CPM technology. Users can track the progress of the project and optimize the project by adjusting the usage of resources, time and cost.

The implement of PMS use the MVCII architecture of J2EE. It can reduce the cost of developing multi-layer applications and their complexity, it can also strongly support the integration of existing applications and is easy for users to package and deploy applications. It enhances security and improves performance. The data access layer, controller layer, and user presentation layer of the PMS are respectively realized by java objects, Servlet components, and JSP components. It designs page layouts using light-weight easyUI component technology from JQuery in JSP pages. EasyUI is a JQuery based plug-in for the front-end user interface: it not only supports multiple page style sheets, and provides activeX controls such as accordion, combobox, menu, dialog, tabs and datagrid components [11], but its library is also composed of three script files (jquery.min.js, jquery.easyui.min.js and easyui-lang-en.js) and two style sheet files (easyui.css and icon.css) which are used on the head of each JSP page.

The main PMS interface is shown in Fig. 7: the home page is divided into three parts: the top of home page is the system menu (area ①) including functions of registration, log-in and log-off: the centre of the main page is divided into the function menu lists of the PMS (area ②) on the left. The system opens the corresponding functions of the JSP page on the right client area (area ③) by clicking a menu item from the menu lists: each JSP page completes the functions of adding, deleting, updating, querying, and statistics. Area ⑤ is a function page for updating project information. The bottom of home page is the copyright statement section of website (area ④).

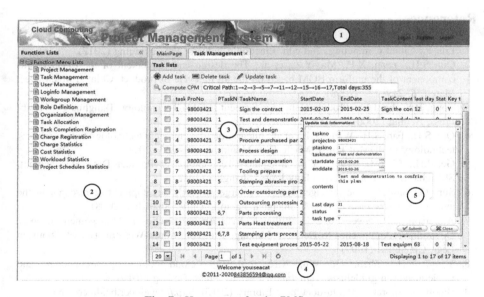

Fig. 7. Home-page for the PMS system

Developers can submit codes using SVN client tools after fulfilling the developing and debugging system requirements in the local environment, before release to the Baidu cloud server platform. The cloud environment is responsible for hosting user systems and Baidu Inc. charge a service fee according to flow occupancy.

6 Conclusion

Cloud computing technology provides an unprecedented opportunity for building an information system for small- and medium-sized manufacturing enterprises (SMME) in the age of big data. Cloud computing shows advantages in reducing development cost, reducing the number of staff, making full use of existing computer and network resources, reducing maintenance cost, providing security, *etc.* Through an example analysis - the design and implementation process of a PDM system - this work validated the feasibility of PMS system development and implementation on the Baidu public cloud computing platform. It also provided a feasible method and reference implementation for the migration of an MIS to a cloud computing platform for SMMEs. It therefore had good application prospects.

Acknowledgments. This work was supported by science research funds of Nanjing Institute of Technology (No. CKJA201401); The humanities and social science research planning funds of the ministry of education, China (No. 12YJCZH209); and college student practice innovation training funds of Jiangsu Province, China (No. 201411276012Z, 201511276009Z).

References

1. Liu, P.: Cloud Computing, 2nd edn. Electronic Industry Press, Beijing (2012)
2. Qiu, S.H., Yuan, X.F., Ma, Y.Z., et al.: The design and implement of the management component of the uniform BOM based on the all lifecycle of product. Chin. J. Mach. Des. Manuf. **6**, 219–221 (2008)
3. Tong, B.S., Li, J.M.: Product Data Management (PDM) Technology. Tsinghua University Press, Beijing (2000)
4. Qiu, S.H., Fan, S.H., Wang, Z.L., et al.: Integration of component technology with workflow technology in PDM. Chin. J. Mach. Des. Manuf. **4**, 251–253 (2011)
5. Ge, J.H., Lv, M., Wang, Y.P.: Integrated Technology of Product Data Management. Shanghai Science and Technology Press, Shanghai (2012)
6. Sun, G.: Studying the management of the large project's development & research in the aviation industry base on PDM. Shanghai Jiaotong Univ., Shanghai (2008)
7. Anderson, D.R., Sweeney, D.J., Williams, T.A.: An Introduction to Management Science: Quantitative Approaches to Decision Making, 10th edn. South-Western, Div of Thomson Lear, Cincinnati (1999)
8. Yang, Z.H., Zhen, Q., Wu, X.H.: Enterprise cloud computing architecture and implementation guidelines. Tsinghua University Press, Beijing (2010)
9. Baidu Incorporated, Baidu document Library (2015). http://developer.baidu.com/bae
10. Baidu Incorporated, Baidu open cloud platform help (2014). http://developer.baidu.com/wiki/index.php?title=docs/cplat/bae
11. jQuery EasyUI Documentation (2014). http://www.jeasyui.com/documentation/index.php

A Ring Signature Based on LDGM Codes

Mingye Liu[1], Yiliang Han[1,2(✉)], and Xiaoyuan Yang[1]

[1] Department of Electronic Technology, Engineering University of Armed Police Force,
Xian, Shanxi, China
yilianghan@hotmail.com
[2] College of Information Science and Technology, Northwestern University,
Xian, Shanxi, China

Abstract. McEliece cryptosystem is a public key cryptosystem that combines channel coding and encryption, and the oldest PKC that is conjectured to be post-quantum secure. To decrease the key size of the original scheme, alternative codes have been adopted to replace Goppa codes. In this paper, we propose a ring signature using low-density generator-matrix codes. Our new scheme satisfies anonymity and existential unforgeability under chosen message attacks (EUF-CMA). As for efficiency, the number of decoding operations has been reduced largely compared with ZLC ring signature, and the size of the public key is about 0.2 % of the ZLC scheme.

Keywords: Post-quantum cryptography · McEliece cryptosystem · Low-density generator-matrix code · Ring signature

1 Introduction

RSA and McEliece cryptosystem are two oldest public-key cryptosystems, and they are based on the large integer factorization and the syndrome decoding problem (SD) respectively. The large integer factorization is a NP problem, while the binary syndrome decoding problem has been proved to be NP-Complete by Berlekamp et al. [1]. The computational assumption in RSA could be broken in a quantum setting by Shor's algorithm, but SD is believed to be secure against quantum attacks. Moreover, MeEliece cryptosystem is safer and faster than RSA. However, its weakness is obvious. Its public key size is too large, and the information rate is low, about 50 % [2].

In order to promote the application of code-based public cryptosystem, many attempts have been made to reduce the public key of the McEliece cryptography. A promising method is to use other code to replace Goppa code to archive the goal. Among these, low-density parity-check (LDPC) codes receives many attentions [3–7].

This work is supported by National Natural Science Foundation of China (61572521, 61103231, 61272492), Project funded by China Postdoctoral Science Foundation (2014M562445, 2015T81047), and Natural Science Basic Research Plan in Shanxi Province of China (2015JM6353, 2014JQ8358, 2014JQ8307).

© Springer Science+Business Media Singapore 2016
W. Chen et al. (Eds.): BDTA 2015, CCIS 590, pp. 155–162, 2016.
DOI: 10.1007/978-981-10-0457-5_15

Nowadays, LDPC-code-based schemes are mainly about encryption, and the signature schemes are much fewer. The first proposal of a digital signature scheme based on low-density generator matrix (LDGM, a special LDPC code) codes and sparse syndromes has appeared in [7]. Before the scheme, code-based signature are mainly CFS signature and some signature schemes with special properties constructed based on CFS [8]. The main idea of CFS is that a signer takes the hash value $h(M)$ of a plaintext M as a syndrome, and then use the private key, the parity check matrix to perform syndrome decoding on $h(M)$, and the resulting error vector e is the digital signature. The main drawback of the CFS scheme is that it is very hard to find a function that quickly transforms an arbitrary hash vector into a correctable syndrome. In paper [7], Marco Baldi uses LDGM codes to replace Goppa codes to construct a digital signature. With the advantage of that LDGM codes are systematic codes, the signer will easily decode a $h(M)$ into a error vector e. Therefore, using LDGM codes to achieve digital signature is a breakthrough.

In order to leak a secret anonymously, Rivest et al. proposed a new digital signature called ring signature in 2001 [9]. This special signature can provide unconditional anonymity for the signer, which is very useful in the particular environment that needs to protect the identity of the signer. In paper [10], Zheng, Li and Chen (ZLC) proposed a code-based ring signature, which extends the CFS signature scheme and is based on the syndrome decoding problem. Because of the inherent weakness of CFS scheme, it is complicated to achieve the ZLC signature.

Inspired by Marco Baldi and ZLC scheme, we propose a ring signature using LDGM codes, and give a security proof of our scheme under random oracle, showing that it satisfies anonymity and existential unforgeability under chosen message attacks (EUF-CMA), and it's more efficient than the ZLC scheme.

2   Ring Signature Using LDGM Codes

In this section, we propose a new ring signature using LDGM codes.

LDGM codes have been used since a long time for transmission application and are recognized to achieve very good error correcting performance. LDGM codes are a special class of codes which are strictly related to LDPC codes: the only difference is in that LDGM codes have a sparse generator matrix and not necessarily a sparse parity-check matrix, while the opposite occurs for LDPC codes.

2.1   Security Model

The anonymity of our scheme is obvious, we only provide the proof of unforgeability.

Definition 1: For any polynomial bound adversary A, if the probability for him to win the following game is negligible, then we say that the signature scheme satisfies existential unforgeability under chosen message attacks (EUF-CMA). The game includes the following three procedures:

(1) **Setup:** Challenger C chooses a system security parameters n, runs this algorithm, sends the public keys set $S = \{PK_i\}$ to adversary A, and preserves the private $\{SK_i\}$.

(2) **Sign Queries:** Adversary A adaptively chooses message $M \in \{0, 1\}^*$. Challenger C runs the algorithm sign, then sends the result x to A.

(3) **Forgery:** Adversary A output a ring signature (R', M, x'), if $Verify(R', M', x')$ output "*Valid*", and (R', M', x') hasn't appeared in the queries, then A wins the game.

2.2 Scheme Description

We use the following notation to describe our scheme. N is the number of potential signers, and l is the number of signers participating in the signature generating. Use S_i and S_r to denote a potential signer and the ring signer, respectively. M is a message and h is a hash function on F_2^{n-k}. $(s_1|s_2)$ means the concatenation of s_1 and s_2. Let $u \xleftarrow{R} U$ represents choosing a vector randomly from a set U. The ring signer runs the following algorithm to produce a ring signature on M.

Key Generation: Potential signers S_i generate their private/public keys as in the [11]:

- The public key: $H_i = Q_i^{-1} \tilde{H} P_i^{-1}$
- The private key: $(P_i, \tilde{H}_i, Q_i)$

The real ring signer is S_r.

Signature: To sign message M

(1) Compute the hash value of M, that is $h(M)$.

(2) Initialization: S_r randomly chooses a vector $v \in F_2^{n-k}$, and chooses a vector $z_i \in F_2^{n-k}$ respectively for the members that participate in the ring signature, $i = r+1, r+2, \ldots, l-1, 0, 1, 2, \ldots, r-1$, their weight is $w(z_i) = t$.

(3) Ring sequences generation:

$$s_{i+1} = h(N \,|h(M)|\, v)$$
$$s_{i+2} = h(N \,|h(M)|\, H_{r+1} z_{r+1}^T \oplus s_{r+1})$$
$$\vdots \qquad\qquad \vdots$$
$$s_l = h(N \,|h(M)|\, H_{l-1} z_{l-1}^T \oplus s_{l-1})$$

Set $s_l = s_0$

$$s_1 = h(N|\, h(M)|\, H_0 z_0^T \oplus s_0)$$
$$\vdots \qquad\qquad \vdots$$
$$s_r = h(N|\, h(M)|\, H_{r-1} z_{r-1}^T \oplus s_{r-1})$$

(4) S_r applies his private key to sign M: computes $s_r' = Q_r \cdot (s_r \oplus v)$, because the parity-check matrix $\tilde{H}_r$ of LDGM is in the systematic form, it is easy for the signer S_r to compute the corresponding error vector e, and $\tilde{H}_r e^T = s_r'$. Randomly chooses a code word $c \in C$, then get the signature $z_r = (e + c)P^T$.

(5) Output the ring signature $(s_0, z_0, z_1, \ldots, z_{l-1})$

Verify: After receiving the message M and the signature $(s_0, z_0, z_1, \ldots, z_{l-1})$, the verifier computes $s_i = h(N| h(M)| H_{i-1}z_{i-1}^T \oplus s_{i-1})$, $i = 1, 2, \ldots, l$. If $s_0 = s_1$, then accept the signature, reject it otherwise.

2.3 Analysis of the Scheme

Correctness Verification

$$H_r z_r^T = Q_r^{-1} \tilde{H}_r P_r^{-1} P(e^T + c^T) = Q_r^{-1} \tilde{H}_r (e^T + c^T) = Q_r^{-1} \tilde{H}_r e^T = Q_r^{-1} Q_r(s_r \oplus v)$$
$$= Q_r^{-1} Q_r(s_r \oplus v) = s_r \oplus v$$

Then $H_r z_r^T \oplus s_r = v$

That means $s_{r+1} = h(N| h(M)| H_r z_r^T \oplus s_r) = h(N| h(M)| v)$

Run the algorithm ahead, we will get that $s_l = s_0$, so we have verified it.

Security Proof
Anonymity:

Any attackers outside a ring of l possible users has probability $1/l$ to guess the real signer, because all z_i but z_r are taken randomly from F_2^n. In fact, $z_r = \bar{z} \oplus \bar{z}_r$, because $\bar{z}$ is distributed uniformly over F_2^n, this result in that z_r is uniformly over F_2^n.

Unforgeability:

We rely its difficulty to two problems issued from coding theory.

Definition 2 (Parameterized Bounded Decoding problem)

Input: A $(n - k) \times n$ binary matrix H and a syndrome $s \in F_2^{n-k}$

Output: A word $e \in F_2^n$ such that $w(e) \leq \frac{n-k}{\log_2 n}$ and $He^T = s$

We also consider the LDGM Code Distinguishing problem (LD).

Definition 3 (LDGM Code Distinguisher). A distinguisher D for a permuted LDGM Code is an algorithm which takes as input a parity check matrix H and outputs a bit. D outputs 1 with probability $\Pr[H \overset{R}{\leftarrow} C(n,k):D(H) = 1]$ if H is a random binary parity check matrix of a LDGM code $C(n, k)$ and outputs 1 with probability $\Pr[H \overset{R}{\leftarrow} B(n,k):D(H) = 1]$ if H is a random binary matrix. We call the advantage of a distinguisher D, denoted by $Adv_{n,k}^{GD}(D)$, the following quantity:

$$\left| \Pr[H \overset{R}{\leftarrow} C(n,k):D(H) = 1] - \Pr[H \overset{R}{\leftarrow} B(n,k):D(H) = 1] \right|$$

$Succ_{n,k}^{LPBD}(\tau)$ denotes the probability of success of the best algorithm that solves the LDGM Parameterized Bounded Decoding problem in time τ and $Adv_{n,k}^{GD}(\tau)$ means the advantage of the best distinguisher for a permuted LDGM code in time τ.

Theorem 1 (Unforgeability). Let A denote a l-adversary against the scheme proposed in Sect. 2.3 that outputs a forgery in time τ with probability p. A makes q_H queries to a hash oracle H, q_ε queries to a cipher oracle ε, and q_Σ queries to a signing oracle Σ. A also can corrupt the private keys of $l-1$ users. We have:

$$p \le q_\varepsilon q_H (N-t+1) \binom{N}{t} Succ_{n,k}^{LPGB}(\tau') + Adv_{n,k}^{LD}(\tau') - \frac{q_\varepsilon}{2^{mt}} \binom{N}{l} \left(\frac{q_\varepsilon}{N-l}\right)^{N-l}$$

And $n = 2^m, k = 2^m - mt, \tau' = Nmt^2 q_\Sigma$

Proof: Let p be the probability of success in time τ of A. We describe how to use it to build an algorithm D that inverses LPBD problem. D receives as inputs a random $mt \times 2^m$ binary matrix $\boldsymbol{H}^*$ and a random vector s^* of F_2^{mt}. Its goal is to output x^* such that $w(x^*) \le t$ and $\boldsymbol{H}^*(x^*)^T = s^*$.

D randomly chooses a user i_0 in the set of users in the ring N and a subset $I_0 \subset N$ and $l-1$ elements such that $i_0 \notin I_0$. D hopes that all corrupted users are in I_0. It sets $i_0's$ public key PK_{i_0} to $\boldsymbol{H}^*$ and for all users $i_0 \in I_0$, sets key pair (PK_i, SK_i) by running the generation algorithm. For all other users, D uses a parity check matrix of random permuted LDGM code as public key, and there is no need to use their private keys. D also randomly picks two indexes $q_{\varepsilon,0}$ and $q_{H,0}$ respectively in $[1, \dots, q_\varepsilon]$ and $[1, \dots, q_H]$ where q_H is the total number queries to the hash oracle H and q_ε is the total number of queries (E or E^{-1}) to the cipher oracle. Then, it picks a random vector $\bar{s} \in F_2^{mt}$. A is initialised with the set of public keys $\{PK_i\}_{i \in N}$.

D classically simulates the oracle H by answering a random value in F_2^{mt} for each new query and maintaining a list of queries. It also simulates the oracle ε according to a permutation by answering a random value for each new query. But to the $q_{H,0}$-th query to the hash oracle, it answers by $\bar{s}$ and to the $q_{\varepsilon,0}$-th query E^{-1} to the cipher oracle, it answers by $s^* + \bar{s}$. Note that D must maintain a list of the queries such that it remembers whether x is the answer to a query $E^{-1}(y)$ or y is the answer to a query $E(x)$. A may corrupt up to $l-1$ users by querying its private key. D trivially answers the queries but fails if A queries the key of user i_0.

Bresson *et al.* showed that, with probability at least $p + \frac{q_\varepsilon}{2^{mt}} \binom{N}{l} \left(\frac{q_\varepsilon}{N-l}\right)^{N-l}$ [11], A produces a forgery such that there are at most $N-l$ cipher queries to ε for $P^*(i)$. Thus there are at least l indexes such that A made a decrypt query to ε for $P^*(i)$; let I^* be the set of those indexes. Since A can corrupt up to l users, there are at least an index i^* such that A did not corrupt its private key. Let q^* be the index of the query to ε for $P^*(i^*)$. With probability at least $\frac{1}{q_\varepsilon}, q^* = q_{\varepsilon,0}$; with probability at least $\frac{1}{n-l+1}, i^* = i_0$ and with probability at least $\frac{1}{q_H}$ the $q_{H,0}$-th query to the hash oracle was $(M\|r_{i_0})$. Then x_{i_0} has weight less than t and satisfies

$$Hx_{i0}^T = E_{H(M),i}^{-1}(P(i_0)) + H(M\|r_{i_0}) = s^* + \bar{s} + \bar{s} = s^*.$$

Thus, x_{i_0} is a t-decoding of s^*.

The running time τ' of D is essentially the running time τ of A and the cost of Nq_Σ syndrome computation, whose cost is mt^2. Note that replacing the public of user i_0 does not alter the probability of success of the simulation more than the advantage $Adv_{2^m,2^m-mt}^{LD}(\tau')$ of the best adversary at solving the permuted LDGM code distinguishing: otherwise D should provide a better distinguisher. We denote $p_{LPBD} = Succ_{2^m,2^m-mt}^{LPBD}(\tau')$ and $p_{LD} = Adv_{2^m,2^m-mt}^{LD}(\tau')$; we obtain:

$$\left(p + \frac{q_\varepsilon}{2^{mt}}\left(\frac{q_\varepsilon}{N-l}\right)^{N-l} + p_{LD}\right)\frac{1}{q_\varepsilon q_H(N-t+1)\binom{N}{t}} \leq p_{LPBD}$$

This concludes the proof.

Efficiency of the Proposed Scheme

(1) **Complexity of Signature**

Based on CFS signature, the ZLC ring signature scheme averagely needs to try about $t!$ times to get the decodable syndrome. As for the LDGM code, we can directly decode any hash value to find the corresponding error vector, so we only need to run 1 decoding operation. So the complexity is much lower.

(2) **Complexity of Verify**

A verifier needs to derive z_i from s_i, run l matrix multiplications $H_i z_i^T$, and 2 hash computation for each s_i. And each matrix multiplication includes n computations of column operation. Therefore, the whole computation is nl column operations and $2(l+1)$ hash computations.

(3) **Size of Public Key**

In the ZLC ring signature, in order to make the computation complexity of forging achieve the level of $O(2^{80})$, we need to use a Goppa code with length $n = 2^{21}$ bits, and the length of redundancy $r = 21 \times 10 = 210$ bits [12]. So every users needs to have a public key with the size 4.4×10^8 bit = 52.5 MB. In our scheme, for the same security level, every user only needs to use a public key of 117 KB [7]. Therefore, the public key size of our scheme is about 0.2 % of the ZLC ring signature, so we can largely reduce the public key size (Table 1).

Table 1. Comparison of the efficiency of the two schemes

Scheme	Decoding operations	Verifying complexity	Public key size
ZLC	$t!$	$nl, 2(l+1)$	52.5 MB
New scheme	1	$nl, 2(l+1)$	117 KB

3   Conclusion

In the post-quantum cryptography age, code-based cryptosystem is safe enough so it can take the place of the cryptosystem based on discrete logarithm and large integer factorization, but the weakness of large public key hinders its extensive use. Inspired by the ZLC ring signature, and use LDGM code, we propose a new ring signature using LDGM code in this paper. We elaborate the procedure of the signature and the verification, and prove that the scheme satisfies existential unforgeability under chosen message attacks (EUF-CMA). As for efficiency, compared with the ZLC ring signature, our scheme has fewer decoding operations and smaller public key size.

References

1. Berlekamp, E., McEliece, R., van Tilborg, H.: On the inherent intractability of certain coding problems. IEEE Trans. Inf. Theory **24**(3), 384–386 (1978)
2. Baldi, M.: QC-LDPC Code-Based Cryptosystem. QC-LDPC Code-Based Cryptography, p. 75. Springer International Publishing, Heidelberg (2014)
3. Monico, C., Rosenthal, J., Shokrollahi, A.: Using low density parity check codes in the McEliece cryptosystem. In: Proceedings of the IEEE ISIT 2000, Sorrento, Italy, p. 215, June 2000
4. Baldi, M: Quasi-cyclic low-density parity-check codes and their application to cryptography. Ph.D thesis, Universita Politecnica delle Marche, Ancona, Italy (2006)
5. Baldi, M., Bambozzi, F., Chiaraluce, F.: On a family of circulant matrices for quasi-cyclic low-density generator matrix codes. IEEE Trans. Inf. Theory **57**(9), 6052–6067 (2011)
6. Baldi, M., Bianchi, M., Chiaraluce, F.: Security and complexity of the McEliece cryptosystem based on QC-LDPC codes. IET Inf. Secur. **7**(3), 212–220 (2013)
7. Baldi, M., Bianchi, M., Chiaraluce, F., Rosenthal, J., Schipani, D.: Using LDGM codes and sparse syndromes to achieve digital signatures. In: Gaborit, P. (ed.) PQCrypto 2013. LNCS, vol. 7932, pp. 1–15. Springer, Heidelberg (2013)
8. Courtois, N., Finiasz, M., Sendrier, N.: How to achieve a McEliece-based digital signature scheme. In: Boyd, C. (ed.) ASIACRYPT 2001. LNCS, vol. 2248, pp. 157–174. Springer, Heidelberg (2001)
9. Rivest, R.L., Shamir, A., Tauman, Y.: How to leak a secret. In: Boyd, C. (ed.) ASIACRYPT 2001. LNCS, vol. 2248, pp. 552–565. Springer, Heidelberg (2001)
10. Zheng, D., Li, X., Chen, K.: Code-based ring signature scheme. Int. J. Netw. Secur. **5**(2), 154–157 (2007)

11. Bresson, E., Stern, J., Szydlo, M.: Threshold ring signatures and applications to ad-hoc groups. In: Yung, M. (ed.) CRYPTO 2002. LNCS, vol. 2442, pp. 465–480. Springer, Heidelberg (2002)
12. Finiasz, M., Sendrier, N.: Security bounds for the design of code-based cryptosystems. In: Proceedings of the ASIACRYPT 2009, Tokyo, Japan, 6–10 December 2009. LNCS, vol. 5912, pp. 88–105. Springer, Heidelberg (2009)

SparkSCAN: A Structure Similarity Clustering Algorithm on Spark

Qijun Zhou and Jingbin Wang[(✉)]

Department of Mathematics and Computer Science,
Fuzhou University, Fuzhou 350108, China
369337098@qq.com, wjbcc@263.net

Abstract. The existing directed graph clustering algorithms are born with some problems such as high latency, resource depletion and poor performance of iterative data processing. A distributed parallel algorithm of structure similarity clustering on Spark (SparkSCAN) is proposed to solve these problems: considering the interaction between nodes in the network, the similar structure of nodes are clustered together; Aiming at the large-scale characteristics of directed graphs, a data structure suitable for distributed graph computing is designed, and a distributed parallel clustering algorithm is proposed based on Spark framework, which improves the processing performance on the premise of the accuracy of clustering results. The experimental results show that the SparkS-CAN have a good performance, and can effectively deal with the problem of clustering algorithm for large-scale directed graph.

Keywords: Directed graph clustering · Parallel algorithm · Spark · RDD

1 Introduction

With the wide application of network data, such as gene regulatory network, social network and other network data in various fields, the scale of directed graph is growing explosively. How to manage and use the massive data has become a hot research topic in recent years [1]. Directed graph contains a wealth of data relationships, such as the behavior of users in social networks. In order to discover the hidden cluster structure in the network, traditional clustering methods are based on the link density, such as Newman [2] algorithms and Kernighan-Lin [3] algorithms, which make the distance between the nodes in the cluster closer, and make the distance between the cluster nodes is far away to achieve the effect of clustering. However, the algorithms above ignore the directed interaction and different functions that nodes may have in the graph data. Based on the link density, Xu Xiao-wei [4] proposed the SCAN algorithm, which is based on the structural similarity. However, the algorithm is only useful to the undirected network of clustering and it doesn't consider the variety of data relationships in the real environment. Zhou Deng-yong [5] proposed a way making the directed edges convert to the undirected edges, but the way ignore the structure information of directed graph. Literature [6] transformed the network clustering problem into the optimization problem of weighted cutting of directed graph for further study. However, literatures [5, 6] did not distinguish the different functions of the nodes.

© Springer Science+Business Media Singapore 2016
W. Chen et al. (Eds.): BDTA 2015, CCIS 590, pp. 163–177, 2016.
DOI: 10.1007/978-981-10-0457-5_16

Chen Jia-jun [7] proposed a directed graph clustering algorithm DirSCAN based on SCAN. Chen Ji-meng [8] proposed a parallel clustering algorithm PDirSCAN which used MapReduce based on literature [7]. Zhao W [9] proposed a clustering algorithm based on MapReduce by looking for connected components, however, there are some problems such as high delay, high I/O operation of HDFS file system, and the poor performance of iterative data processing. In this paper, we propose a structure similarity clustering algorithm based on Spark framework with the advantages of iterative computation. We calculate the similarity of the vertices and build the initial cluster with the distributed framework; we perform cluster label expansion and synchronize operation in parallel to achieve the clustering of the vertices in the graph to reduce the running time and computational cost. The experimental results show that the algorithm can efficiently clustered in the big data environment for directed graph clustering.

The rest of the paper is organized as follows. In Sect. 2, we give a brief introduction to the concept of graph clustering and Spark. In Sect. 3, we introduce our SparkSCAN algorithm in detail. Section 4 presents the experimental results and analysis. Finally we provide our conclusions in Sect. 5.

2 Preliminary

2.1 Spark

Spark [10] is a common parallel framework for the Berkeley AMP Lab UC. Spark has the advantages of MapReduce Job, and intermediate output results can be saved in memory, Job no longer need to read and write HDFS. Thus, Spark can be better applied to data mining and machine learning.

2.2 Rdd

RDD [11] (Resilient Distributed Datasets), is an abstract concept of distributed memory. RDD provides a highly constrained shared memory model, which is a read-only collection of records, and can only be created by performing a set of transformations (such as, join, and map) in the other RDD. These constraints make the cost of achieving fault tolerance very low.

2.3 PDirSCAN Algorithm

A PDirSCAN algorithm based on directed graph is proposed in the paper [8].

Definition 1 (Neighborhood). Given a directed graph $G = \{V, E\}$. The directed edge which from v to u is signed as $< v, u >$, $v, u \in V$. The Neighborhood is a set of nodes and itself which starting from the one step of v, denoted by $\Gamma(v)$.

$$\Gamma(v) = \{u \in V | < v, u > \in E\} \cup \{v\}$$

Definition 2 (Structural Similarity). For two nodes, the more coincident nodes can be reached, the more likely to belong to the same cluster. The definition of structural similarity, denoted by σ, is given by:

$$\sigma(u, v) = \frac{|\Gamma(u) \cap \Gamma(v)|}{\sqrt{|\Gamma(u)| * |\Gamma(v)|}} \qquad (1)$$

Definition 3 (ε Neighborhood-Nodes). The definition of ε Neighborhood-Nodes, denoted by $N_\varepsilon(u)$, is given by:

$$N_\varepsilon(u) = \{v \in \Gamma(u) \mid \sigma(u,v) \geq \varepsilon, \varepsilon \geq 0\}$$

The ε is used to divide the ε neighbor node and non-neighbor node threshold.

Definition 4 (CORE). If a node has enough ε neighborhood-nodes, we called it core. Node u is core if $|N_\varepsilon(u)| \geq \mu$, $u \in V$. μ is a threshold.

Definition 5 (Directly Structure Reachable). If u is one of the v's ε neighborhood-nodes, where v is a core. Then u must belongs to the same cluster with v.

$$DR_{\varepsilon, \mu}(v, u) \Leftrightarrow C_{\varepsilon, \mu}(v) \wedge u \in N_\varepsilon(v) \qquad (2)$$

The cluster is generated from the core in this algorithm. If the v is one of the core u's ε neighborhood-nodes, v is assigned to the same cluster with u. The cluster continued to grow until all the clusters could not be further increased.

Definition 6 (Hub and Outlier). Assume node u does not belong to any cluster. Node u is hub just has node v and w exist in $\Gamma(u)$, which v and w do not belong to the same cluster. Otherwise u is outlier.

3 SparkSCAN

In order to adapt to the large-scale clustering of directed graph, this section we design a parallel algorithm SparkSCAN on Spark.

Definition 7 (Structure Reachable). Given by a directed graph $G = \{V, E\}$. Given a series of vertices $v_1, v_2, ...,v_n$, $v = v_1$, $u = v_n$. We called v and u is structure reachable, which v_i and v_{i-1} is Directly Structure Reachable. If v and u is structure reachable, then u should also belong to the same cluster with v. Any pair node of the same cluster is structure reachable.

In this paper, the operation of the SparkSCAN algorithm is mainly divided into three steps:

1. Parallel recognize ε neighbors nodes and core node, then build the initial clusters;

2. Execute cluster expansion through synchronizing cluster label in parallel, and then, achieve clustering merge;
3. Analysis of clustering results and recognize the hub and outlier node.

In the first step, each node can independently calculate the structural similarity between the other nodes; In the second step, each sub-cluster can be independently calculate the label, cluster labels of vertices can be synchronized according to the intermediate result of the algorithm; In the third step, algorithm can be used to analyze the clustering results of each cluster label. In the process of identifying the hubs and the outliers, each node can do it by itself. SparkSCAN algorithm with the fault tolerance of Spark, so that the whole task will not collapse because of one processing node's paralysis, so as to achieve the purpose of parallel processing.

The overall flow chart of the algorithm is shown in Fig. 1:

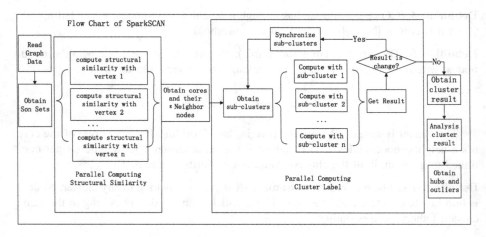

Fig. 1. Flow chart of SparkSCAN algorithm

3.1 Data Structure of SparkSCAN

Data Structure of Graph Storage. There are two kinds of storage methods of giant graph, Edge-Cut and Vertex-Cut. Because per vertex just be stored once by Edge-Cut, so it can save storage space. But if the calculation involves two vertices on the edge are divided into different machines, it communicate and transfer data by crossing machine and cost the network communication traffic. So we use Vertex-Cut way, which each edge only stores one time, and only appears on a machine. This way increases the storage overhead, but it can significantly reduce the amount of network traffic. This method has the advantages especially in large data.

In order to store the graph data effectively, this paper design the data structures to store the nodes and edges with RDD as follows:

Vertex (ID: Long, Arr: String)

Where ID represents the ID of vertex, Arr is a String which represent the properties of vertex, just like "property 1, property 2";

Edge (srcID: Long, dstID: Long, Arr: String)

Where srcID represents ID of the begin node of the edge, dstID represents ID of the end node of the edge, Arr is a String which represents the properties of vertex, just like "property 1, property2";

With the characteristics of Spark, this paper design the data structures to represent the collection of storage points and edges as follows:

VertexRDD = RDD [Vertex (Long, String)];

EdgeRDD = RDD [Edge (Long, Long, String)];

Data Structure of Algorithm. This algorithm involves some important intermediate variables. In order to realize the parallel of the algorithm, we design the data structures as follows:

The data structure of the storage vertices and their son nodes is shown as follows:

neighborRDD = RDD[(Long, Array[Long])]

Each element of the data structure is a key-value pair, denoted by (Long, Array [Long]) where the first Long value represents the ID of vertex, the second Array represents an array of son nodes of vertex.

The data structure of the storage vertices and their ε neighborhood-nodes shown as follows:

eNeighborRDD = RDD[(Long, Array[Long])]

Each element of the data structure is a key-value pair, denoted by (Long, Array [Long]) where the first Long value represents the ID of vertex, the second Array represents an array of the ε neighborhood-nodes of vertex.

The data structure of the storage cores and their ε neighborhood-nodes shown as follows:

uNeighborRDD = RDD[(Long, Array[Long])]

Each element of the data structure is a key-value pair, denoted by (Long, Array [Long]), where the first Long value represents the ID of core, the second Array represents an array of ε neighborhood-nodes of core.

The data structure of the storage the sub-cluster and their cluster label shown as follows:

uAllNeiRDD = RDD[Array[(Long, Long)]]

Each element of the structure is an array, denoted by Array [(vid, label)], storage the relationship between all the vertices of a sub cluster and the cluster labels. Each element of the array is a key-value pair, denoted by (Long, Long), where the first Long value represents ID of vertex, the second Long represents the cluster label of vertex.

The data structure of the storage vertex and the minimum cluster label in all the sub clusters is shown as follows:

minRDD [(Long, Long)]

Each element of the structure is a key-value pair, denoted by (Long, Long), where the first Long value represents the ID of vertex, the second Long represents the minimum cluster label in all the sub clusters of vertex.

3.2   Parallel Recognition ε Neighbors and Core Nodes

Parallel recognize ε neighbor-nodes and cores in three stages:

Stage I, we can get the relationship between node and their son nodes through the calculation of the relationship between the vertices of the graph. Then transform the relationships to key-value pairs and put them into a collection, denoted by neighborRDD. For the convenience of parallel computing, we transform neighborRDD to an array, denoted by neighborArr, and broadcasts the array in all machine.

Stage II, computing structure similarity between vertices in parallel to get all vertices and their ε neighborhood-nodes. Assuming that exist an element, denoted by $(vid_i, Array_i)$, we can get $\Gamma (vid_i) = \{vid_i\} \cup Array_i$. We can get the structural similarity between vertices through the calculation of the element with each other.

The second stage can be described as follows process:

Step.1 Get the current calculation of the element, denoted by $(vid_i, Array_i)$;

Step.2 Assuming that vid_i's has an array to storage the ε neighborhood-nodes, denoted by $eArray_i$. Computing the structure similarity between elements in neighborArr and (vidi, Arrayi). We put the element's ID into $eArray_i$ if the structure similarity is greater than ε.

Step.3 Remove the element from $eArray_i$ which equals vid_i. Then constitute a new element $(vid_i, eArray_i)$ as the result to return.

The second stage of each element is executed independently in parallel. And then all the calculations results of each element are merged to a collection. The collection include all key-value pairs which storage vertex and its ε neighborhood-nodes.

Stage III, filter elements in eNeighborRDD to find the elements, which the size of ε neighborhood-nodes is greater than μ. Merge them to the core collection, denoted by uNeighborRDD.

The process can be described by the following example:

Assuming that exist a directed graph G = {V, E}, its structure is shown in Fig. 2:

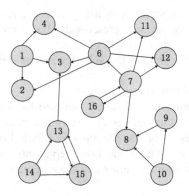

Fig. 2. Directed graph G

We can get a set of vertex and its son nodes which has the out-degree, as show in column "Son Set" in Table 1.

For illustration of purposes, we set the adjustable parameter ε = 0.4, μ = 1;

Take the ID of vertex is 1 as the example, denoted by Vertex 1, its structure similarity as show in Table 2.

According to ε = 0.4 we can get the node's Nε (Vertex 1) is {6}.

Similarly we can get all vertices and their N_ε (Vertex 1) as show in column "ε neighborhood-nodes" in Table 1.

According to μ = 1, we can get set of core is {1,6,7,9,10,13,14,15,16}.

Table 1. Son set and ε neighborhood-nodes of Vertices

Vertex ID	Son set	ε neighborhood-nodes
1	2,3,4	6
6	2,3,4,11,12	1,7
7	6,11,12,16	6,16
9	8	10
10	8,9	9
13	15	14,15
14	13,15	13,15
15	13	13,14
16	7	7

Table 2. The structure similarity of Vertex 1

Vertices pair	Structure similarity
(1,6)	0.45
(1,7)	0
(1,9)	0
(1,10)	0
(1,13)	0.35
(1,14)	0
(1,15)	0
(1,16)	0

3.3   Extension and Synchronization of Cluster Label in Parallel

Extension and synchronization of cluster label in parallel is in two stages:

Stage I. Build and initialize the sub-clusters. Each core and their Nε (u) are transformed to a sub-cluster. The sub-cluster is an array of the key-value pairs with the ID and cluster label of vertex, which in Nε (u) ∪ the ID. The label is initialized to the ID of vertex. The first stage can be described by the following process:

Each element of the uNeighborRDD (core and its ε neighbor-nodes) calculate by the following steps:

Step.1 Get the current calculation of the element, denoted by (vidi, Arrayi);

Step.2 Put vidi into Arrayi;

Step.3 Set labelj = vidj;

Step.4 Transform Array$_i$ to an array which includes all vertices and their label, denoted by Array[(vidj,labelj)], as the result to return.

The stage I of each element of the stage is executed independently in parallel. And then all the calculation results of element are merged a collection, denoted by uAll-NeiRDD, after calculation. The algorithm steps can be described by the following example:

According to the example given by Sect. 4.2, we can build the sub-clusters, results of calculation as show in column "vertices of sub-cluster (initial stage)" in Table 3. Among them, each line represents a sub-cluster, where the key-value pair (id, label) is represents the ID and cluster label of the vertex. Next step, parallel synchronizing the only cluster label of the vertices to all vertices in different sub-cluster.

Table 3. Vertices of sub-clusters

No	Vertices of sub-cluster (initial stage)	Vertices of sub-cluster (after synchronize in sub-cluster)
1	(1,1),(6,6)	(1,1),(6,1)
2	(6,6),(1,1),(7,7)	(6,1),(1,1),(7,1)
3	(7,7),(6,6),(16,16)	(7,6),(6,6),(16,6)
4	(9,9),(10,10)	(9,9),(10,9)
5	(10,10),(9,9)	(10,9),(9,9)
6	(13,13),(14,14),(15,15)	(13,13),(14,13),(15,13)
7	(14,14),(13,13),(15,15)	(14,13),(13,13),(15,13)
8	(15,15),(14,14),(13,13)	(15,13),(14,13),(13,13)
9	(16,16),(7,7)	(16,7),(7,7)

According to Definition 7, if there exists directly structure reachable of any two vertices in the same cluster, then there exists the structure reachable of any two vertices in the sub-clusters if the sub-clusters has same vertices. The vertices of structure reachable should belong to the same cluster. In this paper, vertices are in same cluster if their have same cluster label. Execute cluster expansion through synchronization cluster label in parallel, and then, achieve clustering merge.

Stage II, parallel computing the results of Stage I to synchronize the cluster label in every sub-cluster. Then computing the minimum of the cluster labels of the vertices, which have the same ID. The minimum as the only cluster label of the vertex. Iterative above process until the label of vertices not change any more. Finally we can output the result set. The algorithm process can be described as follows:

Step.1 In each sub-cluster, sort the vertices by their cluster label. We get the minimum as the cluster label of the sub-cluster, and set all vertices's cluster label to this minimum. Then execute Step.2;

Step.2 Merge the elements to a new collection, denoted by allRDD [(vid, label)]. Then execute Step.3;

Step.3 Get the minimum cluster label of vertices, which have the same ID, set the minimum as the only cluster label. Then merge the key-value pair which made up by the ID and only cluster label of vertex to a new collection, denoted by minRDD[(vid, label)]. Then execute Step.4;

Step.4 According to minRDD [(vid, label)], synchronization the only cluster label to all vertices in different sub-cluster in parallel.

Step.5 If the minRDD is the same with the last iterative result, then the iterative is completed, minRDD as the result and return; otherwise execute Step.1.

The process can be described by the following example:

According to the example given by Sect. 4.3, we simulate an iterative process. First, synchronize the cluster label in the sub-cluster, the result as show in column "vertices of sub-cluster (after synchronize in sub-cluster)" in Table 3. Then get every element of all sub-cluster, merge to allRDD as show in column "cluster labels of vertex" in Table 4:

Table 4. Vertices and cluster label

ID of Vertex	Cluster labels of vertex	Only cluster labels of vertex	Only cluster labels of vertex (final)
1	(1,1),(1,1)	(1,1)	(1,1)
6	(6,1),(6,1),(6,6)	(6,1)	(6,1)
7	(7,6),(7,7),(7,1)	(7,1)	(7,1)
9	(9,9),(9,9)	(9,9)	(9,9)
10	(10,9),(10,9)	(10,9)	(10,9)
13	(13,13),(13,13),(13,13)	(13,13)	(13,13)
14	(14,13),(14,13),(14,13)	(14,13)	(14,13)
15	(15,13),(15,13),(15,13)	(15,13)	(15,13)
16	(16,7),(16,6)	(16,6)	(16,1)

Then computing the minimum cluster label of the vertices, which have the same ID. The minimum as their only cluster label. Next set the cluster label to minimum. The result as show in column "only cluster labels of vertex" in Table 4.

According to the result, we synchronize the cluster label in every sub-cluster. The result as show in Table 5:

Table 5. Vertices of sub-cluster (after first iteration)

No	Elements of sub-cluster
1	(1,1),(6,1)
2	(6,1),(1,1),(7,1)
3	(7,1),(6,1),(16,6)
4	(9,9),(10,9)
5	(10,9),(9,9)
6	(13,13),(14,13),(15,13)
7	(14,13),(13,13),(15,13)
8	(15,13),(14,13),(13,13)
9	(16,6),(7,1)

Now we finish an iteration.

In the above example, the final result as show in column "only cluster labels of vertex (final)" in Table 4.

3.4   Clustering Result Analysis

In the result of the parallel clustering algorithm, the vertices of the same cluster label should belong to same cluster.

In this stage, we transform (id, label), which is element in result, to (label, id). Then merge the id of vertices by the same cluster label as the clustering result.

According to example of Sect. 4.3 we get three clusters. They are {1, 6, 7, 16}, {9, 10},{13, 14, 15}.

According to Definition 6, the vertex is hub or outlier, if it is not in any cluster. According to example, hubs are {8, 3}, outliers are {2, 4, 11, 12}.

4   Evaluation

4.1   Data-sets

1. We test and verify the accuracy of algorithm by using binary_networks, which is a tool for generated social network randomly. Datasets generated as follows:
   (1) Dataset binary_networks1K, which includes 1,000 vertices, edges is randomly generated.
   (2) Dataset binary_networks10K, which includes 10,000 vertices, and edges are randomly generated.
   (3) Dataset binary_networks100K, which includes 100,000 vertices, and edges are randomly generated.
2. We test and verify parallel efficiency of algorithm by:
   (1) Random dataset built by binary_networks tool includes 100,000 vertices and 1,532,964 edges.
   (2) Dataset soc-sign-slashdot090216 of Slashdot Zoo signed social network from February 21 2009 includes 82,144 vertices and 549,202 edges.
   (3) Dataset amazon0302 of Amazon product co-purchasing network from March 2 2003 includes 262,111 vertices and 1,234,877 edges.
   (4) Dataset Wiki-vote of Wikipedia who-votes-on-whom network includes 7,115 vertices and 103,689 edges.

Among the above dataset, dataset 1 is simulation and the others are from Stanford University big data network.

4.2   Algorithm Evaluation Index

We used the Precision (P), Recall (R), F1 and Rand Index (RI) to verify the accuracy of algorithm. It is a correct result if two vertices of same real cluster is belong to the same cluster. The greater the value of the four evaluation index, more similar to the real world and better clustering.

We used the speedup verify the parallel efficiency of algorithm, which is the ratio of serial and parallel processing with the shortest time. The greater of speedup, the shorter of parallel time.

4.3   Environment of Experiment

See Table 6.

Table 6. Environment of experiment

CPU	Intel(R) Core(TM) i5-3470 CPU @3.20 GHz
Memory	8 GB
Hard drive	1 TB
OS	Ubuntu14.10
IDE	IntelliJ IDE 14
Programming language	Scala
Spark version	1.4.0

4.4   Parameters of Cluster

This paper select the value of ε and μ by the method of SCAN in literature [4]. This involves making a k-nearest neighbor query for a sample of vertices and noting the nearest structural similarity. The query vertices are then sorted in ascending order of nearest structural similarity. The knee indicated by a vertical line represents a separation of vertices belonging to clusters to the right from hubs and outliers to the left. We recommend a value for μ, of 2.

4.5   Results and Analysis of Experiment

We test and verify the accuracy of algorithm by using binary_networks, which is a tool for generating network randomly.

In binary_networks1K Case, if $\mu = 2$, the result on different values of ε as show in Table 7:

Table 7. The result on different values of ε

ε	P	R	F1	RI
0.2	0.03	1.0	0.06	0.07
0.4	0.99	1.0	0.99	0.99
0.6	1.0	0.83	0.91	0.99
0.8	1.0	0.05	0.10	0.97

From experiments on different values of ε, we can see that value of ε has a significant effect on the accuracy of the clustering results in SparkSCAN. When value of ε is too small, it easily divide different clusters vertices into one cluster; however, when value of ε is too big easily create too much hubs and outlier.

If ε = 0.5, μ = 2, the result on different values of ε as show in Table 8:

Table 8. Indexes of clusteings

Dataset	P	R	F1	RI
binary_networks1K	1.0	0.99	0.99	0.99
binary_networks10K	1.0	0.99	0.99	0.99
binary_networks100K	1.0	0.98	0.99	0.99

The experimental results show that SparkSCAN algorithm with good accuracy performance by selecting P, R, F1, RI reasonably.

Conducting experiments using PDirSCAN and SparkSCAN to verify the parallel efficiency of algorithm:

The speedup of binary_networks100K as show in Fig. 3:

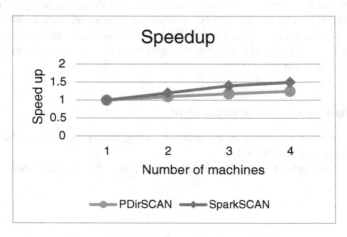

Fig. 3. Speedup of binary_networks100 K

The speedup of soc-sign-slashdot090216 as show in Fig. 4:

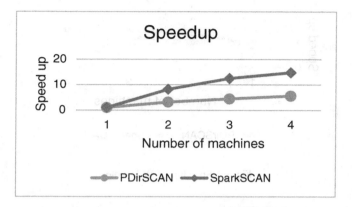

Fig. 4. Speedup of soc-sign-slashdot090216

The speedup of amazon0302 as show in Fig. 5:

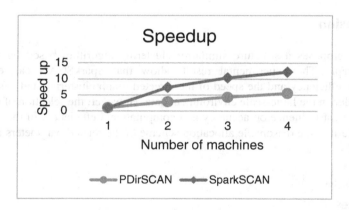

Fig. 5. Speedup of amazon0302

The speedup of Wiki-vote as show in Fig. 6.

Compared with PDirSCAN algorithm, and SparkSCAN has better performance on the above datasets, especially on large datasets. When the number of computer increases, the time algorithm spend is reducing. The result in real world is better than simulation dataset. This is due to the data relationships between simulated data sets is too complex. Synchronizing cluster label increases the network overhead and leading to decrease the effect of parallel. The parallel effect on Wiki-vote dataset is not obvious because of the size of dataset is too small, at this time consuming by Spark framework itself is more apparent.

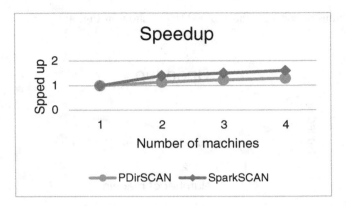

Fig. 6. Speedup of Wiki-vote

In conclusion, SparkSCAN improves the processing speed on the premise of the accuracy of clustering results. SparkSCAN has good performance in large data environment and high practical value. However, the accuracy of the algorithm is more dependent on ε and μ parameters.

5   Conlusion

This paper proposes a structure similarity clustering algorithm based on Spark for directed graph. The experimental results show that SparkSCAN can effectively improve the efficiency and the speed of the directed graph clustering and has a greater practical value in the large-scale environment data. However, the selection of parameter values has great influence on accuracy and computational efficiency. In the future, we will study further the reasonable allocation scheme of the related parameters to achieve better results.

References

1. Ding, Y., Zhang, Y., Li, Z.-H., Wang, Y.: Researach and advances on graph data mining. J. Comput. Appl. **32**(1), 182–190 (2012)
2. Lancichinetti, A., Fortunato, S., Kertész, J.: Detecting the overlapping and hierarchical community structure in complex networks. New J. Phys. **11**(3), 033015-1–033015-18 (2009)
3. Fallani, F.D.V., Nicosia, V., Latora, V., et al.: Nonparametric resampling of random walks for spectral network clustering. Phys. Rev. E **89**(1), 012802-1–012802-5 (2014)
4. Xu, X.-W., Yuruk, N., Feng, Z.-D., et al.: SCAN: a structural clustering algorithm for networks. In: Proceedings of the 13th ACM SIGKDD International Conference on Knowledge Discovery and Data Mining, San Jose, pp. 824–833 (2007)

5. Zhou, D.-Y., Huang, J.-Y., Schölkopf, B.: Learning from labeled and unlabeled data on a directed graph. In: Proceedings of the 22nd International Conference on Machine Learning, Bonn, pp. 1036–1043 (2005)
6. Meila, M., Pentney, W.: Clustering by weighted cuts in directed graphs. In: Proceedings of the 7th SIAM International Conference on Data Mining, Minneapolis, pp. 135–144 (2007)
7. Chen, J.-J.: Research on Clustering Algorithms for Large—Scale Social Networks based on Structural Similarity. Nankai University (2013)
8. Chen, J.-M., Chen, J.-J., Liu, J., Huang, Y.-L., Wang, Y., Feng, X.: Clustering algorithms for large-scale social networks based on structural similarity. J. Electron. Inf. Technol. **02**, 449–454 (2015)
9. Zhao, W., Martha, V., Xu, X.: Pscan: a parallel structural clustering algorithm for big networks in mapreduce. In: 2013 IEEE 27th International Conference on Advanced Information Networking and Applications (AINA), pp. 862–869. IEEE (2013)
10. Zaharia, M.A.: An Architecture for Fast and General Data Processing on Large Clusters. University of California, Berkeley (2013)
11. Zaharia, M., Chowdhury, M., Das, T., et al.: Resilient distributed datasets: a fault-tolerant abstraction for in-memory cluster computing. In: Proceedings of the 9th USENIX Conference on Networked Systems Design and Implementation, p. 2. USENIX Association (2012)

Using Distant Supervision and Paragraph Vector for Large Scale Relation Extraction

Yuming Liu[✉] and Weiran Xu

Beijing University of Posts and Telecommunications, Beijing 100876, China
{yumi,xuweiran}@bupt.edu.cn

Abstract. Distant supervision has the ability to generate a huge amount training data. Recently, the multi-instance multi-label learning is imported to distant supervision to combat noisy data and improve the performance of relation extraction. But multi-instance multi-label learning only uses hidden variables when inference relation between entities, which could not make full use of training data. Besides, traditional lexical and syntactic features are defective reflecting domain knowledge and global information of sentence, which limits the system's performance. This paper presents a novel approach for multi-instance multi-label learning, which takes the idea of fuzzy classification. We use cluster center as train-data and in this way we can adequately utilize sentence-level features. Meanwhile, we extend feature set by paragraph vector, which carries semantic information of sentences. We conduct an extensive empirical study to verify our contributions. The result shows our method is superior to the state-of-the-art distant supervised baseline.

Keywords: Relation extraction · Distant supervision · Paragraph vector

1 Introduction

We are living in information era, still we have difficulty finding knowledge. Relation extraction (RE), the process of generating structural data from plain text, continues to gain attention when PB of natural-language text are readily available. However, most approaches to RE use supervised learning of relation-specific examples, which can achieve high precision and recall. Unfortunately fully supervised methods are limited by the availability of training data and are unlikely to scale to the thousands of relation found on the web.

One of the most promising approaches to RE that addresses this limitation is distant supervision, which generates training data automatically by aligning a knowledge base with text (Bunescu and Mooney [1]; Mintz [2]). For example, taking Fig. 1, we would create a datum for each of the two sentences containing LEBRON JAMES and AKRON, labeled with *city_of_birth*, and likewise with *city_of_residence*, creating 4 training examples overall. Similarly, both sentences involving LEBRON JAMES and PLAYER would be marked as expressing the *title* relation.

© Springer Science+Business Media Singapore 2016
W. Chen et al. (Eds.): BDTA 2015, CCIS 590, pp. 178–190, 2016.
DOI: 10.1007/978-981-10-0457-5_17

Distant supervision introduces two challenges. The first challenge is that some training examples obtained through this hypothesis are not valid, though a sentence contains both entities, it may express no relation on the entity pair. The second challenge is that the same pair of entities may have multiple labels and it is unclear which label is instantiated by any textual mention of the given tuple. To fix these problems Surdeanu [3] cast distant supervision as a form of multi-instance multi-label learning. However, Surdeanu's model only use a latent label of each entity mention when inference relations, which loss too much useful information.

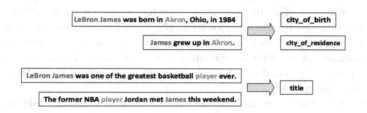

Fig. 1. Examples of distant supervision.

We also find that the features used for RE, consisting of token N-gram, POS N-gram, dependency path between two entities are often sparse, and have drawbacks reflecting meaning of the whole sentence. It causes low performance of system. There are plenty of works on distributed representations of sentences, which can calculate unsupervised like word embedding. Using paragraph vector we could get the global information of each entity mention, which benefits RE system performance.

To improve the performance of RE system, we made two major contributions:

- Propose fuzzy classification based MIML-RE, using both mention level features and latent variable z to train relation level classifiers
- Introduce sentence vector to relation extraction, which could catch global information of a relation mention

2   Related Works

2.1   Distant Supervision

Distant supervision was first proposed by Craven and Kumlien [4] in the bio-medical domain. Since then, it has gain popularity (Mintz et al. [2]; Bunescu and Mooney [1]; Wu and Weld [5]; Yao et al. [6]; Hoffmann et al. [7]; Surdeanu et al. [3]). To tolerate noisy labels in positive examples, Yao et al. [6] use Multiple Instance Learning (MIL), which assumes at-least-one of the relation mentions in each bag of mentions sharing a pair of argument entities which bears a relation, indeed expresses the target relation. MultiR [7] and Multi-Instance Multi-Label

(MIML) learning [3] further improve it to support multiple relations expressed by different sentences in a bag. Angeli et al. [8] combines the distant supervision with partial supervision, they select some mentions from train data and manual label them, then use these labeled data as well as other train data to train the extraction model, which performance well than the original MIML model.

2.2 Paragraph Vector

RE can be seen as a task of text classification, at the heart of RE are machine learning algorithms. These algorithms typically require the text input to be represented as a vector. The most common vector representation for texts is the bag-of-words or bag-of-n-grams. However, the bag-of-words has many disadvantages. The word order is lost, and thus different sentences can have exactly the same representation, as long as the same words are used. Even though bag-of-n-grams captures the word order in short context, it suffers from data sparsity and high dimensionality.

Paragraph vector is inspired by the recent work in learning distributed representations of word embedding (Bengio et al. [9]; Collobert and Weston [10]; Mnih and Hinton [11]; Mikolov et al. [12]). In their formulation, each word is represented by a vector which is concatenated or averaged with other word vectors in a context, and the resulting vector is used to predict other words in the context. Following these successful techniques, researchers have tried to extend the models to go beyond word level to achieve phrase-level or sentence-level representations (Yessenalina and Cardie [13]; Mikolov et al. [14]). Paragraph Vector is capable of constructing representations of input sequences of variable length. Whats more, we can train paragraph vector in an unsupervised way, it suits distant supervision perfect.

Socher et al. [15] present a novel Recursive Neural Networks (RNNs) for relation classification that learns vectors in the syntactic tree path that connects two nominals to determine their semantic relationship. And Zeng et al. [16] present a method using Convolutional Deep Neural Networks (CDNNs) for relation classifier, the input for CDNNs is sentence itself, all other processing like POS, CHUNK is unnecessary. But these works are supervised-learning thus can't be used in distant supervision framework.

3   Task Definition

In this section, we introduce the distant-supervised relation extraction task. In this paper, we define relations are semantic concepts that are true for a give set of entities. An entity is specific person, place, organization, et al. A relation r is a named tuple of the form $r(e_1, e_2)$. An entity mention is a contiguous sequence of textual tokens denoting an entity, and relation mention (for a given relation $r(e_1, e_2)$) as a pair of entity mentions of e_1 and e_2 in the same sentence.

The knowledge-based distant supervision learning problem takes as input

- training corpus
- a set of entities mentioned in that corpus
- a set of pre-defined relations
- a knowledge base about above relations and entities

As output the learner produces a relation extraction model.

4 Modeling Framework

Our model builds on the work of Surdeanu [3], who introduced multi-instance multi-label learning to relation extraction. We extent their work by taking the idea of fuzzy classification. We regard mention level multi-class classifier as a feature generator, using latent variables align with mention level features to form relation level training samples. We will detailedly explain the process above in Sect. 4.1.

Also we introduced text's vector to relation extraction, which could reflect the global information of sentence and overcome sparsity of lexical and syntactic features. Experiment has proved that each of contribution could beat the state-of-the-art, and the combination of them achieves an even higher f1 score.

4.1 Fuzzy Classification Based Multi-instance Multi-label Learning

MIML-RE is a framework for distant supervision model that treats the sentence-level variables z as latent, and uses facts from a knowledge based as supervision for the aggregate-level labels y. MIML-RE model implies that each relation mention involving an entity pair is exactly expressing one relation, but allows the entity pair to exhibit multiple labels across different mentions. Since we do not know the actual relation label of a mention in the distantly supervised setting, we need a latent variable $z^{(m)}$ to indicate mention m presenting which relation, where $z \in \mathbf{R} \bigcup \{NIL\}$.

In fuzzy based MIML-RE model, we create $|R|$ binary variables presenting the known relations for the entity pair. And for a given entity tuple, we generate $|R|$ vectors accordingly, each of them is a positive sample for corresponding relation r. A set of binary classifiers links the entity mentions and each y_j. Figure 2 describes the model. Specifically, in the Fig. 2:

- M is all the mentions for a given entity pair;
- $x^{(m)}$ is an entity mention;
- w_z is the weight vector for the multi-class mention-level classifier;
- $z^{(m)}$ indicates which relation a mention may present;
- a_{y_j} is train data for $j - th$ classifier, which aggregates mention-level features and latent vector z;
- y-classifiers are binary for each relation, y_j is the output for an entity pair as to whether the $j - th$ relation holds;
- w_j is the weight vector for the binary relation-level classifier.

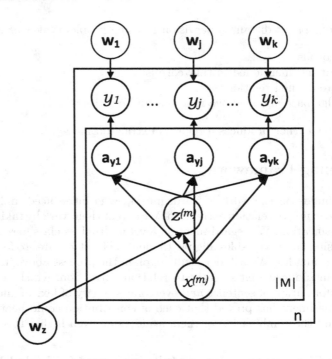

Fig. 2. Framework for FC-MIML. We add a feature-generating layer (between $z^{(m)}$ and y) into MIML to form FC-MIML.

We use a conditional probability model that defines a joint distribution over all of the extraction random variables defined above. Unlike the Surdeanu, our model not only uses latent labels, but also the mention level features to train relation-level classifiers. We add a feature-generating layer into the original MIML framework. Here we take the idea of fuzzy classification, we regard $p\left(z|x_i^{(m)}, \mathbf{w_z}\right)$, i.e., the output of mention-level classifier, as membership function. For each entity pair i, we classify relation mentions into $|R|$ clusters according membership functions. Then each cluster center can be seen as the represent of these mentions, we concatenate latent vector $\mathbf{z_i}$ to cluster center to form $a_i^{y_j}$, which is the train-data for relation-level classifiers.

Formally, given entity tuple i, to generate train data for r-th relation-level classifier, we have:

$$a_i^{y_r} = \frac{1}{M_i} \sum_{m=1}^{M_i} p\left(z|x_i^{(m)}, \mathbf{w_z}\right) x_i^{(m)} + \mathbf{z_i}$$

Where $x_i^{(m)}$ is feature vector of mention m, $\mathbf{z_i}$ is the latent vector.

For each entity pair $i = (e_1, e_2)$, define $\mathbf{x_i}$ to be a vector concatenating the individual sentences, $x_i^{(m)} \in S(e_1, e_2)$, $\mathbf{y_i}$ to be vector of binary $y_i^{(r)}$ random variables, one for each $r \in R$, and $\mathbf{z_i}$ to be the latent vector of $z_i^{(m)}$ variables, one for each sentence $x_i^{(m)}$. Log-likelihood of the train-data is given by:

$$LL\left(\mathbf{w_z}, \mathbf{w_y}\right) = \sum_{i=1}^{n} log\, p\left(\mathbf{y_i} | \mathbf{x_i}, \mathbf{w_z}, \mathbf{w_y}\right)$$

$$= \sum_{i=1}^{n} log \sum_{\mathbf{z_i}} p\left(\mathbf{y_i}, \mathbf{z_i} | \mathbf{x_i}, \mathbf{w_z}, \mathbf{w_y}\right) \quad (1)$$

Which $\mathbf{w_z}$ is the weight vector for the multi-class mention-level classifier, $\mathbf{w_y}$ is the weight vector for the binary top-level classifiers.

4.2 Paragraph Vector

Paragraph vector is inspired by methods for learning word vectors. A well-known framework for learning the word vectors is shown in Fig. 3. The task is to predict a word given the other words in a context.

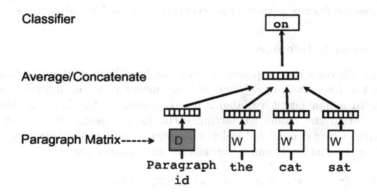

Fig. 3. Framework for word embedding. The paragraph can be taken as a special token.

In this framework, every word is mapped to a vector. The concatenation or sum of these vectors is then used as features for prediction of the next word in a sentence. To learn the distribute presentation for paragraph, we can take paragraph as a special word here, map paragraph to a vector as well. The paragraph vectors are also asked to contribute to the prediction task of the next word given many contexts sampled from the paragraph.

More formally, given a sequence of training words $\mathbf{w_1}, \mathbf{w_2}, \mathbf{w_3}, ..., \mathbf{w_T}$, and the vector of paragraph $\mathbf{w_{para}}$, objective of the word embedding is to maximize the log probability

$$\frac{1}{T} \sum_{t=k}^{T-K} \log p(\mathbf{w_t} | \mathbf{w_{t-k}}, ..., \mathbf{w_{t+k}}, \mathbf{w_{para}})$$

The prediction task is typically done via a multi-class classifier, such as soft-max. There, we have

$$p(\mathbf{w_t}|\mathbf{w_{t-k}}, ..., \mathbf{w_{t+k}}, \mathbf{w_{para}}) = \frac{e^{y_{w_k}}}{\sum_i e^{y_i}}$$

Each of y_i is un-normalized log-probability for each output word i, computed as

$$y = b + Uh(\mathbf{w_{t-k}}, ..., \mathbf{w_{t+k}}, \mathbf{w_{para}})$$

After the training converges, words with similar meaning are mapped to a similar position in the vector space. The vectors of paragraph also express the information that a relation mention contains. These properties make vectors attractive for many natural language processing tasks such as language modeling (Bengio et al. [9]; Mikolov [17]), natural language understanding (Collobert and Weston [10]).

In our case, we can see each sentence as a paragraph, corresponding vector is taking information of whole relation mention. We regard these vectors as semantic features. Using semantic, lexical and syntactic features, we can generate more expressive feature vectors. The experiment proved its advantages.

4.3 Training & Inference

Training. We train the proposed model using hard Expectation Maximization (EM). In the Expectation (E) step, we use current mention and relation level classifiers to assign latent variable z for mentions. In the Maximization (M) step, we retrain the model to maximize the Eq. (1) using the current latent assignments. Using conditional independence, the joint probability in the inner summation of formula (1) can be broken up into simpler parts:

$$p\left(\mathbf{y_i}, \mathbf{z_i}|\mathbf{x_i}, \mathbf{w_y}, \mathbf{w_z}\right) = p\left(\mathbf{z_i}|\mathbf{x_i}, \mathbf{w_z}\right) p\left(\mathbf{y_i}|\mathbf{z_i}, \mathbf{x_i}, \mathbf{w_y}\right)$$

$$= \prod_{m \in M_i} p\left(z_i^{(m)}|x_i^{(m)}, \mathbf{w_z}\right) \prod_{r \in R} p\left(y_i^{(r)}|\mathbf{z_i}, \mathbf{x_i}, w_y^{(r)}\right)$$

And we rewrite the equation above in log form:

$$\log p\left(\mathbf{y_i}, \mathbf{z_i}|\mathbf{x_i}, \mathbf{w_y}, \mathbf{w_z}\right) = \sum_{m \in M_i} \log p\left(z_i^{(m)}|x_i^{(m)}, \mathbf{w_z}\right)$$
$$+ \sum_{r \in R} \log p\left(y_i^{(r)}|\mathbf{z_i}, \mathbf{x_i}, w_y^{(r)}\right) \tag{2}$$

For each entity pair in train data, we need to find suitable latent vector $\mathbf{z_i}$, so we can maximize formula (2). To infer $\mathbf{z_i}$ for a specific entity tuple, given all its mentions $\mathbf{x_i}$, the gold labels $\mathbf{y_i}$, and current model, i.e., $\mathbf{w_z}$ and $\mathbf{w_y}$ weights. We seek to find:

$$\mathbf{z_i^{(*)}} = \arg\max_{\mathbf{z_i}} p\left(\mathbf{z_i}|\mathbf{x_i}, \mathbf{y_i}, \mathbf{z_i}, \mathbf{w_y}\right) \tag{3}$$

As there are exponential numbers of possible assignments for all vectors z_i, it is computationally intractable. So we approximate and consider each mention separately.

$$p\left(z_i^{(m)}|\mathbf{x_i}, \mathbf{y_i}, \mathbf{w_y}, \mathbf{w_z}\right) \propto p\left(z_i^{(m)}, \mathbf{y_i}|\mathbf{x_i}, \mathbf{w_y}, \mathbf{w_z}\right)$$

$$\approx p\left(z_i^{(m)}|x_i^{(m)}, \mathbf{w_z}\right) p\left(\mathbf{y_i}|\mathbf{x_i}, \mathbf{z_i'}, \mathbf{w_y}\right)$$

$$= p\left(z_i^{(m)}|x_i^{(m)}, \mathbf{w_z}\right) \prod_{r \in R} p\left(y_i^{(r)}|\mathbf{x_i}, \mathbf{z_i'}, w_y^{(r)}\right)$$

where $\mathbf{z_i'}$ contains the previously inferred mention labels for entity pair i, with the exception of component m whose label is replaced by $z_i^{(m)}$. So for $i = 1, ..., n$, and for each $m \in M_i$, we calculate:

$$z_i^{(m)*} = \underset{z_i^{(m)}}{\arg\max}\; p\left(z_i^{(m)}|x_i^{(m)}, \mathbf{w_z}\right) \prod_{r \in R} p\left(y_i^{(r)}|\mathbf{x_i}, \mathbf{z_i'}, w_y^{(r)}\right) \quad (4)$$

When we set latent label $z_i^{(m)} = r$, we regard that the corresponding mention is completely belongs to r, i.e. membership function become a binary function, 1 for relation r and 0 for others. So $x_i^{(m)}$ only participates in the generation of $a_i^{y_r}$. Notice that the probability $p\left(y_i^{(r)}|\mathbf{x_i}, \mathbf{z_i'}, w_y^{(r)}\right)$ is actually calculated using $p\left(y_i^{(r)}|a_i^{y_r}, w_y^{(r)}\right)$. Since $\mathbf{w_z}$ and $\mathbf{w_y}$ are fixed in (E) step, $a_i^{y_r}$ is determined by $\mathbf{x_i}$ and $\mathbf{z_i}$, so we use the former here for clarity.

When we get hidden label z, we use it to form $\mathbf{z_i}$ and generate train data $\mathbf{a_i}$ for relation classifiers. Then we retrain $\mathbf{w_y}$, $\mathbf{w_z}$ to maximize the lower bound of the log-likelihood, i.e., the probability in Eq. (2) with the current assignments for $\mathbf{z_i}$ and $\mathbf{a_i}$. From Eq. (2) it is clear that this can be maximized separately with respect to $\mathbf{w_y}$ and $\mathbf{w_z}$. The updates are given by:

$$\mathbf{w_z^{(*)}} = \underset{\mathbf{w_z}}{\arg\max} \sum_{i=1}^{n} \sum_{m \in M_i} \log p\left(z_i^{(m)}|x_i^{(m)}, \mathbf{w_z}\right)$$

$$w_y^{(r)*} = \underset{w_y^{(r)}}{\arg\max} \sum_{i=1}^{n} \sum_{r \in R} \log p\left(y^{(r)}|\mathbf{x_i}, \mathbf{z_i}, w_y^{(r)}\right)$$

To obtain weight vectors, we using a soft-max classifier for mention-level classifier and k binary logistic classifiers for relation classifier. We implement all using the 12-regularized logistic regression with scikit-learn[1]. The implement detail will discuss in Sect. 5.

---

[1] http://scikit-learn.org/stable/.

Inference. Given an entity tuple, we obtain its relation labels as follows.
We first get latent label for each mention to form z_i^*:

$$z_i^{(m)*} = \arg\max_{z_i^{(m)}} p\left(z_i^{(m)} | x_i^{(m)}, \mathbf{w_z}\right) \tag{5}$$

then we generator cluster centers:

$$a_i^{y_r} = \frac{1}{M_i} \sum_{m=1}^{M_i} p\left(z | x_i^{(m)}, \mathbf{w_z}\right) x_i^{(m)} + \mathbf{z_i}$$

finally decide on the final relation labels using the top-level classifiers:

$$y_i^{(r)*} = \arg\max_{\{0,1\}} p\left(y_i^{(r)} | a_i^{y_r}, w_y^{(r)}\right)$$

5   Experiments

To evaluate two major contributions of this work in all directions, we conducted an extensive empirical study.

Firstly, we compare FC-MIML-RE with other distant supervision methods, which is presenting in detail in Sect. 5.1. We find that FC-MIML-RE could improve the performance of RE system, gain about 2.7 % f1 score over a state-of-the-art weakly supervised baseline.

We tested effect of paragraph vector in the second experiment. Results shows that paragraph vector could enhance recall by a large scale, meanwhile only a tiny drop in precision.

Since we have verified each of contribution separately in the first two experiments, we perform the third experiment to test whether the bond of FC-MIML-RE and text's vector make an even bigger difference. Answer is yes, the combination of both two contributions performs the best.

5.1   FC-MIML-RE

To evaluate the performance of FC-MIML-RE, we use corpus created by Surdeanu[2], which is used for the original MIML-RE. This corpus contains approximately 6 million entity mentions, it is generated using mainly resource distributed for the 2010 and 2011 KBP shared tasks. The training relations are from the knowledge base provided by the KBP organizers, which is a subset of Wikipedia infoboxes, these infoboxes contain open-domain relations between named entities.

We adopt the setup of Surdeanu [3] for training MIML-RE model, we randomly select about 5 % samples consisting of entity pairs with no known relations as negative samples for training. Since EM is not guaranteed to converge at the

---

[2] http://nlp.stanford.edu/software/mimlre.shtml.

global maximum of the observed data likelihood, it is important to provide it with good initial values. In our experiment, the starting values are labels assigned to z_i, which are required to compute Eq. (3) in the first iteration. Here we use traditional distant supervision [2] to train mention-level classifier for the first iteration, then use this multi-classifier to generate train-data for relation-level binary classifiers. Then we update z_i and w_z, w_y iteratively.

We do the same as Surdeanu to evaluate MIML-RE models. We adapt official answer provided by KBP 2010 and 2011, then randomly choose 40 queries to set threshold for classifiers, the left 160 used to test performance for MIML-RE. Notice that here we use just the same queries for setting and test in that we can bring in to correspondence with Surdeanu's experiment.

Table 1. Results at the highest F1 point in the precision/recall curve on the KBP dataset

	Precision	Recall	f1
Mintz	0.2192	0.2934	0.2509
AT-LEAST-ONE	0.2637	0.2390	0.2507
MultiR	0.3205	0.2031	0.2486
MIML-RE	0.2530	0.2535	0.2532
FC-MIML-RE	**0.2420**	**0.3308**	**0.2795**

To test the our distant supervision, we compare our FC-MIML-RE model to four algorithm: (1) MIML-RE [3], the Multiple-Instance Multiple Label algorithm which labels the bags directly with the KB; (2) MultiR [7], a Multiple-Instance algorithm that supports overlapping relations (3) AT-LEAST-ONE [6], a Multiple-Instance algorithm assumes that there is at least one sentence presentation a certain relation. (4) Mintz [2], the original distant supervision relation extraction framework.

Table 1 shows that our algorithm consistently outperforms all other algorithms measured by f1. This demonstrates that fuzzy classification based MIML-RE, which generates refined feature for relation-level classifiers, is able to learn a superior model for RE.

5.2 Paragraph Vector

In order to evaluate paragraph vector precisely, we need a manual labeled dataset, so we adapt a corpora provided by Angeli (See footnote 2), which contains about $34k$ sentences. For paragraph vector, we trained 500 dim word vectors using Wikipedia, then we train our paragraph vectors with these word vectors. Notice that when training presentation of sentences, we fix the word vector and only change paragraph vectors. Finally, we select 20000 sentences from the corpus to train a multi-class logistic regression classifier, leave the rest for test.

Table 2. Results at the highest F1 point in the precision/recall curve on Angeli dataset. LEX is for lexical features and SYN is for syntactic.

Feature set	Precision	Recall	f1
Bag-of-word	0.3603	0.2551	0.2987
Paragraph vector	0.3718	0.1637	0.2274
LEX&SYN	0.5903	0.3694	0.4544
LEX&SYN&Paragraph vector	**0.5788**	**0.4104**	**0.4803**

Excluding paragraph vector, lexical and syntactic features we used are the same as [2].

Results are summarized in Table 2, which reports the scores for four kinds of features. We can see that if we use paragraph alone, it can not achieve a high score, but when we combine it with other features, the recall would be improved by a large scale meanwhile only a slightly drop on precision at max f1 point.

5.3   Comprehensive Experiment

The experiments above illustrate that each of our contribution could improve performance measured by f1 score. Since our ultimate goal is to improve the performance on RE system, still we conducted the third experiment, in which we test the combination of FC-MIML-RE and paragraph vector.

Since the Surdeanu corpus only contains extracted features for entity mention, there is no plain text provided, we can not use it to generate paragraph vector. And the second corpus we used is manual labeled, it is not suitable for distant supervision framework. So we employ a corpus generated by Yao[3] to test the performance of FC-MIML-RE align with paragraph vector. This data is developed by aligning Freebase with the New York Times (NYT) corpus. They used the Stanford named entity recognizer to find entity mentions in text and constructed relation mentions only between entity mentions in the same sentence. Since this corpus contains both extracted features and plain text of entity mention, we could use it to evaluate the united performance of two major contributions.

The Yao's dataset contains hundreds of relations, but the majority of them only possess few positive samples. We select the most twenty relations for experiment. We use these data conducting four experiments, the MIML-RE, MIML-RE+paragraph-vector, FC-MIML-RE and FC-MIML-RE+paragraph-vector. We use half of the data as training set, and the rest for test. On the Riedel dataset we evaluate all models using standard precision and recall measures. P/R curve shows in Fig. 4.

From Fig. 4, we can conclude that each of our work could improve system performance, when combing them together, we could achieve an even higher

---

[3] http://people.cs.umass.edu/~lmyao/.

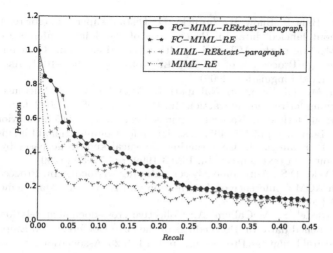

Fig. 4. Result in the Yao dataset

f1 score. When recall is low, FC-MLML-RE beats MIML-RE by a large margin measured by precision. Each of our contribution all performs better than original MIML-RE, and the effect of FC-MLML-RE is slightly better than text's paragraph's according to experiment. With the increment of recall, the advantage broad by text's vector gradually vanishes, still we argue that paragraph-vector yields a smoother curve, which generally means that the model is not over-fitting and achieves a better trade-off between precision and recall.

6   Conclusions

In this paper we presented a new model for distant supervised relation extraction which operates over tuples representing a syntactic relationship between two named entities. We take the idea of fuzzy classification, using cluster center as train-data for relation-level classifiers. Evaluation results on the KBP 2010&2011 (English) Slot-Filling task show that our model outperforms competitive relation extraction approaches by a wide margin. We also examine the effect when taking text's vector as feature. The result shows that paragraph vectors could not achieve high f1 score when it is used alone, but when combining with other features, it improve f1 score by a large margin. Lastly we perform a set of experiments, result demonstrates that the combination of paragraph vector and FC-MIML-RE performs best.

References

1. Bunescu, R., Mooney, R.: Learning to extract relations from the web using minimal supervision. In: Annual Meeting-Association for Computational Linguistics, vol. 45(1), p. 576 (2007)

2. Mintz, M., Bills, S., Snow, R., Jurafsky, D.: Distant supervision for relation extraction without labeled data. In: Proceedings of the Joint Conference of the 47th Annual Meeting of the ACL and the 4th International Joint Conference on Natural Language Processing of the AFNLP, vol. 2, pp. 1003–1011. Association for Computational Linguistics (2009)
3. Surdeanu, M., Tibshirani, J., Nallapati, R., Manning, C.D.: Multi-instance multi-label learning for relation extraction. In: Proceedings of the 2012 Joint Conference on Empirical Methods in Natural Language Processing and Computational Natural Language Learning, pp. 455–465. Association for Computational Linguistics (2012)
4. Craven, M., Kumlien, J.: Constructing biological knowledge bases by extracting information from text sources. In: ISMB 1999, pp. 77–86 (1999)
5. Wu, F., Weld, D.S.: Autonomously semantifying wikipedia. In: Proceedings of the Sixteenth ACM Conference on Information and Knowledge Management, pp. 41–50. ACM (2007)
6. Yao, L., Riedel, S., McCallum, A.: Collective cross-document relation extraction without labelled data. In: Proceedings of the 2010 Conference on Empirical Methods in Natural Language Processing, pp. 1013–1023. Association for Computational Linguistics (2010)
7. Hoffmann, R., Zhang, C., Ling, X., Zettlemoyer, L., Weld, D.S.: Knowledge-based weak supervision for information extraction of overlapping relations. In: Proceedings of the 49th Annual Meeting of the Association for Computational Linguistics: Human Language Technologies, vol. 1, pp. 541–550. Association for Computational Linguistics (2011)
8. Angeli, G., Tibshirani, J., Wu, J.Y., Manning, C.D.: Combining distant and partial supervision for relation extraction. In: Proceedings of the 2014 Conference on Empirical Methods in Natural Language Processing (EMNLP) (2014)
9. Bengio, Y., Schwenk, H., Senécal, J.S., Morin, F., Gauvain, J.L.: Neural probabilistic language models. In: Holmes, D.E., Jain, L.C. (eds.) Innovations in Machine Learning. STUDFUZZ, vol. 194, pp. 137–186. Springer, Heidelberg (2006)
10. Collobert, R., Weston, J.: A unified architecture for natural language processing: deep neural networks with multitask learning. In: Proceedings of the 25th International Conference on Machine Learning, pp. 160–167. ACM (2008)
11. Mnih, A., Hinton, G.E.: A scalable hierarchical distributed language model. In: Advances in Neural Information Processing Systems, pp. 1081–1088 (2009)
12. Mikolov, T., Chen, K., Corrado, G., Dean, J.: Efficient estimation of word representations in vector space. arXiv preprint arXiv:1301.3781 (2013)
13. Yessenalina, A., Cardie, C.: Compositional matrix-space models for sentiment analysis. In: Proceedings of the Conference on Empirical Methods in Natural Language Processing, pp. 172–182. Association for Computational Linguistics (2011)
14. Mikolov, T., Sutskever, I., Chen, K., Corrado, G.S., Dean, J.: Distributed representations of words and phrases and their compositionality. In: Advances in Neural Information Processing Systems, pp. 3111–3119 (2013)
15. Socher, R., Huval, B., Manning, C.D., Ng, A.Y.: Semantic compositionality through recursive matrix-vector spaces. In: Proceedings of the 2012 Joint Conference on Empirical Methods in Natural Language Processing and Computational Natural Language Learning, pp. 1201–1211. Association for Computational Linguistics (2012)
16. Zeng, D., Liu, K., Lai, S., Zhou, G., Zhao, J.: Relation classification via convolutional deep neural network. In: Proceedings of COLING, pp. 2335–2344 (2014)
17. Mikolov, T.: Statistical Language Models Based on Neural Networks. Presentation at Google, Mountain View (2012)

Computer Assisted Language Testing and the Washback Effect on Language Learning

Zhang Hongjun[1(✉)] and Pan Feng[2]

[1] Jilin University of Finance and Economics, Changchun, China
876619502@163.com
[2] Changchun University of Science and Technology, Changchun, China
Z687888@sohu.com

Abstract. Nowadays computer assisted language testing has become an important research both in the field of language teaching and computing test domain. Through the preliminary practice of computer assisted spoken English testing, the feedback of teachers and students, and the washback effect on language learning, we may provide some suggestions in order to improve college students' English abilities and relevant technology of computing test. Therefore, the basis of carrying out this kind of test effectively is provided.

Keywords: Computer · Language testing · Washback effect

1 Introduction

In 2007, the Ministry of Education (2007) awarded new formulation of College English Curriculum Requirements. It put forward a new college English teaching goal of cultivating students' English comprehensive application ability, especially the ability of listening and speaking. Meanwhile it explores ideas that the teaching mode of listening and speaking is set up in the network environment, and the teaching and practicing were directly carried out on the LAN (local area network) or the campus network. From this new requirement we notice that with the rapid development and popularization of computer and multimedia network in China, the teaching environment based on the platform, blackboard and chalk is gradually becoming people's memory. Instead, the teaching and testing mode are inclined to base on multimedia and network.

In the light of these series of reforms, spoken English class has appeared in Jilin University of Finance and Economics for ten years. Students have spoken English test at the end of each semester. We used computer assisted spoken English test in June 2009 for the first time. Till now few references can be found in the field of computer assisted spoken English test. It's necessary for us to analyze and summarize this kind of test and share the experience with others.

This paper will introduce corresponding theory, the preliminary process of computer assisted spoken English testing, feedback from teachers and students, and washback effect on language learning in order to improve college students' English ability and offer some suggestions to perfect computing test system and language learning.

© Springer Science+Business Media Singapore 2016
W. Chen et al. (Eds.): BDTA 2015, CCIS 590, pp. 191–199, 2016.
DOI: 10.1007/978-981-10-0457-5_18

2 Theoretical Framework

2.1 Computer Assisted Language Testing

Computer assisted language testing has appeared for more than 20 years. In 1986, Canale forecast that the age of computerized language assessments would come soon and indicated the revolution of a test pattern—the change from pencil-and-paper tests to internet-based assessments. At present, language testing researchers have been trying to use the existing computer technology to test language skills accurately and effectively. Computer assisted language testing has become the most important development direction in the field of testing. Computer technology is widely used in language testing design, test management and test results analysis and other aspects.

Based on IRT, computer assisted language testing adopts adaptive model, which can use a lot of subjective questions. Thus it has higher validity and authenticity. Computer assisted language testing prompts candidates to improve their language skills by completing authentic language tasks before and after the examination.

Computerized examination system has many characteristics, such as fast running, convenient maintenance and so on. The test database has multi types and wide distribution, so it can test the language level of each examinee. Each function module is powerful with not less than 10 question types, a number of test parameters and a variety of statistical methods. The system is compatible with browsers of IE6, IE7, IE8, FireFox4, Chrome and Opera.

Computer can complete many processes — establishing question base, assigning a topic, developing standards, organizing test, scoring statistics. It not only saves manpower and material resources, but also to reduce the impact of human factors. We can see that this kind of test is efficient and fair.

2.2 Washback Effect

It is generally accepted that testing influences teaching and learning (Alderson and Wall 1993). The research into the testing effect is an important issue in the testing field. Such effect is called washback effect in the language testing. Hughes (1989) defined it as "the effect of testing on teaching and learning". Language testing is used to serve teaching, but without testing, there would be no effective teaching and learning.

Language testing washback effect is said to be either positive or negative. Positive effect of language testing has a constructive impact on language learning which can stimulate teachers and students to do better in the further study. On the contrary, negative effect of language testing plays a bad role in language learning which may lead teachers and students to misunderstand the test or even worse to disturb normal learning. One of the main purposes of the study of washback effect is to explore ways to reduce the negative effect and improve the positive effect. By analyzing computer assisted spoken English testing at length, we will find ways to avoid its bad effect and improve the positive one.

3   Preliminary Practice of Computer Assisted Spoken English Testing

Computer assisted spoken English testing can be divided into three stages: preparation, implementation and evaluation.

3.1   Preparation

In the test experiment, we selected six language labs. Each lab was equipped with 54 sets of excellent computers and related equipment and software. Freshmen and sophomores were respectively arranged in six language labs to take examinations at the same time. There were two invigilators in each language lab, one was mainly responsible for operating the computer terminal and another invigilator was responsible for checking students' admission ticket and organizing students to enter the language lab orderly. Meanwhile two technical staff of Software Company served as the technical consultants of the examination.

Before the examination we had checked the test system and hardware of each computer. Examiners installed the software of examination system to each host computer, which belongs to blue pigeon's science platform. A control system was established which could monitor students' examination all-round. Several sets of spoken English questions were imported from the host of the test console. In the examination system we set up a small library of examination questions. After compression and encryption, examination questions were preserved to test center server.

3.2   Implementation

The examinations were performed for approximately twenty five minutes. At the beginning of the examination, students entered the test detection interface in order to check headphone and microphone and adjust the volume. Then they logged on the system by typing their ID. Confirmed by computer successfully, they would await spoken English test. Subsequently site server requested to distribute questions and decrypted according to the password provided by the test center. After that the server distributed examination questions to each client and examination questions were downloaded from the server to the test. Students could see test interface. All these steps were operated by the main invigilator.

Spoken English test included three parts: reading text, reciting passage and making an impromptu speech by given short subject. Inspection contents were related to usual teaching content. The first part, namely reading text, required students to read clearly and fluently with correct pronunciation and intonation. Students' computer screen would show the paragraphs required to read aloud. They had one minute to prepare and another minute to read the text. The second part, that is reciting passage, was chosen from passages they had learned before. The third part, namely making an impromptu speech by given short subject, aimed to assess the students' English verbal ability. From this part we can know whether the students had obtained the ability to express personal opinions or feelings, state facts or reasons, and describe events.

The test questions were recording test and each test was taped only once. Computer and server could automatically record the whole test process. When the test was over, the computer system submitted papers uniformly, collected and recorded sound package of students' answers. Then the main invigilator played test recording to let students confirm their test recording without any error. After that students left the language lab. At last these answers were packaged to test center server and the backup was stored at the same time.

3.3 Evaluation

Spoken audio files were transferred to the scoring system and marked by all English teachers on computer. The recording used database storage. We can carry out relevant operation to hide students' information. When teachers gave marks they couldn't see students' names or numbers. If they had difficulty in judging, they could consider the recording repeatedly and gave relatively objective scores. At the same time teachers could listen to the recording again and again through the operation of the mouse. Thus the process of scoring is relatively easy. The whole procedure reveals that the evaluation of computer assisted spoken English testing is very impartial and fair.

Evaluation was implemented according to partial marking standard, i.e. marking each test recording document. Marking standards mainly focus on the students' pronunciation and intonation, fluency and coherent expression. Specific criteria are as follows: First is accuracy, which refers to the exact degree of students' pronunciation, intonation, grammar and vocabulary; Second is language range, which refers to the complexity of the use of vocabulary and grammar structure; Third is the length of words, which refers to the number of speech in the entire exam; Fourth is continuity, which investigates whether students have the ability to carry out the coherent speech for a long time; Last is flexibility, which refers to the ability to cope with different situations and topics.

When the marking was over, teachers passed marking files to the teaching secretary. Then the teaching secretary exported all the results as an EXCEL file, and imported them to the management system of college English course performance. So far computer assisted spoken English testing has been over.

4 Result and Discussion

4.1 Result

After the evaluation, we analyzed students' scores. Then teachers received their students' report card. At the same time teachers gave marks according to students' subjective impression. Results showed that there was a high correlation between them. It indicated that computer assisted spoken English testing can reflect students' real English ability to a large extent.

4.2 Discussion

It's just the initial stage of implementing computing spoken English test. We have few relevant experiences and the attitudes of teachers and students towards this kind of test are not sufficient. After the exam, we performed questionnaires and interviews among the teachers and the students to obtain feedback from them.

Feedback from Teachers. The questionnaire and interview among teachers reveal that the teachers were in favor of computer assisted spoken English testing. The computer assisted spoken English testing has improved quality and efficiency of the examination greatly. They thought the results of the test are basically the same as students' class performance. The test can promote students' English learning and stimulate their enthusiasm and confidence in learning English. Therefore the validity and reliability of the test have been proved. This new form of test allowed teachers to pay more attention to the use of multimedia in teaching. In the interview a teacher with thirty years of teaching experience admitted: "I'm too old to know much about computer. But for my students, I use the computer to prepare lessons and make courseware every day now. The use of computer is also more skilled." Most teachers accepted that the test is an effective method for evaluating teaching quality. The normalization of it will make the students concern themselves with spoken language learning.

In the computer assisted examination process, the examinee's original voice data can be retained. Most of teachers believed that these voice data are worth investigating. For each marking standard, they can choose one or tow students' voice data as sample. When the teachers summarized the spoken English test to the whole class after the examination, they would play these typical samples to the students. It's obvious for students to understand the criteria of the test and emulate their strong points. The teachers said it's a good way for students to learn from others and see clearly their own discrepancy to study further.

Nonetheless we need to attach great importance to the problems. For existing question types, more than one-third of the surveyed teachers don't think it can test students' spoken English level. It shows that existing question types should be improved now. In the interview, these teachers thought that the question types of reading, reciting and impromptu speech are models of individual to computer, lacking interaction between individual and individual. They suggested that students would achieve real exchange completely if group discussion could add to question type. Thus the authenticity of the communicative context in the test is insured. After all, the ultimate goal of spoken English test is to evaluate whether the students have grasped the ability of conveying ideas clearly and communicating with others effectively or not.

Feedback from Students. The questionnaire and interview among students show that the students had a favorable attitude to computer assisted spoken English testing. The large-scale computer-based oral test has been accepted and recognized by most of the students. The majority of students believed that computer assisted spoken English testing can reflect their English level objectively and effectively. At the same time it also stimulates their motivation of learning oral English. They thought that it's more efficient than traditional form of test.

Some students whose listening levels were not good enough admitted that it can ease their tension. While in traditional spoken English test they would answer questions asked by the examiner. If they couldn't understand or follow what the examiner had said, they might fail in spoken English test. However, in computer assisted English testing, the examination questions appear on the screen at the same time as tips to make students know the topics which they will talk about. They wouldn't worry about the trouble of listening any more.

For the existing types of questions, the attitudes of students varied. Most of the students thought that making an impromptu speech by given short subject can detect their level of spoken English and is practical. They also believed that reciting passage can urge them to recite sentences, the classic articles and other English expressions. However, some students didn't like to read text, which should arouse our attention. We will guide students to develop a good habit of learning English in order to let them wake up to the importance of reading aloud for the help of training pronunciation and intonation and improving their language sense.

5   Washback Effect on Language Learning

5.1   Positive Effect

Teachers have cultivated the awareness of the use of network technology in teaching because the age of computer assisted language teaching and testing is coming. They realized that to improve the level of their use of the Internet technology is to promote the urgent task of multimedia network teaching. Moreover, computerized testing brings about great help to language teaching. Examination records can be used as teachers' teaching material. By analyzing students' record, teachers will find out what students have mastered and the shortage of their study. Afterwards, teachers may do a further case study or individual counseling for students. It's significant for both teachers and students. For teachers, they can see directly their teaching effect and the focus of teaching on the future. And for students, they can clear their own problems and the target of further effort.

The exam has played a greater role in supervising students' autonomous learning. Computer assisted spoken English testing forced the students to pay attention to listening and speaking. Furthermore, they can make full use of the Internet to assist language learning. It's very good to orient the effect of spoken English. This kind of test has produced a positive effect on students' attitudes of learning English and teachers' teaching method. The exam will not only improve the effect of language teaching, but more importantly make the students adapt to new forms of exam and come into the new world of language learning.

5.2   Negative Effect

Every coin has two sides. Each kind of examination has its advantages and disadvantages, so does computer assisted spoken English testing. We should pay more attention to the negative effect on language teaching and learning and try to figure out ways to solve it.

In view of the fact that current development of software technology is not mature enough, the existing oral exam questions can not fully reflect the characteristics of oral communication. Some students of bright and bubbly personalities reflected that they tended to use some nonverbal communications in the interview, such as eye contact, facial expression and body language to convey information. One student said: "The teacher's smile and approving eyes give me a lot of confidence." It's known that these nonverbal communications take important positions in spoken English test. Yet these have not been achieved in the computer assisted spoken English testing. We expect that engineers and scientists of computer could design the interactive software to represent the real scene of oral communication in real life, in which students may feel role on the scene.

In the evaluation period, we found that for few students we couldn't hear their sound recording. We were confused about the exact reason. We wondered if they didn't answer any question or there was something wrong with the recording process. We need great effort to find out the reason. Although such instances are extremely rare, it should stir up our concern. If students didn't answer any question, teachers have the duty to talk to them in order to find out what's happening on their spoken English test and encourage them to speak English in and after the class. If necessary, teachers may follow up their language learning. The latter is hardly probable, because the invigilator played test recording to let students confirm their test recording before the end of the test. Even so, related department of university and software company should guarantee that the computers and equipment are in normal working order.

6   Suggestions

6.1   To Improve Teachers' Level of Using Network Technology in Teaching

Computer assisted spoken English testing is a new testing format and differs from face to face oral examination. It makes teachers begin to pay more attention to the preparation of lessons by using computer multimedia. In order to improve multimedia English teaching, it is most urgent that teachers should arouse the awareness of making use of the network technology in English teaching so as to enhance their ability of conducting network technology.

Under the network environment, the teaching mode of listening and speaking makes the teacher's teaching contents of spoken language more abundant. Teachers are no longer confined to the contents of the spoken English books. They can focus on teaching contents and then expand and supplement relevant materials. For example, when we discuss the topic of family, the teacher can show the students a short video of foreign family. Through watching their way of life, family values and other aspects, students can be very intuitive to feel some similarities and differences between Chinese and western family. They will have more ideas to exchange with other students. The teacher can ask some deep questions concerning the topic or interact with students.

Colorful teaching contents not only makes oral English class more lively, but also makes the students increase the interest of studying English, especially the ability to speak English. With the aid of computer and network, additional multi sensory information, such as video and audio clips, play a vital role in guiding and stimulating students' expression. Thus, the teachers' teaching enthusiasm will be greatly improved and the students' ability of language use will be exercised.

6.2 To Improve Students' Autonomous Learning Competence

Computer assisted spoken English testing is different from the past. It requires students to express clearly which is dependent on their own expression in order to complete the oral test successfully. This change allows students to be aware of the importance of autonomous learning under the help of computer. In addition to the application in the examination, the examination system should be used appropriately in the students' autonomous listening and speaking training or in the daily teaching. Only when students have increased this kind of training at daily life, can they face the computing test calmly.

Computer assisted language learning is a new autonomous learning mode which is based on computer and network technology. It provides learning equipment, language input materials, learning tasks as well as students' reaction. Therefore, we can say it is one of the most effective tools for English learners to achieve their own learning. Computer assisted language learning emphasizes the importance of students' autonomy which is the ultimate goal of language teaching. Some students are aware that social interaction is no longer limited to the people of their own country by their mother tongue. Under the background of international society and economic globalization, they need to use English to communicate with foreigners sometimes. Therefore, they learn English more consciously at their spare time.

From above we can see that students are given much more freedom for autonomous language learning in computer assisted language learning environment. Students can find ways that are most appropriate to them and develop their autonomy in foreign language study further. Through usual training students will become familiar with the computer-based model so that they can improve the operability of computer assisted language testing. In the language learning platform, students can complete audio-visual network learning by themselves. After a period of practice, students can learn how to adapt to this kind of test. What's more, it also greatly improves students' interest in learning English.

6.3 The Relevant Teaching Management Departments Will Do Their Duties

The smooth implementation of computer assisted spoken English testing can not be separated from the support and cooperation of the relevant teaching management department of the school. Firstly, the daily maintenance and updating of the equipment in multimedia language laboratories demand the support of the school's financial funds. For example, some universities use language discipline platforms of Lange campus network. They need to contact the curriculum development agencies of Lange company regularly to keep pace with the development of program. Relevant departments and leaders will increase the investment of computer hardware and software. Because computer has developed so fast that the system and relevant software and hardware require to be updated in a timely manner. What's more, mouse, keyboard, headset and other parts of computer want daily check. All these are also inseparable from the teachers' hard work of network experimental center.

Secondly, every teacher should understand the test. In the computer era, it is duty for organizers to let language teachers understand the computerized language test. Because there are many functions which can be used in testing procedures, such as the preparation and presentation of questions, statistical analysis of answers, storage, delivery and information extraction, etc. Therefore, there is need to train teachers to operate the computer at regular intervals.

Thirdly, the correlative teaching management departments should pay attention to the development and the use of relevant software of computerized test in time. Take for instance the national college spoken English test, the examination form has transformed from communicating face to face with the examiner to computerized exam. Correspondingly, the examination questions have changed. The question types of self-introduction, answering questions with the picture, group discussion and so on are finished by candidates to interact with other candidates or simulated examiner through the computer. These are worthy of our attention and further research. It is expected that the normalization of the national large-scale computerized examination, and the continuous renewal of the relevant software development technology, which can provide reference for the implementation of this kind of exam in local colleges and universities.

7 Conclusion

From the analysis of computer assisted spoken English testing, its feedback from teachers and students and the washback effect on language learning, we notice that it has been recognized by most of the teachers and students. Furthermore, although it has some shortcomings, it can play a very good positive reaction to oral English teaching and learning. It is sure to rise to urge action in improving students' language ability actively. Therefore, the large-scale implementation of computer assisted language testing by using computer technology will have a brilliant future. It has become the focus of domestic and foreign scholars' attention and research.

With the rapid development of science and technology, we are looking forward to the improvement of computer testing system. Meanwhile, we expect that more and more people will join in the research work to promote the reform of language testing and language learning.

References

Alderson, J.C., Wall, D.: Does washback exist? Applied Linguistics (1993)

Hughes, A.: Testing for Language Teachers. Cambridge University Press, Cambridge (1989)

Ministry of Education. College English Curriculum Requirements 大学英语课程教学要求. Shanghai Foreign Language Education Press, Shanghai (2007)

Using Class Based Document Frequency to Select Features in Text Classification

Baoli Li[(⊠)], Qiuling Yan, and Liping Han

College of Information Science and Engineering,
Henan University of Technology,
Zhengzhou, People's Republic of China
libaoli@gmail.com

Abstract. Document Frequency (DF) is reported to be a simple yet quite effective measure for feature selection in text classification, which is a key step in processing big textual data collections. The calculation is based on how many documents in a collection contain a feature, which can be a word, a phrase, a n-gram, or a specially derived attribute. It is an unsupervised and class independent metric. Features of the same DF value may have quite different distribution over different categories, and thus have different discriminative power over categories. For example, in a binary classification problem, if feature A only appears in one category, but feature B, which has the same DF value as feature A, is evenly distributed in both categories. Then, feature A is obviously more effective than feature B for classification. To overcome this weakness of the original document frequency feature selection metric, we, therefore, propose a class based document frequency strategy to further refine the original DF to some extent. Extensive experiments on three text classification datasets demonstrate the effectiveness of the proposed measures.

1 Introduction

Text classification, which aims at assigning one or more predefined categories to a textual segment, is a key process for dealing with big textual data collections. As there are usually thousands of candidate features in a text collection and these features are not equally effective in text classification, feature selection for finding the most effective feature subset is usually a must step, especially when processing big data in this era. According to the Pareto principle[1] (i.e. the 80–20 rule), using less features may get almost the same results as using all features. Two benefits can be obtained from feature selection: one is better performance, and the other is lower time and space cost. In the past years, a lot of metrics, such as document frequency, information gain, chi-square, bi-normal separation, odds ratio, mutual information, etc., have been proposed to rank features in text classification [1]. Among these metrics, Document Frequency (DF), which counts how many documents a feature appears in, has been recognized as a simple yet quite effective metric in solving different text classification problems, as reported by Yang and Pedersen in [2].

---

[1] https://en.wikipedia.org/wiki/Pareto_principle.

© Springer Science+Business Media Singapore 2016
W. Chen et al. (Eds.): BDTA 2015, CCIS 590, pp. 200–210, 2016.
DOI: 10.1007/978-981-10-0457-5_19

The traditional DF metric is an unsupervised and class independent metric. Features of the same DF value may have quite different distribution over different categories, and thus have different discriminative power over categories. For example, suppose: (1) two features f_i and f_j both have DF value 6; and (2) f_i scatters among 6 categories, but f_j only appears in documents of the same class. Then, f_j is expected to have more discriminative power than f_i, but the original document frequency metric ranks them equally. We, therefore, think over how to revise the traditional DF metric further to make it have finer discriminative power. We propose a class based document frequency strategy to refine the original DF to some extent. For each feature, we count its document frequencies in each category and then choose the maximal class document frequency for ranking. To reduce the negative effects of class imbalance, we design a variant of the simple class based document frequency metric. Experiments on three publicly available datasets show that the proposed metrics perform consistently better than the traditional DF.

The rest of this paper is organized as following: Sect. 2 gives related work; Sect. 3 details the class based document frequency strategy; Sect. 4 presents extensive experiments on three text classification problems and discussion on the results; Sect. 5 concludes the paper with possible future work.

2 Related Work

Feature Selection, which aims at choosing the features that are the most effective ones in solving a supervised or unsupervised learning problem, has been widely studied in machine learning community. Its successful application includes but is not limited to: gene microarray analysis, combinatorial chemistry, image classification, face recognition, text clustering, and text classification. The advent of big data has further raised unprecedented demand for feature selection. Literature [3–6] provides good review of feature selection in different domains.

To obtain the most effective features, we can have two alternatives: choosing a subset from a candidate feature set or deriving a new compact feature set from those candidate features. In this study, we focus on the first type of methods, which are usually further classified into two categories: filter and wrapper. Wrapper methods try to find the ideal feature set by evaluating the performance of each candidate subset, where filter methods try to rank features independently. Filter methods are more popular than wrapper ones, as they are computationally efficient. Forman [1] and Yang and Pedersen [2] conduct empirical comparison of different feature filtering methods for text categorization, including document frequency, information gain, chi-square, bi-normal separation, odds ratio, mutual information, power, and so on. Yang and Pedersen [2] conclude that the DF metric performs as excellent as chi-square and information gain metrics do.

In this study, we concentrate on how to further improve Document Frequency (DF) metric with class information. We propose a class based document frequency feature selection strategy.

3   Class Based Document Frequency for Feature Selection

As we pointed out in Sect. 1, the original document frequency metric neglects the class distribution of a feature over different categories. We expect that features with imbalanced distribution would have more discriminative capacity than those with balanced distribution. We, thus, propose a class based document frequency strategy, which aims at further refining the original DF to some extent. For each feature, we count its document frequencies in each category and then choose the maximal class document frequency for ranking.

To evaluate the discriminative capacity of a feature t for class CLS_i, we usually need to count the following numbers:

A_i: how many documents belong to class CLS_i and contain the feature t;
B_i: how many documents do not belong to class CLS_i but contain the feature t;
C_i: how many documents belong to class CLS_i but do not contain the feature t;
D_i: how many documents do not belong to class CLS_i and do not contain the feature t.

Suppose that there are totally M categories in a classification problem. Then, the original document frequency (DF) measure can be computed as follows:

$$DF = \sum_{i=1}^{M} A_i \tag{1}$$

For performance comparison, in Sect. 4, we experiment with a popular and state-of-the-art feature selection metric, Chi-Square, which is calculated as follows:

$$Chi - Square = \sum_{i=1}^{M} CHI_i$$
$$= \sum_{i=1}^{M} \frac{(A_i + B_i + C_i + D_i) \times (A_i \times D_i - C_i \times B_i)}{(A_i + C_i) \times (B_i + D_i) \times (A_i + B_i) \times (C_i + D_i)} \tag{2}$$

A simple class based document frequency metric (CBDF) can take the following formula:

$$CBDF = \underset{i=1}{\overset{M}{MAX}} A_i \tag{3}$$

CBDF chooses the maximal class document frequency of a feature for ranking. As class imbalance may exist in a text classification problem, a variant of CBDF, which normalizes the CBDF values with class distribution, is designed as follows:

$$CBDF\_N = \underset{i=1}{\overset{M}{MAX}} A_i \bigg/ D(CLS_i) \tag{4}$$

where $D(CLS_i)$ is the total number of documents belonging to class CLS_i in the training dataset.

4 Experiments and Discussion

In order to evaluate the performance of the proposed class based document frequency (*CBDF*) feature selection metrics, we conduct extensive experiments on three single label text classification problems [7].

4.1 Datasets

We experiment with the following three datasets:

20 Newsgroups: this dataset is evenly partitioned into 20 different newsgroups, each corresponding to a specific topic [8]. Its "bydate" version is widely used in literature, as it has a standard training and test split. The training set has 11,293 samples and the test set 7,528 samples.

Reuters52c: it is a single-label dataset derived from Reuters-21578 with 90 classes by Ana Cardoso-Cachopo during her Ph.D. study [9]. Documents with multiple labels in the original Reuters-21578 (90 classes) dataset are discarded and finally the Reuters52c dataset contains 52 categories, 6,532 documents for training, and 2,568 documents for test. The dataset is imbalanced and some categories only have a few documents, e.g. classes *cpu* and *potato*. We use the all-terms version without stemming.

Sector: this dataset is a collection of web pages belonging to companies from various economic sectors. It has 105 categories, 6,412 training samples, and 3,207 test samples. In the training dataset, the largest categories have 80 samples, while the smallest category only has 10 samples. Most categories have around 40-80 samples. The class imbalance is not serious. This dataset was first used by McCallum and Nigam in their paper [10]. We used a version of this dataset from the LIBSVM data collection[2]. The stop words and rare words (DF = 1) are removed in this version.

4.2 Experimental Settings

We use the vector space model (VSM) for data representation, in which the dimension is determined by the size of the dataset's vocabulary. Each document is then represented as a space vector where the words in the document are mapped onto the corresponding coordinates. The weight of a feature is given as follows:

$$x_i = \frac{(1 + \log(TF(w_i, d))) \cdot \log(\frac{|D|}{DF(w_i)})}{\sqrt{\sum_j ((1 + \log(TF(w_j, d))) \cdot \log(\frac{|D|}{DF(w_j)}))^2}}, \tag{5}$$

which is the same as the standard representation "ltc" in Manning and Schutze [11]. Here, D is the document collection, and TF and DF are a term's frequency in a document d and its document frequency in the collection D respectively. In classification,

---

[2] http://www.csie.ntu.edu.tw/~cjlin/libsvmtools/datasets/multiclass.html#sector.

we use four widely used algorithms: Centroid, Multinomial Naive Bayes [10], Linear (Liblinear [12]) and SVM (Libsvm [13]).

We evaluate the performance of different classification algorithms with the original DF and our proposed CBDF feature selection metrics.

4.3 Evaluation Metric

To evaluate the effectiveness of category assignments to documents by classifiers, the harmonic average of the standard precision and recall, F1 measure, is used as follows:

$$F1 = \frac{2recall * precision}{recall + precision} \tag{6}$$

The overall performance on all categories can be computed either by the micro-averaging method or by the macro-averaging method. In micro-averaging, the MicF1 score is computed globally over all the binary decisions. In macro-averaging, the MacF1 score is computed for the binary decisions on each individual category first and then averaged over the categories. The micro-averaged score tends to be dominated by the classifier's performance on common categories, while the macro-averaged score is more influenced by the performance on rare categories.

4.4 Results and Discussion

Figure 1 shows the averaging Micro-F1 and Macro-F1 results of the four algorithms with DF and CBDF as feature selection metrics. We experiment with top N features, where N varies from 500 to 10,000 with interval of 500. Liblinear achieves the highest scores, especially for averaging Micro-F1 (around 84), where Multinomial Naive Bayes performs the poorest. SVM demonstrates advantage over Centroid. With fewer features, SVM can obtain the best results, which demonstrates its excellent discriminative capacity with fewer features. Liblinear can successfully utilize more features to get better results.

CBDF constantly performs much better than the original DF. The difference is evident when using fewer features, but tends to be narrower when using more features. With Liblinear, DF looks approximately good as others when we use more than 9,000 features (the total number of candidate features is 73,712), although CBDF does beat DF (at least 0.3).

As Liblinear achieves the best results, we will only report the results of this algorithm in the next experiments, but we do get similar trends with other three algorithms.

Figure 2 compares the normalized and un-normalized versions of CBDF on the three datasets. Normalized version does not constantly show better performance than its un-normalized variant. On the 20 newsgroups dataset and the sector dataset, CBDF_N performs almost the same as CBDF on Micro-F1 and Macro-F1. It's not out of expectation as these two datasets are almost balanced, especially for the 20 newsgroups

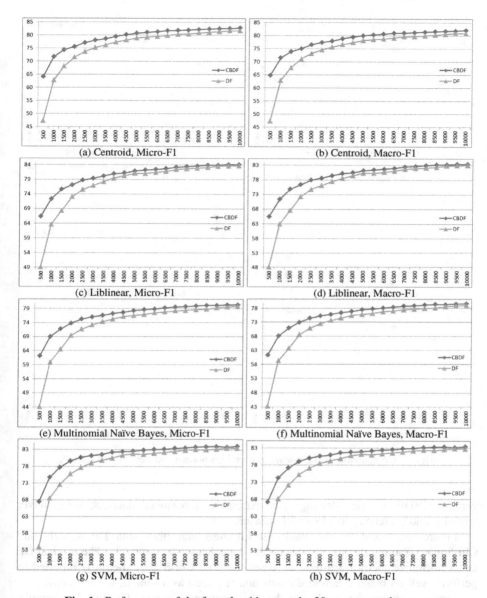

Fig. 1. Performance of the four algorithms on the 20 newsgroup dataset.

dataset. On the imbalanced dataset reuters52c, for Micro-F1, CBDF_N shows small advantage over CBDF with fewer features. For Macro-F1, on this seriously imbalanced dataset, CBDF_N clearly beats it counterpart when using fewer features. The normalized version can alleviate the negative effects of extremely imbalanced categories and obtain better Macro-averaging performance. We should choose the normalized

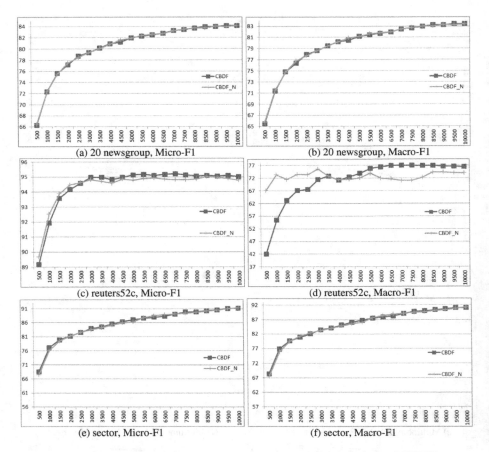

(a) 20 newsgroup, Micro-F1

(b) 20 newsgroup, Macro-F1

(c) reuters52c, Micro-F1

(d) reuters52c, Macro-F1

(e) sector, Micro-F1

(f) sector, Macro-F1

Fig. 2. Comparison of the normalized and un-normalized versions of CBDF.

version CBDF_N when dealing with extremely imbalanced datasets, using fewer features, and preferring to Macro-F1 measure.

Figures 3, 4 and 5 give results of Liblinear algorithm with DF, CBDF, and Chi-Square feature selection metrics on the three datasets respectively. The original DF and Chi-Square metrics are used as baselines. The Chi-Square metric is reported to perform well on many different datasets and regarded as a state-of-the-art metric.

Figure 3 shows the results on the 20 newsgroup dataset. The top 2 subgraphs (a) and (b) illustrate the Micro-averaging and Macro-averaging F1 scores respectively when the ratio of the selected features to the total features ranges from 0.05 to 1 with an interval of 0.05. They give us overall pictures with different feature selection metrics. The bottom 2 subgraphs (c) and (d) give the results when we choose top N features, where N varies from 500 to 10,000 with an interval of 500. We can observe from the top 2 subgraphs that CBDF performs better than DF when the ratio is less than 0.3. When we consider the top 10,000 features, it's clear that it achieves better results than

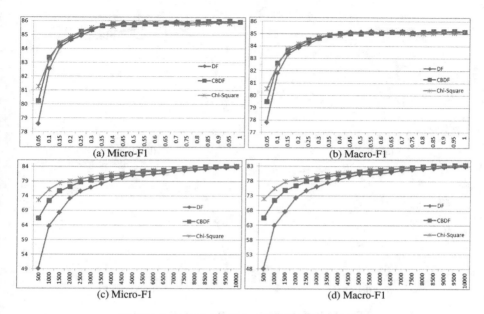

Fig. 3. Performance on the 20 newsgroup dataset.

DF. The Chi-square metric is the best one among all the three metrics on this dataset, but CBDF performs equally well as the Chi-Square metric, especially using more than 10 % features.

Figure 4 shows the results on the Reuters52C dataset. There are totally 22,274 candidate features. In the left 2 subgraphs on Micro-F1 scores, CBDF performs better than DF overall when we use less than top 3,500 features. CBDF exhibits better performance than Chi-Square when using top 10 % to 35 % features. Even for Chi-Square, it cannot beat DF all the time, e.g. when using 6,500 to 9,000 features. On Macro-F1, Chi-square achieves much better results when using less than 3,000 features, but fails to beat CBDF wen using top 30 % to 50 % features.

Figure 5 shows the results on the sector dataset. There are totally 48,988 candidate features in this dataset. CBDF obtains much better results than DF when we use less than 45 % features. The difference is quite distinct when we use less than 30 % features. The Chi-Square metric performs the poorest when we use less than 3,300 features. It tells us again that feature selection is a process dependent on the data to deal with. For a text classification problem, it's necessary to do some cross validation to find a most suitable feature selection metric.

Overall, class based document frequency metrics demonstrate advantage over the original DF metric. Normalized metric CBDF_N does show advantage over its un-normalized variant CBDF especially with extremely imbalanced datasets for Macro-averaging scores.

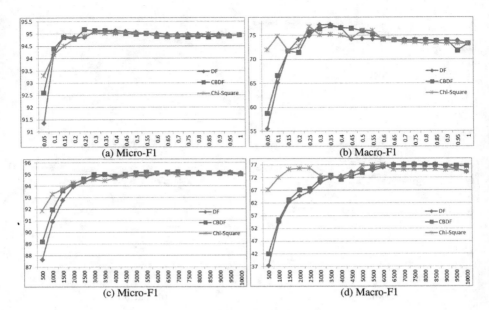

Fig. 4. Performance on the Reuters52C dataset.

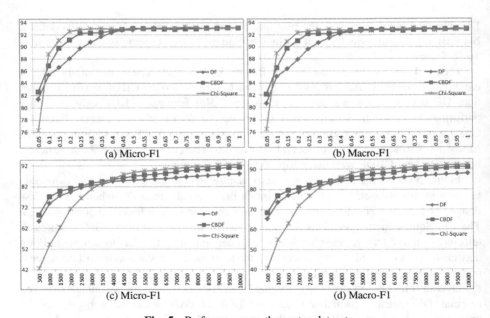

Fig. 5. Performance on the sector dataset.

5 Conclusions and Future Work

Document frequency, as an unsupervised and class independent metric, is reported as a simple yet quite effective feature selection metric in text classification. Features having the same DF value may have quite different distribution over different categories, and thus have different discriminative capacity over categories. Targeting at overcoming this drawback of DF, we propose a class based document frequency strategy and design a series of metrics for selecting features in text classification. Experiments on three publicly available datasets demonstrate that the proposed metrics does perform better than the traditional DF metric.

In the future, we would like to experiment with more datasets and apply the proposed feature selection metric into other text mining applications, e.g. text clustering, sentiment analysis, and so on.

Acknowledgments. This work was supported by the Henan Provincial Research Program on Fundamental and Cutting-Edge Technologies (No. 112300410007), and the High-level Talent Foundation of Henan University of Technology (No. 2012BS027).

References

1. Forman, G.: An extensive empirical study of feature selection metrics for text classification. J. Mach. Learn. Res. **3**, 1289–1305 (2003)
2. Yang, Y., Pedersen J.O.: A comparative study on feature selection in text categorization. In: Proceedings of Fourteenth International Conference on Machine Learning, pp. 412–420 (1997)
3. Guyon, I., Elisseeff, A.: An introduction to variable and feature selection. J. Mach. Learn. Res. **3**, 1157–1182 (2003)
4. Chandrashekar, G., Sahin, F.: A survey on feature selection methods. Comput. Electr. Eng. **40**(1), 16–28 (2014)
5. Tang, J., Alelyani, S., Liu, H.: Feature selection for classification: A review. In: Aggarwal, C. (ed.) Data Classification: Algorithms and Applications. Chapman & Hall/CRC Data Mining and Knowledge Discovery Series. CRC Press, Boca Raton (2014)
6. Lazar, C., Taminau, J., Meganck, S., Steenhoff, D., Coletta, A., Molter, C., Nowe, A.: A survey on filter techniques for feature selection in gene expression microarray analysis. IEEE/ACM Trans. Comput. Biol. Bioinfor. (TCBB) **9**(4), 1106–1119 (2012)
7. Sebastiani, F.: Machine learning in automated text categorization. ACM Comput. Surv. **34**(1), 1–47 (2002)
8. Lang, K.: Newsweeder: learning to filter netnews. In: Proceedings of the Twelfth International Conference on Machine Learning, 9–12 July 1995, pp. 331–339 (1995)
9. Cardoso-Cachopo, A.: Improving Methods for Single-label Text Categorization. Ph.D. thesis, Instituto Superior Técnico, Portugal (2007)
10. McCallum, A., Nigam, K.: A comparison of event models for naive bayes text classification. In: Proceedings of AAAI 1998 Workshop on Learning for Text Categorization (1998)

11. Manning, C.D., Schutze, H.: Foundations of Statistical Natural Language Processing. MIT Press, Cambridge (1999)
12. Fan, R.-E., Chang, K.-W., Hsieh, C.-J., Wang, X.-R., Lin, C.-J.: LIBLINEAR: A library for large linear classification. J. Mach. Learn. Res. **9**, 1871–1874 (2008)
13. Chang, C.-C., Lin, C.-J.: LIBSVM: a library for support vector machines. ACM Trans. Intell. Syst. Technol. **2**, 27:1–27:27 (2011)

A Generalized Location Privacy Protection Scheme in Location Based Services

Jing-Jing Wang[1(✉)], Yi-Liang Han[1], and Jia-Yong Chen[2]

[1] Key Laboratory of Network and Information Security of APF, Engineering College of APF,
Xi'an 710086, China
344505421@qq.com, yilianghan@hotmail.com
[2] Laboratory of Information and Network of APF, Engineering College of APF,
Xi'an 710086, China
260840527@qq.com

Abstract. When the user getting location based services by the traditional technology, his location information of region is always be exposed. However, in modern mobile networks, even the current geographical region is a part of privacy information. To solve this problem, a new generalized k-anonymity location privacy protection scheme in location based services (LPPS-GKA) with the third trust servicer is proposed. And it can guarantee the users get good location-based services (LBS) without leaking the information of the geo-location region, which has protected the perfect privacy. Analysis shows that LPPS-GKA is more secure in protecting location privacy, including region information, and is more efficient than other similar schemes in computational and communicational aspects. It is suitable for dynamic environment for different user's various privacy protection requests.

Keywords: Location privacy protection · Generalized k-anonymity · Location based service

1 Introduction

Nowadays, social communication method has evolved dramatically along with the big data era and the development of data networks. So the new secure schemes or systems should be 'application-oriented' with strong reality backgrounds. And more and more people join in the social networks to share sources and information, and generate a lot of data, namely big data, at the same time. What's more, the users are always not willing to let other people know these data. And the location information is one of the most important contents of the user's privacy data, which should be protected properly. However, most of the new-type applications of the mobile social networks, especially the location based services, are always publishing the user's location information at any time [1], and the traditional opinion of anonymous privacy protection is not valid any more in modern society.

As a result, some newer methods for location data preservation are proposed successively for protecting location privacy comprehensively, such as the technology of

© Springer Science+Business Media Singapore 2016
W. Chen et al. (Eds.): BDTA 2015, CCIS 590, pp. 211–217, 2015.
DOI: 10.1007/978-981-10-0457-5_20

randomization [2], space-vagueness [3] and time-vagueness [4]. In these methods mentioned above, space-vagueness technology has its dominant position in real applications in big data and social networks relatively because of its moderate computation cost and easy realization.

2 Related Works

In the modern social networks, the researches on location-preserving have gotten lots of outcomes. Just as described in paper [5], the location privacy protection technology based on k-anonymity has made some remarkable developments in these outcomes.

k-anonymity is one of the specific ways to realize space-vagueness. When publishing data, the real data should be disposed first, then published together with other $(k - 1)$ data simultaneously [6]. k-anonymity technology is first used in location-privacy preserving by Gruteser in 2003 [7], which means to fuzzy the user into k adjacent access points in one region. But it cannot protect the located region from leaking when each space-vagueness. From then on, it has been used widely in the location privacy protection and gotten great developments.

However, the research about how to protect location data in location based services has a rather late start. Since 2011, Huang Yi, etc. proposed a method for location privacy preserving in location services, but the time for user to get location services increases greatly when the privacy request k increasing [8]. And it is also supported by querying adjacent points in one region, so the geo-location region is exposed too.

In 2013, Damiani and Cuijpers first pointed out the issue that the specific information of user's region is also an important part of location privacy [9]. And they proposed one protection scheme based on privacy policy, but its efficiency is not high enough for mobile Internet. In 2014, Peng etc. proposed a method to judge the location privacy attacks according with the region located [10]. Then in 2015, Wang etc. proposed a location privacy protection approach named KAP, which aimed at the privacy issue of location service under the mobile Internet combing the concept of k-anonymity [5]. The approach had stronger security, but it needs a number of other access points' data around the user during every location service, which may leak the user's real located region again.

To achieve the perfect location privacy, we must protect the specific location. On the other hand, we must pay attention to the content about the located region including the user too. And the method realizing this idea should be feasible and efficient.

3 Our Scheme

In the existing outcomes in researches on the issue, most of schemes about location privacy protection are based on the center server and distributed P2P structure [11]. So in this paper, we will discuss about this type protection schemes to realize perfect privacy.

3.1 Basic Definition

Definition 1: CRT (Chinese Remainder Theorem): Suppose $m_1, m_2 \cdots m_k$ are positive integers that are pairwise co-prime. Then, for any given sequence of integers $a_1, a_2 \cdots a_k$, there exists an integer x solving the following system of simultaneous congruences.

$$\begin{cases} x \equiv a_1 \bmod m_1 \\ x \equiv a_2 \bmod m_2 \\ \cdots \\ x \equiv a_k \bmod m_k \end{cases} \tag{1}$$

If $M = m_1 m_2 \cdots m_k$, $M_i = \frac{M}{m_i}$ and $M_i M_i^{-1} = 1 \bmod m_i$, we can computer the solution set of the equations above:

$$X = M_1 M_1^{-1} a_1 + M_2 M_2^{-1} a_2 + \cdots + M_k M_k^{-1} a_k + K * M, \quad K \in Z \tag{2}$$

Obviously, the original number x is included as one element in the solution set X.

K is an integer in Eq. (2). Choose K as different values, we can get different solutions to satisfy all the equations in Eq. (1).

Definition 2: Generalized k-Anonymity (GKA): In location based services, if one user's accurate location data is extend to k access points, and: (a) It is not necessary that these k access points are adjacent neighbours in one region and interact with the user during location. (b) These k access points must belong to a equivalence class, which means for the attacker, he can not tell which one is the user's real location data.

It can be defined as generalized *k-anonymity*.

From the definition above, we can see that GKA is more scientific and practical than the traditional definition in the background of big data and complex social networks.

3.2 LPPS-GKA

At present, most of the researches based on *k-anonymity* used the trusted third party, namely center servicer [8]. The main function of the center servicer is anonymity and agency query. As shown in Fig. 1, we begin with step 1:

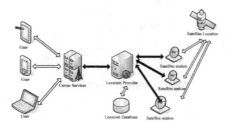

Fig. 1. Illustration for scheme with center servicer

Here,

x: The longitude-coordinate of the user's location;
y: The latitude-coordinate of the use's location;
t: The time when the user asks for service;
v: The velocity when the user asks for service;
$con.$: The query content of the user;
k: The privacy protection request parameter wanted by the user;
$K, K':K, K' \in Z$ It is the integer chosen in the process of CRT computation.

Step 1:
Query: The user sends his location information, queries contents and privacy request to the center anonymous servicer;

$$Q = (x, y, t, v, con., k)$$

Step 2:
GKA: The center servicer first extends the accurate location coordinate of user into an equivalence set including k different elements through CRT. Then it sends the query content and these k elements as a query set to the location provider.

– Choose $m_1, m_2, \cdots m_k$ randomly, satisfying the refine condition:
  $m_1, m_2, \cdots m_k \in Z^+, gcd\left(m_i, m_j\right) = 1, i, j \in Z^+$, and if the user doesn't want to leak his region information, make sure $m_1 * m_2 * \cdots m_k \gg 10^4$ (because we take the precision of coordinates up to four decimal places), then the results returned are far apart from each other; Or else, if the user doesn't care about his current area information, $m_1, m_2, \cdots m_k$ can be chosen randomly even its multiply is very small. So as to realize generalized *k-anonymity* defined in our paper, which is more scientific than the traditional definition. We give a toy example for the results from GKA, which is illustrated by Fig. 2.
– Send the query message *KAC* to the location provider

$$KAC = \left((x_1, y_1), (x_2, y_2), (x, y), con., t\right)$$

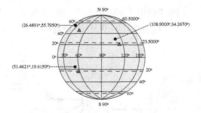

Fig. 2. Illustration for GKA

Step 3:
LBS: The location provider offers the query results set QC to the center servicer:

$$QC = \left(con_1., con_2., con., t'\right)$$

The center servicer finds the accurate result $con.$ in the set, and computes r according the time interval and velocity. Through $r = v(t' - t)$ and the original location of the user to judge the current area with radius r, then sends $con.$ to the user by locating this area. In this way, LPSS-GKA is completed with a center servicer.

4   Discussion and Analysis

The privacy protection technology of location big data not only needs to protect the user's location data, but also to balance the feasibility of services and overheads [12]. In this chapter, we will discuss the features and performance of our schemes from 3 aspects respectively: (1) Geo-indistinguishability; (2) Survivability of services; (3) Overheads in computation and communication.

4.1   Geo-Indistinguishability

Theorem 1. LPSS-GKA - has realized geo-indistinguishability.

Proof. As the user asks for service every time, we get an equivalence class including his real location data, and in this equivalence class, there are k different elements with equivalent relationship with each other. And in theory, these k elements have the uniform probability to be chosen for the attacker. Even in the continual services during a period of time, the attacker can't identify with greater advantage. Because in fact, the different regions located successively form an equivalence class too. So we can say it realized the objective of geo-indistinguishability.

4.2   Feasibility of Services

Theorem 2. LPPS-GKA is feasible. It supports high quality of services, reliability of query results.
We prove its feasibility mainly through two standards used universally: (a) Quality of services; (b) Reliability of query results.

(a)   *Quality of services*

Definition 3. The quality of service can be judged by the proportion of the number of the successful privacy requests n' as k changed in the whole number of all the privacy requests n, namely the success rate of anonymity. Its math expression is:

$$R_{SA} = \frac{n'}{n} \times 100\% \ (n' \leq n)$$

Proof. Because we used the classic math tool CRT to realize *k-anonymity*, when k is changed by user, the only additional thing need to do is choose more or less K values in the solution set according to different k.

As we all known, the number of integers is countless, which means $R_{SA} \approx 1$ if the servicer and the terminal devices run normally.

(b) *Reliability of query results*

Definition 4. The reliability of query results can be judged by the relationship of the distance between the user's original location p and the current location p_1 when receiving query result after a time interval Δt, and the radius of user's location region.

- For LPSS-GKA $r = v(t' - t) = v * \Delta t$. If $\frac{|pp_1|}{v * \Delta t} \leq 1$, we say the query result is reliable and accurate. Or else, the query result can't be returned to the user's hand, because the user's location has gone beyond the communication range accepted by the center servicer. In this case, the location based service is invalid, but this case can be neglected if we choose proper parameters.

4.3 Performance

High efficiency is the main advantage for our scheme to real application in social networks or mobile networks.

(a) In LPSS-GKA, the space-vagueness degree, namely the privacy request k can be changed easily and smoothly without adding more computation overheads. Shown as Fig. 3, even k becomes larger, the overhead of computation and communication of both center servicer and user's terminal devices is almost fixed because the high efficiency of CRT.

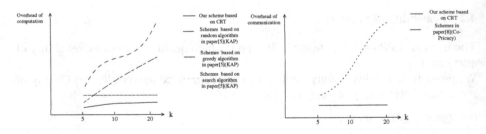

Fig. 3. Illustration for the overheads comparison of our scheme and classic schemes

When asking for location based services, the users in our scheme needn't to interact with other adjacent users for several times any more like most of the existing schemes to get other fuzzy access points. Therefore, it can spare more communication time and network source.

5 Conclusion and Future Work

A new LPPS-GKA for space-vagueness with center servicer in location services based on CRT is proposed in this paper. The scheme can be proved to get good LBS for users without leaking the information of the user's location region. Besides, it can meet different location privacy preserving requests of users with high efficient. When the privacy request is higher, namely k is larger, the time increased can be neglected.

Since the third trust servicer is not so trustable, so the future work is to design a better protection scheme without third center servicer.

Acknowledgment. This work was supported in part by the National Natural Science Foundation of China (Nos. 61272492 and 61572521), Shaanxi Province Natural Science Foundation of China (No. 2015JM6353) and the Basic Foundation of Engineering University of CAPF (No. WJY201521).

References

1. Jabeur, N., Zeadally, S., Sayed, B.: Mobile social networking applications. Commun. ACM **56**(3), 71–79 (2013). doi:10.1145/2428556.2428573
2. Suzuki, A., Iwata, M., Arase, Y., Hara, T., Xie, X., Nishio, S.: A user location anonymization method for location based services in a real environment. In: Proceedings of the 18th ACM SIGSPATIAL International Symposium on Advances in Geographic Information Systems, San Jose, pp. 398–401 (2010). doi:10.1145/1869790.1869846
3. Gredik, B., Liu, L.: Protecing location privacy with personalized k-anonymity: architecture and algorithms. IEEE Trans. Mob. Comput. **7**(1), 1–18 (2008). doi:10.1109/TMC.2007.1062
4. Pan, X., Xu, J., Meng, X.: Protecing location privacy against location-dependent attacks in mobile services. IEEE Trans. Knowl. Data Eng. **24**(8), 1506–1519 (2012). doi:10.1109/TKDE.2011.105
5. Wang, Y., Zhang, H., Yu, X.: KAP: location privacy-preserving approach in location services. J. Commun. **35**(11), 182–190 (2014). (in Chinese)
6. Wernke, M., Skvortsov, P., Durr, F., et al.: A classification of location privacy attacks and approaches. Pers. Ubiquit. Comput. **18**(1), 163–175 (2012)
7. Gruteser, M., Grunwald, D.: Anonymous usage of location-based services through spatial and temporal cloaking. In: Proceedings of the 1st International Conference on Mobile Systems, Applications and Services (MOBISYS 2003), San Francisco, California, pp. 31–42 (2003)
8. Huang, Y., Huo, Z., Meng, X.F.: Coprivacy: a collaborative location privacy preserving method without cloaking region. Chin. J. Comput. **34**(10), 1977–1985 (2011). (in Chinese)
9. Damiani, M.L., Cuijpers, C.: Privacy challenges in third-party location services. In: IEEE 14th International Conference on Mobile Data Management (MDM 2013), Milan, Italy, pp. 213–225 (2013)
10. Peng, Z.T., Kaji, K., Kawaguchi, N.: Privacy protection in Wi-Fi based location estimation. In: The 7th International Conference on Mobile Computing and Ubiquitous Networking (ICMU 2014), Singapore, pp. 62–67 (2014)
11. Yang, S.-T., Ma, C.-G., Zhou, C.-L.: LBS-oriented location privacy protection model and scheme. Chin. J. Commun. **35**(8), 116–127 (2014)
12. Wang, L., Meng, X.F.: Location privacy preservation in big data era: a survey. J. Softw. **25**(4), 693–712 (2014). doi:10.13328/j.cnki.jos.004551. (in Chinese)

Cooperation Oriented Computing: A Computing Model Based on Emergent Dynamics of Group Cooperation

Jiaoling Zheng$^{(\boxtimes)}$, Hongpin Shu, Yuanping Xu, Shaojie Qiao, and Liyu Wen

Department of Software Engineering, Chengdu University of Information Technology, No. 24, Block 1, Xuefu Road, Chengdu 610225, China
zjl9191@163.com

Abstract. Traditional way of problem solving tries to deliver data to program. But when the problem's complexity exponentially increases as the data scale increases, to obtain the solution is difficult. Group cooperation computing model works in an inverse way by delivering program to data. It first models each single data as individual and data unit as group of individuals. Then, different cooperation rules are designed for individuals to cooperate with each other. Finally, the solution of the problem emerges through individuals' cooperation process. This study applies group cooperation computing model to solve Hamilton Path problem which has NP-complete time complexity. Experiment results show that the cooperation model works much better than genetic algorithm. More importantly, the following properties of group cooperation computing are found which may be different from the traditional computing theory. (1) By using different cooperation rules, the same problem with the same scale may exhibit different complexities, such as liner or exponent. (2) By using the same cooperation rule, when the problem scale is less than a specific threshold, the problem's time complexity is liner. Otherwise, the problem complexity may be exponent.

Keywords: Group cooperation · Cooperation rules · Social computing · Hamilton path · NP-hard

1 Introduction

There are many research findings in biological and social science which demonstrates the power of group cooperation. The famous Drachten traffic management experiment [1] shows that automobiles can cooperate with each other without the management of traffic police and traffic signals. The Wikipedia website shows that individuals can

This work was supported by National Science Foundation of China (61202250, 61203172); Key Project of Sichuan Provincial Department of Education (15ZA0184); Youth Foundation Project of Sichuan Provincial Department of Education (11ZB088); Scientific Research Foundation of CUIT (KYTZ201110, KYTZ201111, J201208, J201101).

W. Chen et al. (Eds.): BDTA 2015, CCIS 590, pp. 218–233, 2016.
DOI: 10.1007/978-981-10-0457-5_21

write encyclopedia together without strict auditing process [2]. Moreover, researchers of computer science try to solve NP-hard problems with methods of group cooperation. Scott Aaronson tries to solve Steiner tree problem through the autonomous connection process of soap bubbles [3]. Group cooperation game Foldit can produce accurate models of the protein by letting game user cooperate with each other [4]. The main contribution of this study is not to introduce algorithms solving one single problem. More importantly, it tries to design cooperation rules to solve computing problems and to explore the inner relation between micro individual dynamics and macro group dynamics. Finally, this study tries to understand the relation between the algorithm's complexity and the algorithm's micro, macro dynamics.

2 Motivation

Traditional way of problem solving tries to deliver data to program. But when the problem's complexity exponentially increases as the data scale increases, to obtain the solution is difficult. Cooperation computing model works in an inverse way by delivering program to data. It is motivated by DNA computing and Artificial Chemistry Programming [5]. The architecture of Artificial Chemistry Programming can be defined by the triple (S, R, A) as illustrated in Fig. 1. S denotes a set of objects and R a set of interaction rules constituting the reaction schemes applied whenever molecules collide. The system's dynamics is controlled by an algorithm A describing how the rules are applied to a population of molecules. Accordingly, the architecture of cooperation computing can also be defined by a triple (S, R, A) as illustrated in Fig. 2. Here, S denotes a set of sub-solutions of the problem and R a set of interaction rules according to which the sub-solutions may cooperate with each other. The system's dynamics is controlled by an evolution algorithm A describing how the rules are applied to those sub-solutions. The following illustrates the triple (S, R, A) of Cooperation oriented computing to solve Hamilton Path problem.

(1) **S(Sub-solutions).** Data unit is a group of individuals or sub-solutions.

Definition 1 SHP (SHP) and SHP-set (SHP-set): Let $G(V,E)$ be an undirected graph. V is the node set of G, $V = (v_1, v_2, \ldots, v_n)$. E is the edge set of G. An **SHP** is a

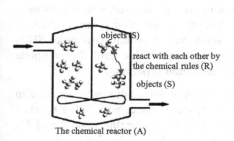

The chemical reactor (A)

Fig. 1. Architecture of cooperation oriented computing

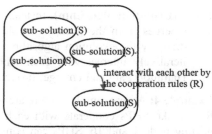

The evolution algorithm (A)

Fig. 2. Architecture of artificial chemistry programming

path which contains a sub set of V and visits each vertex exactly once. An **SHP-set** of graph G is a set of SHPs, SHP-set = {$SHP_1, SHP_2,...,SHP_m$}. SHP-set satisfies the following constraints: (1) $V(SHP_1) \cup ... \cup V(SHP_m)$ = V; (2) $V(SHP_1) \cap ... \cap V(SHP_m)$ = Ø. $V(SHP_i)$ = {$v|v \in SHP_i$ ($1 \le i \le m$) and $v \in V$}. Suppose there is a graph G (V,E) as illustrated in Fig. 3. The following gives an example of G's SHP-set.

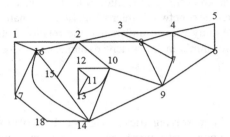

Fig. 3. A graph example

Example 1. SHP-set = {SHP_1, SHP_2, SHP_3}. SHP_1 = <1-2-3-4-5-6-9-7-8>, SHP_1's marginal nodes are node 1, 8; SHP_2 = <17-18-14-10-12-13-11>, SHP_2's marginal nodes are node 17, 11; SHP_3 = <16-15> SHP_3's marginal nodes are node 16, 15;

(2) **R(rules).** Different rules are designed for sub-solutions to cooperate.

Connect rule: If there exists an edge between the end nodes of two different SHPs, the two SHP can be connected.
Example 2: Assume graph G in figure has 3 SHPs, its SHP-set = {SHP_1, SHP_2, SHP_3}. SHP_1 = <1-2-3-4-5-6-9-7-8>, SHP_2 = <17-18-14-10-12-13-11>, SHP_3 = <16-15>. First, SHP_1 can connect with SHP_2 by connecting node 1 and 17. The new SHP-set = {SHP_1, SHP_2}. SHP_1 = <11-13-12-10-14-18-17-1-2-3-4-5-6-9-7-8>, SHP_2 = <16-15>.
Disconnect rule: One SHP can be cut off into two shorter SHPs.
Example 3: Cut off SHP_1 by disconnect node 1,2. The new SHP-set = {SHP_1, SHP_2, SHP_3}. SHP_1 = <11-13-12-10-14-18-17-1> SHP_2 = <2-3-4-5-6-9-7-8>, SHP_3 = <16-15>.

(3) **A(Evolution algorithm).** Instead of searching for the final solution, the solution emerges from the group sub-solution's dynamic cooperation process which is drive by the evolution algorithm. In this study, the algorithm is to make SHPs iteratively interact with each other according to the rules until the complete Hamilton Path can emerge through the collaboration process.

Example 4: After example 3, there are 3 SHPs. The algorithm continues to use rules in R to make SHPs cooperate with each other. SHP_1 can connect with SHP_3 by connecting node 1 and 16. SHP_2 can connect with SHP_3 by connecting node 2 and 15. A complete Hamilton Path is <11-13-12-10-14-18-17-1-16-15-2-3-4-5-6-9-7-8>.

3 Related Works

Firstly, a lot of researchers believe that group intelligence may emerge through the cooperation of individuals. Nicholas Stern's research report shows that by increase society's communication and trustiness, the global climate change problem can be solved much better compared with current solution [6]. Nicholas A. Christakis of Harvard University believes that the evolution of society will let individuals develop new skills which can make them cooperate with each other better when social network become more and more complex [7]. The famous American reporter James Surowiecki argues in his book "The Wisdom of Crowds" that the information of a group can lead to better decisions than the information of an individual [8]. Socialist Robert Axelrod's research results show that cooperation can emerge in a group in which individuals do not have any friendship with each other [9]. Dunbar's social evolutionary theory points out that the number of relationship an individual can maintain is less that 150 [10]. J. Bruggeman and other socialists believe that diverse social relationship can bring more opportunity to an individual [11–14]. Secondly, many researchers try to find the mechanism of group cooperation. J.P. Eckmann finds that different interactions in a social group will strongly affect the productivity of individuals in it [15, 16]. In another aspects, in order to analyze human mobile and behavior, Alex Pentland proposes several methods of how to collect the immense data generated by individuals' daily life [17–19]. By using data collected by mobile phon, Gonzales et al. find that human mobility follows simple and reproducible patterns [20–22]. Thirdly, many computer scientists try to solve computing problems through group cooperation. Scott Aaronson tries to solve Steiner tree problem which is NP-hard by through the connection of soap bubbles [3]. Firas Khatib tries to produce accurate models of the protein by challenging players with a protein folding game Foldit. Surprisingly, through the cooperation of playing the game Foldit, players are able to generate models of sufficient quality for successful molecular replacement [4]. Liu Jiming et al. tries to solve the N-queen problem by autonomy oriented computing which obtain good experiment results [23]. As to solving Hamilton Path problem, there are several heuristic methods. Most of these heuristic rules are applicable for specific graphs.

4 Cooperation Rules

Duncan Watts finds that a giant social network with long circle may emerge from the composition of short circles [24]. Kim finds that to connect nodes with larger node degree may improve the probability of quickly finding a path from the starting node to the target node [25]. And according to the above intuition, we design 6 types of rules. Table 1 illustrates the notations that will be used to describes those rules.

Rule 1 (Connect Strategy): Minimum Degree Node Connect First. Assume there are three SHPs of G as shown in Fig. 4(a). Node a is the marginal node of SHP_3 and Node d is the marginal node of SHP_2, and $E_{original}(a,d) = 1$. According to rule 1, node a can connect with node d and SHP_2 can join with SHP_3 to form a longer SHP. Moreover, if there exist more than one nodes similar to node d, then node a should connect with the

Table 1. Notation

Notation	Meaning of notation
Original Neighbors	If node i, j are neighbors in the original graph G, then node i, j are original neighbors.
$E_{original}$	If node i, j are original neighbors, $E_{original}(i,j) = 1$, Otherwise, $E_{original}(i,j) = 0$
Original_Neighbors (node i)	$\{v\mid E_{original}(i,v) = 1\}$
SHP Neighbors	If node i, j are neighbors in the one of G's SHP, then node i, j are SHP neighbors.
E_{SHP}	If node i, j are SHP neighbors, $E_{SHP}(i,j) = 1$. Otherwise, $E_{SHP}(i,j) = 0$.
SHP_Neighbors (node i)	$\{v\mid E_{SHP}(i,v) = 1\}$
marginal node	Node which has only one neighbor in an SHP.
marginal node set	$\{v\mid v$ is marginal node$\}$
SHP's non-marginal node	Node which has two neighbors in an SHP.

node that has the minimum degree. For example, assume $E_{original}(a, c) = E_{original}(a, d) = 1$, Degree(d) = 5, Degree(c) = 2. Then node a should connect with node c, because Degree(c) < Degree(d).

```
Rule1(Node i)
Input:G(V,E),node i (node i is the marginal node of one
of G' s SHP)
Output:SHP_Neighbors(node i)
1 Candidate_node_set = Original_Neighbors(node i) ∩ Mar-
                                      ginal_Node_set(G);
2 If (Candidate_node_set ≠ null)
3 {Node j = one of nodes in Candidate_node_set which has
                              the minimum node degree;
4 Set E_SHP(i,j) = 1;}
```

Example 5: Suppose there is a graph G(V,E) as illustrated in Fig. 3. SHP-set = {SHP_1, SHP_2, SHP_3}. SHP_1 = <1-2-3-4-5-6-9-7-8>, SHP_2 = <17-18-14-10-12-13-11>, SHP_3 = <16-15>. Then node 1 could connect with node 16 or node 17. Because degree (16) = 5 and degree(17) = 3, node 1 should connect with node 17 according to rule 1. Consequently, SHP-set = {SHP_1, SHP_2}. SHP_1 = <11-13-12-10-14-18-17-1-2-3-4-5-6-9-7-8>, SHP_2 = <16-15>.

Rule 2(Connect Strategy): Random Connect. Rule 2 is similar with rule 1. But rule 2 randomly chooses one marginal node of other SHPs to connect instead of connecting the node with minimum degree. For example, as in Fig. 4(a), according to rule 1, node a should connect with node c. By rule 2, node a can connect either with node c or node d randomly. The following gives the algorithm of rule 2.

```
Rule2(Node i)
Input:G(V,E),node i, node i is the marginal node of one
of G's SHP
Output:SHP_Neighbors(node i)
1 Candidate_node_set = Original_Neighbors(node i) ∩ Mar-
                                       ginal_Node_set(G);
2 If ( Candidate_node_set ≠ null)
3 {    Node j = randomly choose one node in candi-
                                       date_node_set;
4      Set E_SHP(i,j) = 1;}
```

Example 6: Suppose there is a graph G(V,E) as illustrated in Fig. 3. SHP-set = {SHP$_1$, SHP$_2$,SHP$_3$}. SHP$_1$ = <1-2-3-4-5-6-9-7-8>, SHP$_2$ = <17-18-14-10-12-13-11>, SHP$_3$ = <16-15>. Then node 1 could either connect with node 17 or 16 according to rule 2. If node 1 connect with node 16, then SHP-set = {SHP$_1$,SHP$_2$}. SHP$_1$ = <15-16-1-2-3-4-5-6-9-7-8>, SHP$_2$ = <17-18-14-10-12-13-11>.

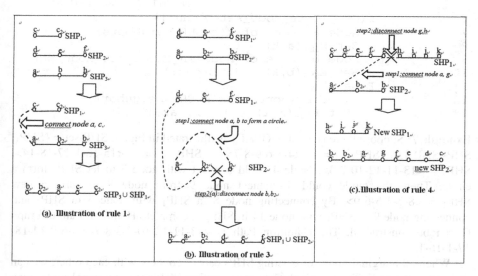

Fig. 4. Illustration of rule 3

Rule 3(Connect Strategy): Circle Composition. Assume there are two SHPs as shown in Fig. 4(b) and E$_{original}$(a,b) = 1. Then rule 3 can be used and it contains 2 steps. **Step 1:** Connect node a, b to let the SHP$_2$ form a circle. **Step 2:** Disconnect two of the circle's neighboring nodes to form new SHP. The following describes two types of situations when executing step 2.

Situation 1. Assume $E_{original}(d,b_2) = 1$, b_2 belongs to the circle and d is marginal node. Then, **(a)** disconnect node b_2 with any one of its two SHP-neighbors, say node b, in the circle; **(b)** connect node b_2 with node d to combine SHP_1 and SHP_2.

Situation 2. If situation 1 does not exist, randomly disconnect any two neighboring nodes of the circle.

Situation 1 can join the circle with another SHP to form longer SHP. Situation 2 cannot form longer SHP but can generate new marginal nodes. Figure 4(b) illustrate situation 1. The following gives the algorithm of rule 3. Line $5 \sim 10$ represents situation 1 of step 2. Line $11 \sim 13$ represents situation 2 of step 2.

```
Rule3(Node i)
Input:G(V,E),node i, node i is the marginal node of one
of G's SHP
Output:SHP_Neighbors(node i)
1 Let Node j is the other one marginal node which are in
the same SHP with node i;
2 If (E_original(i,j) = 1)
3 {    Set E_SHP(i,j) = 1;
4        Foreach node k in the circle
5        {    Candidate_node_set = Orginal_Neighbors(k)∩
                                    Marginal_nodes(G);
6            If(Candidate_node_set ≠ null)
7            {    Node a = one of k's SHP neighbors;
8                 Set E_SHP(a,k) = 0;
9                 Node b = one of node in Candidate_node_set;
10                Set E_SHP(b,k) = 1; break;}
11            Else
12            {    Node a = one of k's SHP neighbors;
13                 Set E_SHP(a,k) = 0;  break;} }}
```

Example 7: Suppose there is a graph G(V,E) as illustrated in Fig. 3. SHP-set = {SHP_1, SHP_2,SHP_3}. SHP_1 = <4-5-6-9-8-7>, SHP_2 = <15-16-1-17-18-14>, SHP_3 = <13-11-12-10-2-3>. Node 4 could connect with node 7 to let SHP_1 form a circle. Then, the circle could disconnect node 9 from node 8 to form a new SHP_1 = <8-7-4-5-6-9>. By connecting node 8 in SHP_1 with node 3 in SHP_3 and connecting node 9 in SHP_1 with node 14 in SHP_2, a complete Hamilton Path of graph G can be constructed. The Hamilton Path is <13-11-12-10-2-3-8-7-4-5-6-9-14-18-17-1-16-15>.

When all marginal nodes of existing SHPs cannot connect with each other, we can first connect one SHP's marginal node with another SHP's non-marginal node, then disconnect the non-marginal node with any one of its two SHP neighbors. These operations can generate new marginal nodes and are represented by Rules 4 and 5.

Rule 4 (Disconnect Strategy): Maximum Degree Node Disconnect First. Assume there are two SHPs of G as shown in Fig. 4(c). None of SHP_1 and SHP_2's marginal node can connect with each other. Node a is SHP_1's marginal node and node g is SHP_2's non-marginal nodes, $E_{original}(a,g) = 1$. In order to generate new marginal nodes to create new connection opportunities, there are two steps to take. **Step 1**, connect node a with node g. **Step 2**, disconnect node g with node h or node f. Node h, f are node g's SHP-neighbors. The previously marginal nodes are {a,b,c,k}. After the two steps, the marginal nodes are {b,c,h,k}. Figure 4(c) illustrates the two steps.

If there exist more than one node in SHP_1 that can connect with node a, node a should connect with the node whose SHP-neighbors has the largest degree. For example, in Fig. 4(c), assume $E_{original}(a,g) = 1$, $E_{original}(a,j) = 1$. Degree(f) = 5, Degree (h) = 3, Degree(i) = 1, Degree(k) = 3. Then node a should connect with node g. And node g should disconnect with node k, as the newly generated marginal node k has the largest degree. The following gives the algorithm of rule 4. Line $1 \sim 4$ shows when node a in SHP_2 given, how to choose node g and node k. Line 5 represents step 1 and step 2.

```
Rule4(Node i)
Input:G(V,E), node i (node i is the marginal node of one
of G's SHP)
Output:SHP_Neighbors(node i)
1 Candidate_node_set = Original_Neighbors(node i);
2 Candidate_neighbor_set = SHP_Neighbors (Origi-
                             nal_Neighbors(node i));
3 Node k = one-of node in Candidate_neighbor_set with the
                             largest node degree;
4 Node j = {E_SHP(j,k) = 1 and E_original(i,j) = 1}
5 Set E_SHP(i,j) = 1; Set E_SHP(j,k) = 0;
```

Example 8: Suppose there is a graph G(V,E) as illustrated in Fig. 3. SHP-set = {SHP_1, SHP_2}. SHP_1 = <11-13-12-10-14-18-17-1-2-3-4-5-6-9-7-8>, SHP_2 = <16-15>. Node 16 can connect with node 1,2,15,14,17. According to rule 4, node 16 should connect with node 1 and node 1 should disconnect with node 2. Because node 2 has the maximum degree in the neighbor set of node 1,2,15,14,17. The new SHPs is <11-13-12-10-14-18-17-1-16-15>, <2-3-4-5-6-9-7-8>.

Rule 5(Disconnect Strategy): Random Disconnect. Rule 5 is similar with rule 4. The difference between rules 5 and 4 is that if SHP's marginal node can connect with more than one non-marginal node of other SHPs, rule 5 will let the marginal node randomly choose a non-marginal node to connect with. In Fig. 3(c), assume $E_{original}(a,g) = 1$ and $E_{original}(a,j) = 1$, then node a can randomly connect with g or j and the node g or j can randomly disconnect with one of its two SHP neighbors.

Rule 6(Disconnect Strategy): Massive Disconnect. Rule 6 randomly choose some non-marginal nodes and disconnect these nodes with one of their two SHP neighbors. Rule 6 can be described by the following. Parameter T controls the ratio of how many nodes will execute rule 6.

```
Rule 6
Randomly choose M nodes which are not marginal nodes
1 For each of the M nodes
2 { rand = random(0,1)
3    if(rand < T){Disconnect with one of SHP neighbors}}
```

5   Evolution Algorithm Based on Cooperation Rules

5.1   Algorithm Framework

```
Algorithm Framework
Input: Let graph G(V,E) be an undirected graph, |V|=N;
Output: Hamilton path of graph G
1   Initialize SHP-set
2   while( number of SHP-set > 1 )
3        { Evolution(Combination of Rules);   };
4   End While
```

Line 1 is to initialize SHP set. Initialize SHP set = $\{SHP_1, SHP_2, \ldots, SHP_n\}$ in which $SHP_i (1 \le i \le n)$ contains only one node, i.e. node i. Line 2 ~ 4 is to let the graph nodes cooperate with each other by using the rule in the rule set. We design four combinations of rules. Combination 2 C_2 is similar with Combination 1 C_1 except that C_2 does not use Rule 3 after Rule 1. Combination 4 C_4 is similar with Combination 3 C_3 except that C_4 does not use Rule 3 after Rule 1.

```
Evolution(Rule 1, Rule 3,        Evolution(Rule 2, Rule 3,
Rule 4, Rule 6)//C1              Rule 5, Rule 6)// 3
1Foreach  SHP_i in SHP Set      1 Foreach  SHPi in SHP Set
2     Foreach Marginal Node     2     Foreach Marginal Nodei
              i in SHP_i                       in SHP_i
3        Rule 1( Node i);       3        Rule 2( Node i);
4        Rule 3( Node i);       4        Rule 3(Node i);
5      End For                  5      End For
6   End For                     6   End For
7   Foreach  SHP_i in SHP Set   7   Foreach  SHPi in SHP Set
8     Foreach Marginal Node     8     Foreach Marginal Node
              i in SHP_i                       i in SHP_i
9        Rule 4( Node i);       9        Rule 5( Node i);
10     End For                  10     End For
11 End For                      11 End For
12 Rule 6;                      12 Rule 6;
```

5.2   Theoretical Analyses

Lemma 1. Let graph G(V,E) be an undirected graph, |V| = N. G's average node degree is D and G's maximum node degree is Max_D. Assume G is divided into k SHPs, i.e. SHP-set = $\{SHP_1, SHP_2, \ldots, SHP_k\}$. The probability that there exist at least one original edge between any two of the 2 k marginal nodes is $1 - ((N - 2 K)/N)^D$.

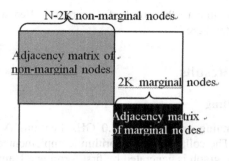

Fig. 5. Analogy of adjacency matrix

Proof. First, the probability that there exist no original edge between any two marginal nodes is $((N - 2K)/N)^D$. Assume Fig. 5 is the analogy of graph G's adjacency matrix M. $M(i,j) = 1$ means $E_{original}(i,j) = 1$ and $M(i,j) = 0$ means $E_{original}(i,j) = 0$. The black part is the adjacency matrix of marginal nodes and the gray part is the adjacency matrix of the non-marginal nodes. Each row contains 2k marginal nodes and N − 2k non-marginal nodes. Therefore, in each row, the probability that one 1 falls into the white part I is $(N - 2k)/N$. Moreover, since G's average node degree is D, so each row of the adjacency matrix contains D 1s on average. And according to the hypothesis that there exist no original edge between any two marginal nodes, so the adjacency matrix of the marginal nodes (the black part) does not contain 1s and the adjacency matrix of the non-marginal nodes must contain D 1s. Therefore, the probability that there exist no original edge between any two marginal nodes is $((N - 2K)/N)^D$. Accordingly, the probability that there exist at least one original edge between any marginal nodes is $1 - ((N - k)/N)^D$, q.e.d.

Let $p_{dc}(k - 1, rule_i | k, rule_j)$ denotes that after using $rule_j$ to divide the k SHPs into k + 1 SHPs, the probability of using $rule_i$ to connect the k + 1 SHPs into k − 1 SHPs. Let $p_c(k - 1, rule_i | k, rule_j)$ denotes that after using $rule_j$ to connect the k SHPs into k − 1 SHPs, the probability of using $rule_i$ to connect the k − 1 SHPs into k − 2 SHPs. The following demonstrates that nodes with larger degree are important.

Theorem 1. $p_{dc}(k - 1, rule1 | k, rule4) > p_{dc}(k - 1, rule1 | k, rule5)$.

Proof. After using rule4 to divide the k SHPs into k + 1 SHPs, the k + 1 SHPs must be able to connect into k SHPs. Let node a denotes the newly generated marginal node. The key point is whether there is an edge between node a and the other marginal nodes. According to Lemma 1, the probability that there exist at least one original edge between two of the 2 k marginal nodes is $1 - ((N - 2k)/N)^D$. If the new marginal node a is generated by rule 5, then node a's degree is D on average. Thus $p(k - 1, rule1 | k, rule4) = 1 - ((N - 2k)/N)^D$. However, if the new marginal node a is generated by rule 4, then node a's degree is Max_D on average. Thus $p_{dc}(k - 1, rule1 | k, rule4) = 1 - ((N - 2 k)/N)^{Max\_D}$. Because Max_D > D, so $p_{dc}(k - 1, rule1 | k, rule 4) > p_{dc}(k - 1, rule1 | k, rule 5)$.

Theorem 2. $p_{cc}(k - 2, rule1 | k, rule1) > p_{cc}(k - 2, rule2 | k, rule2)$.

Proof. The proof is similar to the proof of Theorem 1. Because limited space, the detail proof is eliminated here.

6   Experimental Results

6.1   Experiment Setting

The experiment is running on dual-core 3.0 GHz Pentium IV, windows XP, with 2 Gbytes of memory. The collaboration algorithm is implemented on the Netlogo [26] platform. The synthetic graph is generated by first forming a Hamilton path through all nodes and then let each node randomly connect with other K nodes. In the experiment K is 3. The real data is obtained from [27]. It contains 1589 nodes and 2742 edges. In case that this network does not contain a complete Hamilton Path, we construct a Hamilton Path by sequentially connecting each node.

6.2   Experiments on Synthetic Data

(1)   *Rule capacity*

Tables 2, 3, 4 and 5 describe the time complexity of the four types of rule combinations. In order to test the capacity of different combinations of rules, the graph scale increases from $|V| = 200$ to $|V| = 500$. The experiment is repeated 5 times. The success rate is the percentage of successful experiments that find out one of the network's Hamilton Path in total 5 times of experiments. The M threshold in rule 5 is set to 50 and the T threshold is set to 0.002.

Table 2. Capacity of C1

Scale	200	300	400	500
Time (s)	4.29	5.86	29	1800
Success rate	100 %	100 %	60 %	20 %

Table 3. Capacity of C_2

Scale	200	300	400	500
Time (s)	6.8	21.5	145.8	-
Success rate	100 %	100 %	20 %	-

Table 4. Capacity of C3

Scale	200	300	400	500
Time (s)	8.8	74.1	-	-
Success rate	100 %	60 %	-	-

Table 5. Capacity of C4

Scale	200	300	400	500
Time (s)	57.4	117.313	-	-
Success rate	100 %	60 %	-	-

According to Tables 2, 3, 4 and 5, when the graph scale is within 200 to 400, C_1 can solve the problem in a linear time complexity. But when problem scale increases to 500, C_1's time complexity suddenly increases. When the graph scale is from 200 to 300, C_2 can solve the problem in linear time. But when problem scale is larger than 300, C_2's time complexity also suddenly increases. C_3 and C_4 can hardly work out the solution when the graph scale is larger than 300. It is reasonable to say the complexity of the problem is not decided by the problem itself, but also by the way that we model the problem. If the rules are good enough, the problem complexity may decrease.

We try to find out Hamilton Path of the synthetic graph using Genetic Algorithm (GA). Each chromosome contains one gene and the length of gene is the number of nodes in the graph. The gene contains the number of each node. Gene(i) is the i-th position of the gene and the value of gene(i) is the number of one of G's nodes. The fitness measurement is in Eq. (1). The scale of population is 100. We use two types of genetic operators, the mutation operator and the crossover operator. The mutation rate and the crossover rate is both 0.01. The selection method is roulette wheel.

$$
\begin{aligned}
Fitness = \sum gene(i, i+1) \quad & gene(i, i+1) = 1 \, (if \, E_{original}(gene(i), gene(i+1)) = 1) \\
(0 \leq i \leq gene\_length - 1) \quad & gene(i, i+1) = 0 \, (if \, E_{original}(gene(i), gene(i+1)) = 0)
\end{aligned}
\tag{1}
$$

Since GA can not find the Hamilton path of any of the synthetic data, the experiment results is not listed here. Comparing with the cooperation rules, the genetic operators work in a macro level to adjust the system's evolution process. It is not subtle enough to detect the difference between individual nodes.

(2) Algorithm dynamics of different rules

This experiment investigates how the graph's SHP number changes during the evolution process. The horizontal axis represents the time stamp. The vertical axis represents the number of SHPs at time stamp t. The curve shows whether the system can converge to the final solution in linear or sub-linear time complexity. Figure 6 (a $\sim$ l) is the SHP number change history using C_1 strategy with $|V| = 200,300,400,500$. There are several fluctuations in each curve. The fluctuation means the number of SHP suddenly increases. This phenomenon is caused by the using of rule 6 which tries to lead the group jump out of the local optima by cutting off several existing SHPs. By using C_1, When $|V| = 200 \sim 400$, rule 6 is used quite infrequently. When $|V| = 200$, rule 6 is used only once. When $|V| = 300$, rule 6 is used twice. But when $|V| = 500$, rule 6 is used a lot of times. This means C_1 can easily jump out of the local optima when the problem scale is less than 300. But when the graph scale is larger than 300, it is difficult for C_1 to jump out of the local optima. Thus, $|V| = 300$ seems a complexity threshold. Below the threshold, C_1's time complexity is linear or sub-linear. Beyond the threshold, C_1's time complexity suddenly becomes exponent. $C_2 \sim C_4$'s SHP evolutionary process exhibit the similar phenomenon. The only difference is the complexity threshold. C_2, C_3 and C_4's complexity threshold is about 200, 80 and 50. According to experimental results, if the collaboration rule is good enough, its complexity threshold could be high. The algorithm may find Hamilton path of more complex graphs.

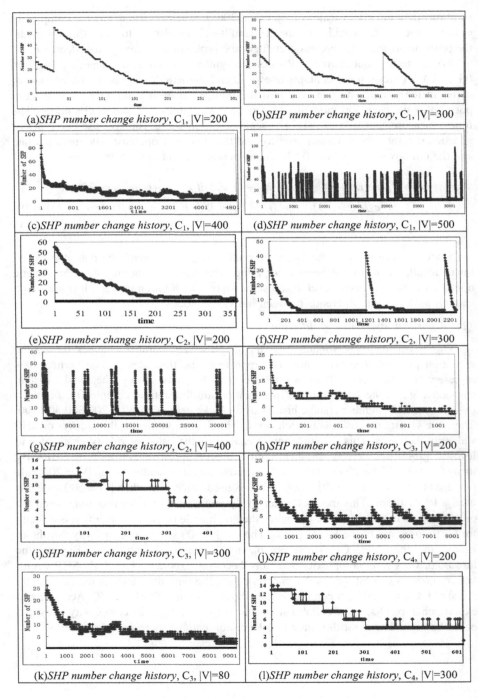

(a)*SHP number change history*, C_1, $|V|$=200

(b)*SHP number change history*, C_1, $|V|$=300

(c)*SHP number change history*, C_1, $|V|$=400

(d)*SHP number change history*, C_1, $|V|$=500

(e)*SHP number change history*, C_2, $|V|$=200

(f)*SHP number change history*, C_2, $|V|$=300

(g)*SHP number change history*, C_2, $|V|$=400

(h)*SHP number change history*, C_3, $|V|$=200

(i)*SHP number change history*, C_3, $|V|$=300

(j)*SHP number change history*, C_4, $|V|$=200

(k)*SHP number change history*, C_3, $|V|$=80

(l)*SHP number change history*, C_4, $|V|$=300

Fig. 6. SHP number change history

This experiment investigates the relationship between node's aliveness and degree. Equation (2) defines $Node_i$'s aliveness. T is the evolution time of the algorithm.

$$Aliveness(node_i) = \sum C(t)\,(1 \leq t \leq T) \quad C(t) = 0,\, if\ SHP\_neighbors(node_i, t)$$
$$= SHP\_neighbors(node_i, t-1)$$
$$C(t) = 1,\ otherwise$$

$$(2)$$

Reference [25] points out that node with larger degree are more active. In Fig. 7, the vertical axis represents node's aliveness. The horizontal axis represents node's degree. The dotted line divides the figure into several sections. Nodes in the same section have the same degree. Node degree in each section from left to right is 8,7,6,5,4,3,2. By C_1, the average aliveness of nodes with larger degree is higher than those with smaller degree. However, by C_2, the average aliveness doesn't have significant difference. Since C_1 performs better than C_2, this experiment verifies Theorems 1 and 2.

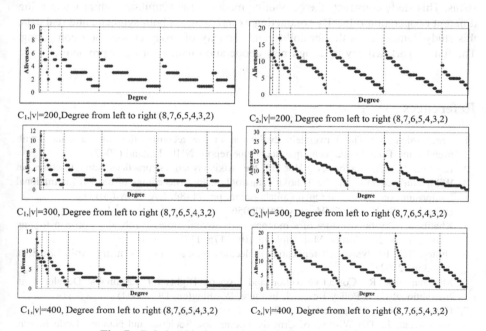

C_1,|v|=200,Degree from left to right (8,7,6,5,4,3,2) C_2,|v|=200, Degree from left to right (8,7,6,5,4,3,2)

C_1,|v|=300, Degree from left to right (8,7,6,5,4,3,2) C_2,|v|=300, Degree from left to right (8,7,6,5,4,3,2)

C_1,|v|=400, Degree from left to right (8,7,6,5,4,3,2) C_2,|v|=400, Degree from left to right (8,7,6,5,4,3,2)

Fig. 7. Relationship between nodes' aliveness and degree

6.3 Experiments on Real Data

This section conducts experiments on real data as described in Sect. 6.1. Two other algorithms are selected to compare with C_1. In Table 6, Mathematica 9 only contain algorithm to find a graph's Hamilton Cycle. We add an additional node which has an edge to every other node in the graph. If the algorithm can find a Hamilton Cycle in the new graph, it finds the Hamilton Path in the original graph. ACOTSP is a well known ant colony optimization package. If there is an edge between two nodes, the distance between them is 0. Otherwise, the distance is 1. If ACOTSP can find a path in the graph whose length is 0, it means the algorithm finds the Hamilton Path in the graph. Two methods cannot find the Hamilton Path within two hours.

Table 6. Capacity comparison between C_1 and two existing methods

	C_1	Mathematica 9	ACOTSP
Time (s)	<500	>7200	>7200

7 Conclusion and Future Work

The traditional computing method is to deliver data to algorithm. The group cooperation model is to deliver algorithm to data. Therefore, the cooperation computing model can adjust its evolution process according to each individual data's real-time status. This study constructs the computing model of the Hamilton Path problem using different cooperation rules. By conducting experiments on both synthetic and real data, this study demonstrates the feasibility and capacity of group cooperation computing. The future work will try to design new cooperation model for other problems.

References

1. Leeuwarden, N.: The Laweiplein evaluation of the reconstruction into a square with roundabout. Bouwkunde technique, technique report: NHL, Finland (2007)
2. Keegan, B., Contractor, N.: Hot off the wiki: dynamics practices and structures in Wikipedia's coverage of the Tōhoku catastrophes. In: Proceedings of the 7th International Symposium on Wikis and Open Collaboration, pp. 71–84, CA, USA (2011)
3. Bringsjord, S.: P = NP.2004. cs. CC/0406056
4. Khatib, F.: Crystal structure of a monomeric retroviral protease solved by protein folding game players. Nat. Struct. Mol. Biol. **18**(10), 1175–1177 (2011)
5. Hutton, T.J.: Evolvable self-replicating molecules in an artificial chemistry. Artif. Life **8**(4), 341–356 (2002)
6. Meinzen-Dick, R.: Collective Action and Property Rights for Sustainable Development. International Food Policy Research Institute, Washington, DC (2004)
7. Christakis, N.A.: Connected. Little Brown, New York (2009)
8. Surowiecki, J.: The Wisdom of Crowds, Economies, Societies and Nations. Little Brown, New York (2004)
9. Robert, A.: The evolution of cooperation. Science **211**, 1390–1396 (1981)
10. Dunbar, R.: Co-evolution of size. Behav. Brain Sci. **16**(4), 681–735 (1993)
11. Eagle, N., Macy, M., Claxton, R.: Network diversity and economic development. Science **328**(5981), 1029–1031 (2010)
12. Page, S.E.: The Difference: how the Power of Diversity Creates Better Groups. Princeton University Press, New York (2008)
13. Granovetter, M.S.: The strength of weak ties. Am. J. Sociol. **1973**, 1360–1380 (1973)

14. Burt, R.S.: Structural Holes: the Social Structure of Competition. Harvard University Press, New York (2009)
15. Sinan, A.: Network structure and information advantage. In: Proceedings of the Academy of Management, vol. 3, PA (2007)
16. Dawber, T.: The Framingham Study: the Epidemiology of Atherosclerotic Disease. Harvard University Press, New York (1980)
17. Pentland, A.: Computational social science. Science 323(5915), 721–723 (2009)
18. Pentland, A., Madan, A., Gatica-Perez, D., Farrahi, K.: Pervasive sensing to model political opinions in face-to-face networks. In: Lyons, K., Hightower, J., Huang, E.M. (eds.) Pervasive 2011. LNCS, vol. 6696, pp. 214–231. Springer, Heidelberg (2011)
19. Olguín, D.O.: Sensible organizations: technology and methodology for automatically measuring organizational behavior. IEEE Trans. Syst. Man Cybern. Part B: Cybern. 39(1), 43–55 (2009)
20. Gonzalez, M.: Understanding individual human mobility pattern. Nature 453, 779–782 (2008)
21. Raul, M.: Discovering human places of interest from multimodal mobile phone data. In: Proceedings of the 9th International Conference on Mobile and Ubiquitous Multimedia, vol. 12, New York, USA (2010)
22. Lu, H.: The jigsaw continuous sensing engine for mobile phone applications. In: Proceedings of the 8th Conference on Embedded Networked Sensor Systems, pp. 71–84, CA, USA (2010)
23. Liu, J., Hu, B.: On complex emergent behavior, self-organized criticality, and phase transition in multi-agent systems: the autonomy oriented computing perspectives. Model. Ident. Control 3(1), 3–16 (2008)
24. Watts, D.: Small Worlds: the Dynamics of Networks Between Order and Randomness. Princeton University Press, Princeton (1999)
25. Kim, B.J.: Path finding strategies in scale-free networks. Phys. Rev. E 65(2), 027103 (2002)
26. Railsback, F.: Agent-based individual-based modeling. Princeton University, NJ (2011)
27. Boccaletti, S., Latora, V., Moreno, Y.: Complex networks: structure and dynamics. Phys. Rep. 424(4), 175–308 (2006)

SVR Recommendation Algorithm of Civil Aviation Auxiliary Service Based on Context-Awareness

Haohan Liu, Hongli Zhang$^{(\boxtimes)}$, Kanghua Hui, and Huaiqing He

College of Computer Science and Technology,
Civil Aviation University of China, Tianjin, China
{hhliu,khhui}@cauc.edu.cn, hlzhang1006@163.com,
huaiqinghe@aliyun.com

Abstract. Analysis of the particularity of the civil aviation passenger auxiliary service recommendation scenario. As application of the traditional recommendation algorithm has certain limitation in civil aviation auxiliary services recommendation, a SVR recommendation algorithm of auxiliary service of civil aviation based on context-awareness was proposed. Analysis of the civil aviation passenger travel data, construct the civil aviation passenger preference model, then recommend auxiliary service for passengers. Based on the traditional two-dimensional user-item recommendation, considering the user characteristics, item attributes and user contextual information in the process of recommendation, which can effectively reduce the data sparseness in some degree. In addition, when there is a new user or a new item, whose similar users or items can be found according to the user or item attributes, to some extent, which can solve the problem of cold start. The experimental results show that the algorithm can recommend auxiliary service for passengers more accurately, which can provide convenience for passengers as well as increase the quality of airlines' services.

Keywords: Context-aware recommendation · SVR · Auxiliary service of civil aviation recommendation · Collaborative filtering

1 Introduction

Recommendation system has been widely used in every field of e-commerce. It recommends items and information which the user may be interested in while improving the service quality of the e-commerce business. Nowadays, most of the recommendation systems rely on the user's rating data to the item, which brings about the problems of sparsity and cold-start. In reality, there are many factors that impact the determination of the user when they choose to buy something or depend on some information. And the user requirements also change with the contextual information. For example, the user preference will alter with the time [1, 2], the user attributes information and the item characteristic attributes also influence the user preference [3]. In recent years, many scholars have pointed out the concept of context-aware recommendation system, which brought in the contextual information in the process of recommendation and had become the research hot-spot in the domain of recommendation system.

© Springer Science+Business Media Singapore 2016
W. Chen et al. (Eds.): BDTA 2015, CCIS 590, pp. 234–246, 2016.
DOI: 10.1007/978-981-10-0457-5_22

From the point of civil aviation industry, more and more types of civil aviation auxiliary services have appeared. How to recommend the service which the passenger may interest from various of services appropriately is much more important. As the passenger trip will be influenced by many factors, a SVR recommendation algorithm of auxiliary service of civil aviation based on context-awareness is proposed. Firstly, the algorithm constructed the model of the user and item characteristic attributes respectively, and then the preference model was built based on the user ratings to the item. Finally, we added the contextual information into the user-item preference model, thus constructing the model of the user-item-context, which is the preference model of the user to item under certain contextual information.

2   Construct the User Preference Model

Based on the traditional two dimensional of the user-item recommendation, the algorithm took the user and the item characteristic attributes information and the contextual information to construct the user preference model.

2.1   Construct the User Attributes Model

Construct the user characteristic attributes model with the attribute information (such as: gender age occupation).

$$U_{ij} = \begin{bmatrix} u_{11} & \cdots & u_{1n} \\ \vdots & \ddots & \vdots \\ u_{m1} & \cdots & u_{mn} \end{bmatrix} \tag{1}$$

Where, u_{ij} represent the characteristic attributes of the user:

$$u_{ij} = \begin{cases} 0, user\ i \text{ doesn't belong to attribute } j \\ 1, user\ i \text{ belongs to attribute } j \end{cases} \tag{2}$$

Such that m denotes the number of the user, n denotes the number of the characteristic attributes of the user.

2.2   Construct the User Preference Model

(1) Construct the item characteristic attributes information model, which is given by:

$$Item_{ij} = \begin{bmatrix} item_{11} & \cdots & item_{1n} \\ \vdots & \ddots & \vdots \\ item_{m1} & \cdots & item_{mn} \end{bmatrix} \tag{3}$$

Where, $item_{ij}$ denotes the characteristic attributes of item I, which is defined as follows:

$$item_{ij} = \begin{cases} 0, item\ i\ doesn't\ possess\ to\ attribute\ j \\ 1, item\ i\ posesses\ to\ attribute\ j \end{cases} \tag{4}$$

Such that m denotes the number of the item, n denotes the number of the characteristic attributes of the item.

(2) Construct the preference model according to the characteristic attributes, which is denoted as follows:

$$P_{ij} = \begin{bmatrix} p_{11} & \cdots & p_{1n} \\ \vdots & \ddots & \vdots \\ p_{m1} & \cdots & p_{mn} \end{bmatrix} \tag{5}$$

Where, p_{ij} denotes the preference value to the item of the user, m represents the number of the user, n represents the number of the characteristic attributes of the item. The preference value to the item of the user is defined as follows:

$$p_{ij} = \frac{n_j}{m} \tag{6}$$

Such that, n_j represents the number of ratings to the item j of the user i, m denotes the number of all items which are rated by the user ($n_j <= m$). The range of the value of p_{ij} is between 0 and 1.

(3) Combine the model which is built in (1) and (2), we can define the user-item characteristic attributes preference model M_u as follows:

$$M_u = Item \times P_u \tag{7}$$

2.3   Construct the Contextual User Preference Model

As the choice of the user will change under different contexts, we add the context into the model constructed above. Considering these contexts which influence the user behavior in the process of recommendation can recommend better, and the accuracy of the recommendation has been improved.

(1) According to the context which the user is under to construct the contextual matrix model:

Where, the context c_{ij} is defined as follows:

$$C_{ij} = \begin{bmatrix} c_{11} & \cdots & c_{1n} \\ \vdots & \ddots & \vdots \\ c_{m1} & \cdots & c_{mn} \end{bmatrix} \tag{8}$$

$$c_{ij} = \begin{cases} 0, \text{user } i \text{ doesn't under the context j} \\ 1, \text{user } i \text{ under the context j} \end{cases} \qquad (9)$$

Such that, m denotes the number of users, n denotes the number of contexts.
(2) Based on the user-item preference model constructed above, we can construct the user-item-context preference model [4]:

$$y : (U, M_u, C) \rightarrow R \qquad (10)$$

Where, C is the context, R is the item rating from the user under the context C.

3 Nonlinear SVR Algorithm

The basic idea of Nonlinear SVR is mapping the input vector into a high-dimensional feature space through pre-determined nonlinear mapping, then create a optimal separating hyperplane which has the maximum distance in the feature space. From the point of geometry, support vector is the minimum number of sample vector to determine the optimal separating hyperplane. And then in the high-dimensional space to conduct linear regression, so as to obtain the nonlinear regression effect in the original space [5]. The specific algorithm is described as follows:

(1) Firstly, the training set is given as $T = \{(x1, y1),...,(xl, yl)\} \in (R_n \times Y)^l$, where, $xi \in R_n$, $yi \in Y = R$, $i = 1,...l$;
(2) Then, choose the kernel function $K(x,x')$ and the precision $\varepsilon > 0$ and the penalty parameter $C > 0$. The most common used kernel functions are as follows: Radial Basis Kernel, Polynomial kernel, Fourier kernel and B-spline kernel. Penalty parameter is used to measure the error during the process of learning of the learning model, which has defined before the learning. The Loss Function is different for different learning model. And for the same learning problem, the learning model is different under different Loss Function.
(3) Construct and solve the quadratic programming problem

$$\min_{a^{(*)} \in R^{2l}} \frac{1}{2} \sum_{i,j=1}^{l} (a_i^* - a_i)(a_j^* - a_j)K(x_i, x_j) +$$

$$\varepsilon \sum_{i=1}^{l} (a_i^* + a_i) - \sum_{i=1}^{l} y_i(a_i^* - a_i), \qquad (11)$$

$$\text{s.t.} \sum_{i=1}^{l} (a_i - a_i^*) = 0,$$

$$0 \le a_i^{(*)} \le C, i = 1, \cdots, l,$$

(4) Compute $\bar{b}$: choose the component $\bar{a}_j$ and $\bar{a}_k^*$ of $\bar{a}^*$ between the interval (0,C) (penalty factor can be defined as 1).

If $\bar{a}_j$ is selected, then:

$$\bar{b} = y_j - \sum_{i=1}^{l} (\bar{a}_i^* - \bar{a}_i)K(x_i, x_j) + \varepsilon \tag{12}$$

if $\bar{a}_k^*$ is selected, then:

$$\bar{b} = y_k - \sum_{i=1}^{l} (\bar{a}_i^* - \bar{a}_i)K(x_i, x_k) - \varepsilon \tag{13}$$

(5) Construct the decision function

$$y = g(x) = \sum_{i=1}^{l} (\bar{a}_i^* - \bar{a}_i)K(x_i, x) + \bar{b} \tag{14}$$

4 SVR Recommendation Algorithm Based on Context

The main idea of Context-aware recommendation system based on SVR is to add the context which may influence the user preference based on the traditional user-item two-dimension recommendation, then construct the user-item-context preference model, and adopt the SVR algorithm to predict the rating and finally make recommendations for users. The algorithm is described as follows:

Input: User-Item-Context matrix, N recommendation items set;
Output: Top-N recommendation set for the user u

(1) Firstly, make sure the range of all the rating vector;
(2) Standardized pre-processing the user-item-context matrix (U,Mu,C) and the user rating matrix about the item under different contexts;
(3) Choose the kernel function K(x,x') and the precision $\varepsilon > 0$ and the penalty parameter $C > 0$. Without the condition of prior knowledge to guide, we often choose RBF kernel function, which can obtain the better nonlinear fitting results, the calculation formula is as follows:

$$k(x, x') = \exp(-\frac{||x - x'||^2}{\sigma^2}) \tag{15}$$

(4) Use the training set to train the model, pre-processing the input and output parameter data of the sample. Then regard the value under relevant attribute as the input of SVM (Support Vector Machines), take the rating as the output of SVM. Thus establish a nonlinear map relationship between attributes and rating. MAE (mean absolute error) is used to measure the difference between the actual rating

and the prediction rating. Adjust the value of precision ε and the penalty parameter C, and optimize the model constantly. During continuous learning and testing, the model can be taken as the effective non-linear prediction model when it achieves a higher precision.

(5) Adopt the model constructed above to predict the user rating for items those haven't been rated by the user. And then recommend the top-N rating items to the user, realizing the recommendation for the target user.

5   The Application of the Passenger Auxiliary Service of Civil Aviation

The chapter is mainly to analysis and arrangement the civil aviation passenger data set, then apply the recommendation algorithm to the civil aviation passenger data set. And finally provide appropriate auxiliary service to the passenger.

5.1   Analysis of the Passenger Ticket Booking Data Set

In this experiment, we adopt the passenger ticket booking data set of civil aviation, the data set is from the passenger ticket booking data from the airlines during the year of 2011. The number of the data set is 1048753. Considering the security, we deal with some of the book ID digits, thus the data set doesn't involve the user's privacy. As the amounts of data, we filter the user data those who take plane less 3 times a year. Then choose the 80 % of the data as the training set, the other 20 % as the testing set. The training set is used to construct the contextual user preference model, and the testing set is used to predict and measure the accuracy of the recommendation algorithm.

5.2   Pre-processing of the Passenger Ticket Booking Data Set of Civil Aviation

The algorithm proposed is mainly used to predict the passenger preference for the auxiliary service, and then recommend the auxiliary service that the user may interest. Firstly, we need to transform the original data set into a rating data set of the passenger about different auxiliary services under different contexts.

We choose the sex and age as the user attributes. Regard the start time of the flight and the outing weather and company (the person with the passenger) as the contexts that influence the passenger preference for auxiliary services. Select airport services, in-flight services and luggage services as the characteristics attributes of the item.

Where, sex include M(man) and F(female). Through the statistics we found that the range of the age is between 21 and 77. According to the age division of the United Nations world health organization, we divide the age into four groups: 21–34, 35–44, 45–59 and 60–77. The start time is divided into four groups: 0:00–6:00, 6:00–12:00, 12:00–18:00 和 18:00–24:00. The weather includes sunny and rainy. Company includes family, friends and colleague. The auxiliary service attributes are described as Fig. 1:

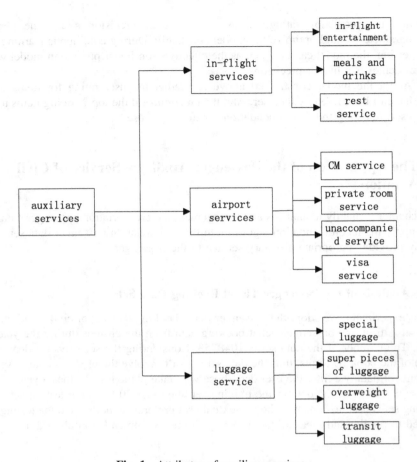

Fig. 1. Attributes of auxiliary services

After dividing attribute categories for every attribute, we consider the marking strategy. For example, Table 1 is the book information segment of the user ID 63**9 during one quarterly.

Table 1. Booking data of a user

Booking ID	Age	Gender	Start time	Weather	In-flight services	Airport services	Luggage service
19**2	48	F	16:05:00	1	A1	B2	C2
19**2	48	F	21:15:00	0	A1	B2	C4
19**2	48	F	11:50:00	1	A2	B2	C4
19**2	48	F	19:30:00	0	A3	B4	C1

Where, 0 and 1 in the weather column represent the rainy and sunny. A1, A2, A3 in the in-flight service column represent separately: in-flight entertainment, meals and drinks and rest service; B1, B2, B3, B4 in the luggage service column represent separately: CM service, private room service, unaccompanied service and visa service. C1, C2, C3, C4 represent separately: special luggage, super pieces of luggage, over-weight luggage and transit luggage. For example, if the user whose ID is 19**2 selected in-flight entertainment twice, while select meals and drinks and rest service once among the four bookings. In-flight entertainment accounts for 50 % while meals and drinks and rest service just accounts for 25 %. We define the marking criterion as Table 2:

Table 2. Marking criterion

Proportion	Rating
X ≤ 25 %	1
25 % < X ≤ 50 %	2
50 % < X ≤ 65 %	3
65 % < X ≤ 85 %	4
others	5

According to the Table 2, we can obtain the user's rating for in-flight entertainment is 2, the rating for meals and drinks and rest service is 1. Use the same method, we can achieve the ratings for other auxiliary services. Then we obtain the rating data set of the passenger for the auxiliary services. According to the Sect. 2 to construct the model use this rating data set. Finally we obtain a rating data set for each auxiliary service for users with different attributes under different context.

5.3   Evaluation Metric

The performance of the recommendation system is mainly measured from the respect of accuracy, diversity, novelty and coverage rate. The most commonly used is the accuracy of recommendation. Accuracy measurements include: predictive accuracy, Classification Accuracy and rank accuracy. As the measurement for regression effect usually measure the error between the predict rating and the real rating [6]. Therefore, we combine the metric of these two. We adopt MAE to evaluate the model. MAE is defined as follows [7]:

$$MAE = \frac{1}{N}\sum_{1}^{N}|y'_{ij} - y_{ij}| \qquad (16)$$

Where, N is the number of the user, and y_{ij} denotes the prediction rating and y'_{ij} denotes the real rating.

5.4   Experimental Results and Analyses

The algorithm chooses the RBF kernel, we need to set the value of the precision ε and the penalty parameter C and the radial basis kernel parameter σ. According to literature [8], we set C = 100. After a series of experiments, consider the algorithm scalability and prediction accuracy comprehensively, we set $\varepsilon = 0.01$, C = 100 and $\sigma = 1$.

Experimental operation environment and implementation:

Hardware Environment: Intel Core(TM) i5-2400 CPU, 4G memory; operating system: Windows7 32 bit;

Programming software: Matlab 7.0 and VC++6.0.

5.4.1   Analysis of the Influence of SVR Parameters' Selection on MAE

(1) The influence of the precision ε on MAE

Literature [9] indicates the selection of ε independent of the selection of (σ, 2, C). For different (σ, 2, C), less prediction error was achieved approximately in the same value of ε [8]. Therefore, the value of ε can be confirmed, and then change the value of (σ, 2, C) in order to find the minimum prediction error. The experimental selected a group value of ε according to literature [8], Fig. 2 shows the change trade of the value of MAE under different ε. Figure 2 shows the proportion of the number of support vector under different values of ε.

The experimental result indicated that when the value of ε is in a range, the value of MAE is smaller, and in this range MAE changes smaller as the change of the value of ε. Then we choose the maximum of ε, because the bigger the value of ε, the less number of the support vector is, in such case, the more simple regression function, the faster calculation speed is. Combine the Figs. 2 and 3, when the value of ε exceeds a certain value, the value of MAE increase very fast, and the support vector is very little and close to zero, this phenomenon belongs to under-fitting obviously.

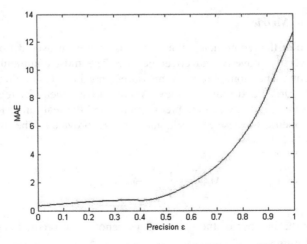

Fig. 2. Change trade of the value of MAE under different ε

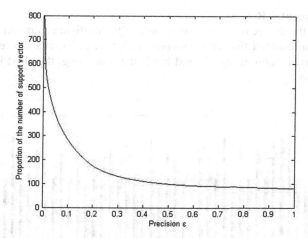

Fig. 3. Proportion of the number of support vector under different values of ε

(2) The influence of radial basis kernel parameter σ on MAE

In the experimental, we select the value of σ between 0 and 1 according to the method to select the value of radial basis kernel parameter in literature [10]. With the change of σ, the change trade of MAE is as Fig. 4:

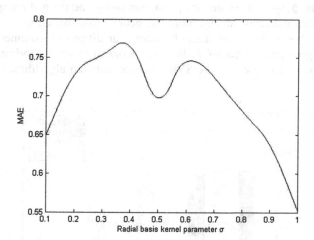

Fig. 4. Change trade of the value of MAE under different σ

The experimental result indicates that, when the value of radial basis kernel parameter σ reach a certain value, MAE shows a trend of decrease, and when the value of σ is the maximum, the value of MAE is the minimum.

5.4.2   Experimental Results

Figure 5 shows the degree of rating consistency of the in-flight entertainment. From the bar chart we can contrast the error between prediction rating and the real rating (the blue bar is the prediction rating, the red bar is the real rating, the chart just shows 50 samples' results).

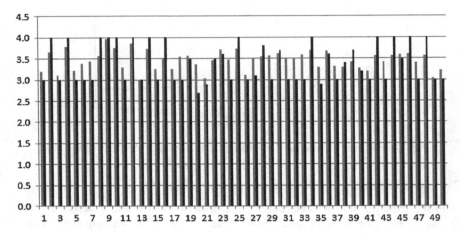

Fig. 5. Consistency between the prediction rating and the real rating (Color figure online)

From the Fig. 5, we can see that the prediction rating and the real rating are in good agreement. Which indicates the algorithm has better predictive effect.

The Fig. 6 shows the different result under four different recommendation algorithms. Four algorithms are User-CF, Item-CF, Context-aware Factorization Machine [4] and our proposed Context-aware SVR Recommendation algorithm.

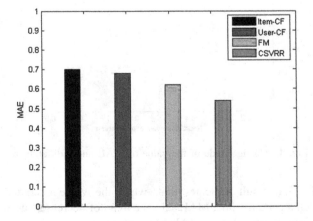

Fig. 6. Value of MAE in four recommendation algorithms

From the Fig. 6, compared to the traditional user-based and item-based Collaborative Filtering and the Context-aware Factorization Machine, our proposed method has the minimum mean absolute error value.

6 Conclusion

The algorithm proposed in this paper apply the machine learning into the recommendation system. Based on the traditional two-dimension of user-item algorithm, this algorithm introduces the context, and construct user-item-context preference prediction model to provide auxiliary services recommendation for the passenger, which can induce the data sparsity effectively. When there is a new user or a new item, we can rely on the attribute of the user and the item to find the similar users and items, which can resolve the cold-start problem. Through the experimental, the result indicated this algorithm can predict the passenger preference well and more accurately. Based on the work in this paper, the next thing we try to do is to consider the user dependence on the recommendation system and the rating fair of the user about the item. And combine the recommendation accuracy and diversity to measure the performance of the recommendation system.

Acknowledgment. This work is supported by Key project plan of Tianjin applied basic and frontier technology research (14JCZDJC32500), Project supported by Civil aviation science and technology fund (MHRDZ201207), China's civil aviation college preparatory major projects (3122013P003) and China civil aviation university research start-up fund (2010QD10X).

References

1. Liu, X., Aberer, K.: SoCo: a social network aided context-aware recommender system. In: Proceedings of the 22nd International Conference on World Wide Web. International World Wide Web Conferences Steering Committee, pp. 781–802 (2013)
2. Wang, L., Meng, X., Zhang, Y.: Context-Aware Recommender Systems. J. Softw. **23**(1), 1–20 (2012)
3. Zhuang, Y.: A collaborative filtering recommendation algorithm based on the model of items' features. Comput. Appl. Softw. **26**(3), 260–262 (2009)
4. Rendle, S., Gantner, Z., Freudenthaler, C., et al.: Fast context-aware recommendations with factorization machines. In: Proceedings of the 34th International ACM SIGIR Conference on Research and Development in Information Retrieval, pp. 635–644. ACM (2011)
5. Cristianini, N., Shawe-Taylor, J., Li, G.: An Introduction to Support Vector Machine (SVM). Electron. Industry Press, Beijing (2004)
6. Feng, X., Liu, Y., Wang, D., et al.: UV spectrometry and support vector regression for simultaneous determination of paracetamol, aspirin and caffeine. ACAD J GCP **22**(1) (2006)
7. Zhu, Y.X., Lv, L.Y.: Evaluation metrics for recommender systems. J. Univ. Electron. Sci. Technol. China **41**(2), 163–175 (2012)

8. Yan, G., Zhu, Y.: Parameters selection method for support vector machine regression. Comput. Eng. **35**(14), 218–220 (2009)
9. Liu, J., Cai, H., Tan, Y.: Heuristic algorithm for tuning hyper parameters in support vector regression. J. Syst. Simul. **19**(7) 2007
10. Su, G., Deng, F.: Introduction to model selection of SVM regression. Bus. Sci. Technol. **22**(2), 154–158 (2006)

Crystal MD: Molecular Dynamic Simulation Software for Metal with BCC Structure

He Bai[1(✉)], Changjun Hu[1], Xinfu He[2], Boyao Zhang[3], and Jue Wang[3]

[1] University of Science and Technology Beijing, Beijing, China
baihe@xs.ustb.edu.cn
[2] China Institute of Atomic Energy, Beijing, China
hexinfu@ciae.ac.cn
[3] Supercomputing Center, Computer Network Information Center,
Chinese Academy of Science, Beijing, China

Abstract. Material irradiation effect plays an important role in material science. However, it is lack of high-throughput irradiation facility and process of evolution and development, which lead to lack of basic scientific theory about atomic scale materials design and development guidance. High-performance computing for simulation makes deeply understanding of micro-level-material possible. In this paper, a new data structure is proposed for the parallel simulation of metal materials evolution with crystal structure under irradiation defects. Compared with LAMMPS and IMD, which are two popular molecular dynamic simulation versions, our method takes much less memory on multi-core clusters.

Keywords: Irradiation effect · Molecular dynamics · Crystal structure · High-performance computing

1 Introduction

Materials irradiation effect mainly refers to the interaction of neutrons, charged particles or electromagnetic rays and other radiation with solid materials product. Study on materials irradiation effect involves many areas, such as nuclear reactor structural material, ion probe, plasma processing and ion modification. However, in addition to building a nuclear reactor systems theory, material problem is a major factor restricting the development of nuclear power. Especially for the economics and safety of nuclear power plant, material irradiation effect plays an important role.

Macroscale material radiation damage are 9 orders of magnitude, from the atomic scale to the macroscopic scale: from bond breaking processes in time scale of picosecond to nonlinear process of engineering structural failure and destruction in time scale of decades. Because of lack of high-throughput irradiation facility and difficulty in observing and discovering the process of evolution and development, we cannot understand the internal microstructure evolution process and mechanism of radiation effect on materials deeply. This situation leads to the lack of basic scientific theory about atomic scale materials design and development guidance.

W. Chen et al. (Eds.): BDTA 2015, CCIS 590, pp. 247–258, 2016.
DOI: 10.1007/978-981-10-0457-5_23

The development of high-performance computer technology makes deeply understanding of micro-level-material possible and provides an important contribution to reveal of microstructure evolution of material failure, quantitative relationship between macro properties and microscopic processes. Computational material science has become one of the most important research fields of materials science. Since the computer simulation can offer us insight into the details of materials and reduce development costs, the combination of radiation experiments and computer simulations can not only improve research efficiency, but also reduce a lot of human and financial resources cost.

Molecular Dynamics (MD) simulates the trajectory of particles by solving the equations of motion of all the particles in the system. Thermodynamic quantities and other macroscopic properties of system are calculated by simulating the interaction and motion of microscopic particles. MD can be used in the ensemble calculation such as NPT, NVE, NVT, it is a thermodynamic calculation method based on the theory of Newtonian mechanics, which has been widely used in various fields of physical, chemical, biological, materials, medicine, etc.

At present, the scales of molecular dynamics simulations are usually from tens of thousands to tens of millions of atoms, due to the storage capacity constraints. However, typical material microstructure and defect size are much larger than the calculated scale that existing storage capacity can achieve, which makes a great difference between computer simulation results and the actual material microscopic processes, as well as macroscopic properties.

Recently, with the rapid development of high performance computer technology, MD has further development. Scientists developed new MD software based on hybrid architecture high performance computer. LLNL and IBM firstly developed MD software facing high performance computer—ddcMD [1]. Sandia national laboratory developed massive parallel MD software—LAMMPS [2], which is being widely used today. University of Stuttgart developed MD software in 1997, named IMD [3–5], and in 1999, it achieved the world biggest record with $5*10^9$ atoms. Based on IMD, University of Stuttgart developed new MD software—ls1 Mardyn [6]. It holds the world biggest MD simulation record with $4*10^{12}$ molecules.

In this paper, a new data structure has been designed for parallel MD simulation. It focuses on the crystal structure characteristics of BCC metallic materials without neighbor list or linked cell. Based on this data structure, we developed molecular dynamics simulation software named Crystal MD. To take advantage of characteristics of BCC structure, array elements' location in Crystal MD can be used directly to reflect the atoms' spatial arrangement. Compared with the traditional MD data structure, this new data structure can reduce memory usage per atom, which results in larger scale MD simulations on the same memory computer. Compared with the current mainstream large-scale material MD open source software IMD and LAMMPS, the Crystal MD can significantly expand the scale of simulation for crystal with BCC metal materials. Besides, we are now expanding the code to be able to simulate the crystal with FCC metal materials.

The main contributions of our work lie in three aspects:

(1) We propose a new data structure of MD simulation for metal with BCC structure, which efficiently reduces memory usage, and expands the scale of MD simulation.

(2) Based on the data structure, we propose a new method to find neighbor atoms' during the simulation. We call it lattice neighbor list. Besides, we propose a new method of communication, which efficiently reduces the pack/unpack calculation during the communication.

(3) We compare the memory usage among LAMMPS, IMD and Crystal MD, and analyze the constraint of the simulation scale.

The rest of this paper is organized as follows: Sect. 2 generally introduces Crystal MD. Section 3 proposes the new MD simulation data structure. Section 4 presents a new communication method based on the new data structure. Performance analysis and discussion are given in Sect. 5. Conclusions and future work are drawn in Sect. 6.

2 Crystal Structure and MD Calculation

2.1 Crystal Structure

Mechanical properties of materials are dedicated by its' microstructure [7], let's take radiation effect of steels for example. Structural materials in reactor systems are predominantly crystalline, metallic alloys. Neutron radiation has the capability to displace atoms from their lattice sites and the formation of point defects, the migration and clustering of point defects will lead to microstructure evolution under service condition [8]. In Crystal MD, we want to study the radiation effect of alloy metal with BCC structure, thus we focus on MD simulation with BCC structure. The body-centered cubic (BCC) system has one lattice point in the center of the unit cell in addition to the eight corner points. It has a net total of 2 lattice points per unit cell ($1/8 \times 8 + 1$). Figure 1 shows its physical structure.

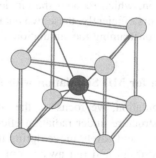

Fig. 1. Physical structure of body-centered cubic (BCC)

2.2 MD Calculation

MD program's computational cost depends on the calculation of forces acting on the molecules (atoms), which are based on molecular models for the physical interactions. Usually, we choose different potential functions for force calculation according to different types of molecules (atoms).

Here, in Crystal MD, we use embedded-atom method (EAM) [9] potentials as physical interaction because we are aiming at studying metals and metal alloys. The basic idea of EAM potential is to divide the potential energy into two parts: pair potentials; embedding potentials.

After force calculation, Newton's equations of motion are solved numerically for all atoms to obtain the configuration in the next time step. Usually, the algorithm based on the (Størmer-) Verlet method [10, 11] are used. But here, we employ the velocity Verlet method [12], which is more efficient.

3   Data Structure in Crystal MD

The design of data structure in MD simulation determines the efficiency of finding neighbors during the force calculation; furthermore, it determines the scale that computer simulation can reach. There are two data structures which are commonly used in MD simulations: Linked cell [13–15]; Neighbor list [16]. Linked cell data structure divides the simulation volume into a grid of equally sized cubic cells, which have an edge length equals to the cut-off radius. Therefore, this data structure ensures that all interaction partners for any given molecule are situated either within the cell of the molecule itself or the 26 surrounding cells (3d condition). Linked cell is easy to build, and has great scalability with little memory cost, but it has to be built every time step. Neighbor list data structure is to build a list, which all atom pairs within a neighbor cutoff distance equal to the force cutoff plus the skin distance are stored in the list [17]. Typically, the larger the skin distance, the less often neighbor lists need to be built, but more pairs must be checked for possible force interactions every time step. Neighbor list is more complicated to be built than Linked cell, and uses much more memory to store the list. However, Linked cell iterates much more atoms beyond the cutoff than neighbor list during the force calculation, which reduces the efficiency of computation significantly. In order to find neighbors efficiently, these two data structures not only result in additional memory, but also constraining the simulation scale.

3.1   Data Structure Design for MD Simulation with BCC Structure

For bigger scale of MD simulation, we focus on the special physical evolutionary process of metal with BCC structure under radiation effect: atoms are regularly constrained near every lattice point, and barely move during the whole simulation. Only a few atoms will break the constraint and run away from the lattice point where they originally located. Based on this physical feature, we propose a new data structure to store the information of atoms. According to the atoms' spatial distribution, Crystal MD put the information of atoms, such as position, velocity, etc. into the corresponding arrays in order, which makes a row of array corresponding to a lattice point, as shown in Fig. 2.

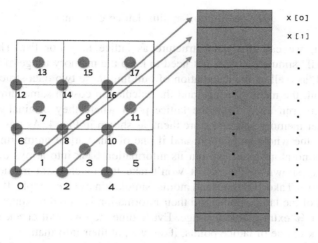

Fig. 2. Data structure facing metal with BCC structure in 2-dimension

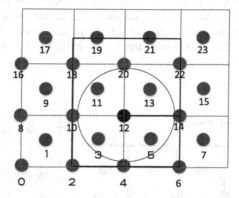

Fig. 3. Lattice neighbor list for BCC structure in 2-dimension; take atom with red as example, all its possible neighbors are in the black box (the radius of circle is cutoff), the offsets are −10, −9, −8..., all other atoms can use the same offset to find all their possible neighbors (Color figure online)

The information of atoms is stored in arrays regularly according to the atoms' spatial distribution. Meanwhile, atoms that interact with each other are also distributed in space regularly with a cutoff radius. Atoms that interact in space are distributed in arrays regularly as well. Therefore, using cutoff and lattice constant, we can calculate the offsets of neighbor atoms where they are stored in arrays apart from center atom. In this way, it is easy to locate the position where atoms' neighbors' information is stored, as Fig. 3 shows. To find the neighbors of atom with red, a black rectangle whose width is N (calculated using Eq. 1) times of lattice constant is used.

$$N = \text{ceil(cutoff radius/lattice constant)} \qquad (1)$$

At present, we call this data structure as lattice neighbor list. This new data structure for MD simulation does not need to keep the memory usage of Neighbor list or Linked cell, as well as the calculation of building these two data structure. Thus, it will reduce both the memory usage and the calculation cost in same simulation scale.

When atoms run away from the lattice point where they originally located, we allocate another memory space to store them, as shown in Fig. 4. As this part of atoms is very few in the whole simulation, and if one of this part of atoms runs to a lattice point which is no atom fixed, we put its information back into arrays of lattice point (the right array shown in Fig. 2). It won't take too much extra memory and extra computation cost. Take 100 thousand atoms' simulation as an example. If there are 1 % atoms run out of the lattice point, and their information is stored in xinter array, so that there will be 1 % extra memory usage. Every time step we will check if this part of atoms run back to one of lattice points, if so, we put their information back into arrays of lattice point; if not, we use array to tag which lattice point each of these atoms is closed to. The extra cost is focused on the calculation of which lattice point this part of atoms are closed to, thus, it is proportionate to the scale of this part of atoms. During the force calculation, Crystal MD first judges which lattice point the atoms in xinter array is closed to according to the array of tag, then computes the forces between them and other atoms in lattice points with the lattice neighbor list, thirdly computes the forces between atoms in xinter array.

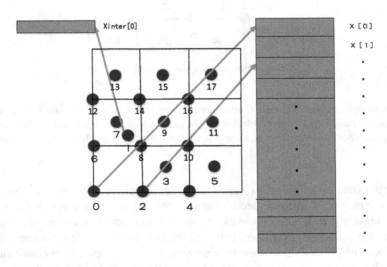

Fig. 4. Data structure for metal with BCC structure in 2-dimension, when the atom with number 1 run away from the lattice point where it is fixed at the beginning, move its information into xinter array

4 Communication Scheme in Crystal MD

In Crystal MD, as we are aiming at massively parallel MD simulation, we use standard domain decomposition to divide the simulation box, which means every process gets same volume of box. As the scenario that Crystal MD can simulate is crystal structure of BCC, and the feature of BCC structure determines its spatial distribution is almost fixed during the whole simulation, so, in program, the atoms in lattice points which are needed to send to neighbor processors as their ghost area are fixed during the whole simulation. Thus, in the first time step of simulation, we calculate which atoms in lattice points are needed to be sent (received) to (from) neighbor processors, and store their index in arrays, as Fig. 5 shows. We don't need to calculate which atoms to send and receive every time step, we only need to pack (unpack) and send (receive) the information to (from) neighbor processors according to the index we stored. In this way, we can increase the efficiency of communication.

As for the atoms which run away from the lattice point where they originally located, they also need to communicate with neighbor processors. Because of the BCC structure, these atoms are very few during the whole simulation. We choose the same way as other MD programs to handle the communication of these atoms. When atom run out of the local box, we pack the information of this atom, and send it to the

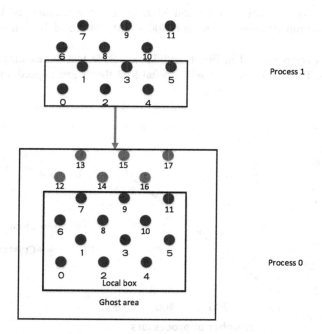

Fig. 5. When process 1 send atoms (in the square circle) to the ghost area of process 0 (atoms in grey), we store the index (process 1:0, 1, 2...) (process 0:12, 13, 14...) in the first step of simulation, after that, we know exactly which atoms to be sent and received at every time step later

corresponding neighbor process. If the atom is needed for neighbor's ghost area, its information will be sent to neighbor processors.

5 Performance Analysis and Discussion

In order to test the performance of Crystal MD on large-scale parallel computers, the experiments include two aspects: scalability, memory usage. In order to prove that there is less memory with Crystal MD, we compared the memory usage with LAMMPS (version 20150320) which uses neighbor list data structure and IMD (version 20140121) which uses Linked cell data structure.

All tests in this section were finished at Supercomputing cluster "Era" in Supercomputing Center of Computer Network Information Center.

The experimental environment is infiniband switched cluster with Intel Xeon E5-2680. Each node has 64 GB memory and run RedHat Linux version 4.4.7-3, with kernel 2.6.32-358.el6. Time is measured with the function "MPI_Wtime()". A single-precision is used for the experiments.

5.1 Performance Test and Discussion

For the scalability evaluation of Crystal MD, we use the scenario of Fe at 600 K temperature, and run 10 steps. The scale is $800 \times 800 \times 800$, and the number of atom is 1.024×10^9.

The results are presented in Fig. 6 and Table 1. We test the execution time with different cores. Take 100 cores as basis. Crystal MD shows great speed up in this case.

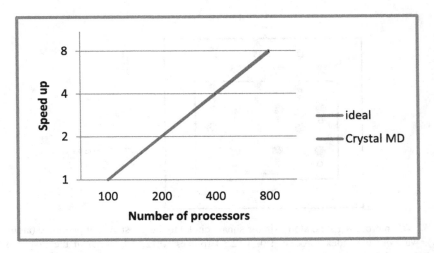

Fig. 6. Scalability of Crystal MD with $800 \times 800 \times 800$ Fe example

Table 1. Execution time of Crystal MD

Name	Type of potential function	Calculation scale (bcc length)	CPU cores	Computation time (s)
Crystal MD	EAM	1.024×10^9 (800)	100	1480.01221
		1.024×10^9 (800)	200	729.889124
		1.024×10^9 (800)	400	375.261747
		1.024×10^9 (800)	800	189.048273
		1.024×10^9 (800)	1000	148.758292

Table 2. Memory usage test of three MD simulation versions, which fully fill the memory that each CPU core can reach

Name	CPU cores	Calculation scale (bcc length)	Memory usage (G/core)	Memory usage per atom (B/atom)
Crystal MD	100	1024000000(800)	2.1	220.20
LAMMPS	100	351232000(560)	2.0	611.41
IMD	100	686000000(700)	2.0	313.04

5.2 Memory Usage Test

We compared the memory usage among Crystal MD, LAMMPS and IMD to prove Crystal MD performs the best at memory usage. Firstly, we fully fill the memory size that each node can reach. As Table 2 shows, Crystal MD uses the least memory per atom, while LAMMPS uses the most. In the same memory size, Crystal MD can simulate much bigger scale of BCC structure simulation. To figure out the detailed difference of memory usage among these three versions, we test the memory usage of these three versions in same scale. And we analyzed the code of these three versions to estimate memory usage per atom. We start from the memory usage that stores atom information, record the variables associated with atom information, and estimated memory usage. The record is shown in Table 3.

According to Table 3, Crystal MD uses the least memory to store atoms' information, LAMMPS uses the most. The difference in memory usage to store atoms' information among three versions is 10–60 bytes, but the total memory usage is quite different as shown in Table 2. Therefore, we made a further analysis of these three versions' memory usage in data structure. We use the same scale of BCC structure simulation to test memory usage of three versions, as shown in Table 4.

LAMMPS uses neighbor list to store atoms' neighbors' index. In the test case, the average number of each atom's neighbor is 84. Neighbor list stores the index of neighbors. Each neighbor index takes 4 bytes to store, therefore, the list takes 84*4 = 336 bytes per atom to store neighbor index. The total memory of neighbor list is 78.231 GB, which occupies 65.2 % of total memory usage.

IMD uses Linked cell to find neighbor atoms. During the simulation, IMD's cells allocate memory space to store atoms' information. Because atoms move among cells,

Table 3. Memory usage for each atom of three versions

Crystal MD		LAMMPS		IMD	
Coordinate (x)	3 × double	Coordinate (**x)	3 × double	Coordinate (ort)	3 × double
velocity (v)	3 × double	velocity (**v)	3 × double	electron cloud density (eam_rho)	1 × double
force (f)	3 × double	force (**f)	3 × double	derivative of embedding energy (eam_dF)	1 × double
Atom Id (id)	1 × unsigned long	Atom Id (*tag)	1 × int64_t	Id	1 × integer
Atom type (type)	1 × int	Atom type (*type)	1 × int	force (craft)	3 × double
electron cloud density (rho)	1 × double	*image	1 × int64_t	momentum (impuls)	3 × double
derivative of embedding energy (df)	1 × double	*mask	1 × int	Atom type (sorte)	1 × short int
		electron cloud density (*rho)	1 × double	Vsorte	1 × short int
		derivative of embedding energy (fp[i])	1 × double	mass (masse)	1 × double
		which element each atom type maps to (*map)	1 × int	Number	1 × integer
		Index of neighbors (*jlist)	1 × int	pot_eng	1 × double
		Index of lattice point neighbors (*ilist)	1 × int		
		First neighbor (*firstneigh)	1 × int		
		Number of neighbors (*numneigh)	1 × int		
		whether each i, j has been set (**setflag)	1 × int		
		cutoff sq for each atom pair (**cutsq)	1 × double		
		accumulated per-atom energy (*eatom)	1 × double		
		Accumulated per-atom virial (**vatom)	1 × double		
Memory usage per atom	100 bytes	Memory usage per atom	160 bytes	Memory usage per atom	116 bytes

Table 4. Memory usage test of three versions in the same scale of BCC structure

Name	Cpu cores	Calculation scale (bcc length)	Memory usage (G/core)
Crystal MD	100	250000000(500)	0.5664
LAMMPS	100	250000000(500)	1.2
IMD	100	250000000(500)	0.7656

the memory space of cells will larger than the memory space to store information for the actual number of atoms in cells. In the test case, the total number of atoms including ghost area is: 267635200, but the memory space in cells is available for 502574200 atoms. Thus, the memory of linked cell is 25.38 GB, which occupies 33.15 % of total memory usage.

Crystal MD uses lattice neighbor list to store neighbor offsets. In the test case, it only needs 112 neighbor offsets, and each neighbor offset takes 4 bytes to store. Thus, it takes only 448 bytes to store lattice neighbor list. This new data structure saves much more memory usage than neighbor list or Linked cell do. This is the reason why Crystal MD uses much less memory in the same scale of BCC structure simulation than LAMMPS and IMD.

6 Conclusions

We presented the new molecular dynamic simulation software, Crystal MD. We compared memory usage with other MD versions to analyze the limitation of MD simulation scale. Focusing on metal with BCC structure, in the same memory capacity, Crystal MD saves the memory to improve the scalability of MD simulation. Crystal can simulate much bigger scale with BCC structure than other MD software on 100 CPU cores.

Other crystal structure MD simulation, such as FCC structure, will be done in future works.

Acknowledgements. The research is partially supported by Natural Science Foundation of China under Grant No. 61303050, the Hi-Tech Research and Development Program (863) of China No. 2015AA01A303 and the Youth Innovation Promotion Association, CAS (2015375).

References

1. Streitz, F.H., Glosli, J.N., Patel, M.V., et al.: 100+ TFlop solidification simulations on BlueGene/L, SC05 (2005)
2. LAMMPS manual. http://LAMMPS.sandia.gov/
3. Stadler, J., Mikulla, R., Trebin, H.-R.: IMD: a software package for molecular dynamics studies on parallel computers. Int. J. Mod. Phys. C **8**(5), 1131–1140 (1997)
4. www.itap.physik.uni-stuttgart.de/~johannes/IMD-home.html
5. Roth, J.: IMD - a molecular dynamics program and applications. In: Attig, N., Esser, R. (eds.) Proceedings of the Workshop: MD on Parallel Computers (to be published by World Scientific, Singapore)

6. Large systems 1: molecular dynamics. http://www.ls1-mardyn.de/. Accessed 19 August 2014
7. Was, G.S.: Fundamentals of Radiation Materials Science. Springer, Heidelberg (2007)
8. Wirth, B.D., Odette, G.R., Marian, J.: J. Nucl. Mater. **329–333**, 103 (2004)
9. Daw, M.S., Baskes, M.T.: Embedded atom method: derivation and application to impurities, surfaces and other defects in metals. Phys. Rev. B **29**, 6444–7991 (1984)
10. Störmer, C.: Radium (Paris) **9**, 395–399 (1912)
11. Verlet, L.: Phys. Rev. **159**, 98–103 (1967)
12. Swope, W.C., Anderson, H.C., Berens, P.H., Wilson, K.R.: A computer simulation method for the calculation of equilibrium constrains for the formation of physical clusters of molecules: application to small water clusters. J. Chem. Phys. **76**, 637–649 (1982)
13. Quentrec, R., Brot, C.: J. Comput. Phys. **13**, 430–432 (1973)
14. Hockney, R.W., Eastwood, J.W.: Computer Simulation Using Particles. McGraw-Hill, New York (1981)
15. Schamberger, S., Wierum, J.-M.: Graph partitioning in scientific simulations: multilevel schemes versus space-filling curves. In: Malyshkin, V.E. (ed.) PaCT 2003. LNCS, vol. 2763, pp. 165–179. Springer, Heidelberg (2003)
16. Rapaport, D.C.: The Art of Molecular Dynamic Simulation, pp. 70–93. Cambridge University Press, Cambridge (2004)
17. Sandia National Laboratories: LAMMPS user manual, p. 996 (2014)

GPU Acceleration of the Locally Selfconsistent Multiple Scattering Code for First Principles Calculation of the Ground State and Statistical Physics of Materials

Markus Eisenbach[1(✉)], Jeff Larkin[2], Justin Lutjens[2], Steven Rennich[2], and James H. Rogers[1]

[1] Oak Ridge National Laboratory, Oak Ridge, TN 37831, USA
eisenbachm@ornl.gov
[2] NVIDIA Corporation, Santa Clara, CA 95050, USA

Abstract. The Locally Self-consistent Multiple Scattering (LSMS) code solves the first principles Density Functional theory Kohn-Sham equation for a wide range of materials with a special focus on metals, alloys and metallic nano-structures. It has traditionally exhibited near perfect scalability on massively parallel high performance computer architectures. We present our efforts to exploit GPUs to accelerate the LSMS code to enable first principles calculations of O(100,000) atoms and statistical physics sampling of finite temperature properties. Using the Cray XK7 system Titan at the Oak Ridge Leadership Computing Facility we achieve a sustained performance of 14.5PFlop/s and a speedup of 8.6 compared to the CPU only code.

1 Multiple Scattering Theory

Density Functional Theory [4], especially in the Kohn-Sham formulation [6] represents a major, well established, methodology for investigating materials from first principles. Most computational approaches to solving the Kohn-Sham equation for electrons in materials attempt to solve the eigenvalue problem for periodic systems directly. The solution of the eigenvalue problem for dense matrices results in cubic scaling in the system size. Additionally these spectral methods require approximations such as the use of pseudopotentials or linearized basis sets for all electron methods to make the size of the basis set manageable.

This manuscript has been authored by UT-Battelle, LLC under Contract No. DE-AC05-00OR22725 with the U.S. Department of Energy. The United States Government retains and the publisher, by accepting the article for publication, acknowledges that the United States Government retains a non-exclusive, paid-up, irrevocable, world-wide license to publish or reproduce the published form of this manuscript, or allow others to do so, for United States Government purposes. The Department of Energy will provide public access to these results of federally sponsored research in accordance with the DOE Public Access Plan (http://energy.gov/downloads/doe-public-access-plan).

© Springer Science+Business Media Singapore 2016
W. Chen et al. (Eds.): BDTA 2015, CCIS 590, pp. 259–268, 2016.
DOI: 10.1007/978-981-10-0457-5_24

In this paper we utilize a different approach to solving the Kohn-Sham equation using multiple scattering theory in real space. The basis of this method is formed by the Kohn-Korringa-Rostocker (KKR) method [5, 7], that allows for the solution of the all electron DFT equations without the need for linearization.

1.1 The LSMS Algorithm

For the energy evaluation, we employ the first principles framework of density functional theory (DFT) in the local density approximation (LDA). To solve the Kohn-Sham equations arising in this context, we use a real space implementation of the multiple scattering formalism. The details of this method for calculating the Green function and the total ground state energy $E[n(r), m(r)]$ are described elsewhere [3, 14]. Linear scaling is achieved by limiting the scattering distance of electrons in the solution of the multiple scattering problem. Additionally the LSMS code allows the constraint of the magnetic moment directions [11] which enables the sampling of the excited magnetic states in the Wang-Landau procedure described below.

For the present discussion it is important to note that the computationally most intensive part is the calculation of the scattering path matrix τ for each atom in the system by inverting the multiple scattering matrix.

$$\tau = [I - tG_0]^{-1} t \tag{1}$$

While the rank of the scattering path matrix τ is proportional to the number of sites in the local interaction zone and to $(l_{max} + 1)^2$ (typically a few thousand, e.g. for $l_{max} = 3$ and 113 atoms in the local interaction zone, the rank is 3616.), the only part of τ that will be required in the subsequent calculations of site diagonal observables (i.e. magnetic moments, charge densities, and total energy) is a small (typically 32×32) diagonal block of this matrix. This will allow us to employ the algorithm described in the next section for maximum utilization of the on node floating point compute capabilities (Fig. 1).

From benchmarking a typical calculation of 1024 iron atoms with a local interaction zone radius of $12.5a_0$ on the AMD processors on Titan, we find that the majority of time (95 %) and floating point operations are spent inside the inversion of the multiple scattering matrix to obtain the τ-matrix. About half of the remaining time is used to construct this matrix. Thus our approach to accelerate the LSMS code for GPUs concentrated on these two routines that will be presented in detail in the following two sections.

1.2 Scattering Matrix Construction

To calculate the τ-matrix (Eq. 1) the first step involves constructing the scattering matrix

$$m = I - tG_0 \tag{2}$$

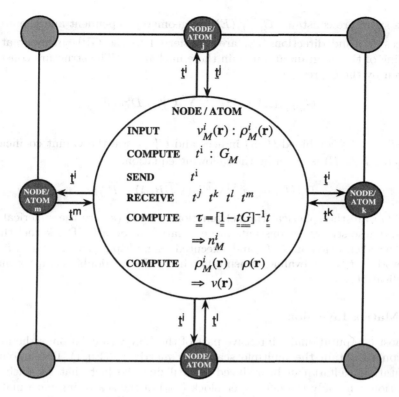

Fig. 1. Parallelization scheme of the LSMS method [14].

to be inverted. The m matrix is constructed from blocks that are associated with the sites i and j in the local interaction zone.

$$m_{ij} = I\delta_{ij} - t_i G_0^{ij} \tag{3}$$

Each of these block can in principle be evaluated in parallel. The size of these individual blocks is given by the cut-off in the l expansion of the scattering expansion. For a spin-canted calculation the size of a block is $2(l_{max} + 1)^2$, *i.e.* for a typical $l_{max} = 3$ each of the m_{ij} blocks has rank 32. The indices inside each block label the angular momentum l, m. For a typical number of $O(100)$ atoms in the local interaction zone, there are $O(10,000)$ blocks of the m matrix that need to be calculated, thus providing significant parallelism that can be exploited on accelerators. On the accelerator the m matrix is first initialized to a unit matrix to account for the $I\delta_{ij}$ part. The single site scattering matrices are currently calculated on the CPU, as this involves only the numerical evaluation of ordinary differential equations for initial values determined by the energy and the angular momentum l and are communicated as needed to remote nodes and transferred to the GPU memory.

The structure constants $G_{0,LL'}^{ij}(E)$ are geometry dependent and the atomic distances R_{ij} and directions $\hat{R}_{ij}$ are transferred to the GPU memory at the beginning of the program and remain there unchanged. The structure constants are given by the expression

$$G_{0,LL'}^{ij}(E) = 4\pi i^{l-l'} \sum_{L''} C_{L'L''}^{L} D_{L''}^{ij}(E) \tag{4}$$

where L are the combined (l, m) indices and $C_{L'L''}^{L}$ are the Gaunt coefficients. The factor $D_{L''}^{ij}(E)$ is given by the following equation.

$$D_{L}^{ij}(E) = -i^{l+1}\sqrt{E}h_l(\sqrt{E}R_{ij})Y_L^*(\hat{R}_{ij}) \tag{5}$$

Here $h_l(x)$ are the spherical Hankel functions and $Y_L(\hat{r})$ are the spherical harmonics. These structure constants are evaluated inside a CUDA kernel that is allows parallelization in L, L' and executed in multiple streams for i, j. The final product $t_i G_{ij}$ is evaluated using batched cuBLAS double complex matrix multiplications.

1.3   Matrix Inversion

The most computationally intensive part of the LSMS calculation is the matrix inversion to obtain the multiple scattering matrix τ. (Eq. 1) The amount of computational effort can be reduced by utilizing the fact that for each local interaction zone only the left upper block (τ_{00}) of the scattering path matrix τ is required. LSMS uses an algorithm that reduces the amount of work needed while providing excellent performance due to its reliance on dense matrix-matrix multiplications that are available in highly optimized form in vendor or third party provided implementations (*i.e.* ZGEMM in the BLAS library).

The method employed in LSMS to calculate the required block of the inverse relies on the well known expression for writing the inverse of a matrix in term of inverses and products of subblocks:

$$\begin{pmatrix} A & B \\ C & D \end{pmatrix}^{-1} = \begin{pmatrix} U & V \\ W & Y \end{pmatrix}$$

where

$$U = (A - BD^{-1}C)^{-1}$$

and similar expressions for V, W, and Y. This method can be applied multiple times to the subblock U until the desired block τ_{00} of the scattering path matrix is obtained.

The operations needed to obtain the τ_{00} thus are matrix multiplications and the inversion of the diagonal subblocks. For the matrix multiplication on the GPUs we can exploit the optimized version of these routines that are readily available in the cuBLAS library. The size of the intermediate block sizes used in the matrix inversion serves as a tuning parameter to optimize the performance of our block inversion algorithm. As the block size becomes larger, the resulting matrices entering the matrix multiplication usually result in significant improvements of the matrix multiplication performance at the cost of performing more floating-point operations then are strictly needed to obtain the final τ_{00} block. Thus there exists a optimum intermediate block size that minimizes the runtime of the block inversion. For the CPU only code this is achieve at a rank of the intermediate blocks of approximately 1000. For the GPU version we employ an optimized matrix inversion algorithm written in CUDA that executes the whole inversion in a single kernel in GPU memory, thus avoiding costly memory transfers and kernel launches. The maximal rank of double complex matrices that can be handled by this algorithm is 175, thus providing the size limit for the intermediate blocks and the block size for which we observe the maximum performance of the GPU version reported in this paper.

2   Wang-Landau Monte-Carlo Sampling

The LSMS method allows the calculation of energies for a set of parameters or constraints $\{\xi_i\}$ that specify a state of the system that is not the global ground state. Examples of this include arbitrary orientations of the magnetic moments or chemical occupations of the lattice sites. Thus we can calculate the energy $E(\{\xi_i\})$ associated with these sets of parameters. Evaluating the partition function

$$Z(\beta) = \sum_{\{\xi_i\}} e^{-\beta E(\{\xi_i\})}, \tag{6}$$

where $\beta = 1/k_B T$ is the inverse temperature and the sum is over all possible configurations $\{\xi_i\}$, allows the investigation of the statistical physics of the system and the evaluation of its finite temperature properties. In all but the smallest most simple systems (e.g. for a few Ising spins), it is computational intractable to perform this summation directly. Monte-Carlo methods have been used successfully to evaluate these very high dimensional sums or integrals using statistical importance sampling. The most widely used method is the Metropolis method [8], which generates samples in phase space with a probability that is given by the Boltzmann factor $e^{-\beta E(\{\xi_i\})}$.

For our work we have chosen to employ the Wang-Landau Monte-Carlo method [12,13], which is a method to calculate the density of states $g(E)$ of the system for the phase space spanned by the set of classical parameters that describe the system.

We have parallelized the Wang-Landau procedure by employing multiple, parallel walkers that update the same histogram and density of states [2] as illustrated in Fig. 2.

3   Scaling and Performance

The WL-LSMS code has been known for its performance on scalability, as has been shown on CPU only architectures, such as the previous Jaguar system at

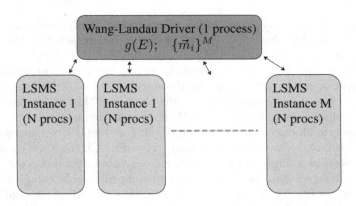

Fig. 2. Parallelization strategy of the combined Wang-Landau/LSMS algorithm. The Wang-Landau process generates random spin configurations for M walkers and updates a single density of states $g(E)$. The energies for these N atom systems are calculated by independent LSMS processes. This results in two levels of communication, between the Wang-Landau driver and the LSMS instances, and the internal communication inside the individual LSMS instances spanning N processes each.

the Oak Ridge Leadership Computing Facility [2]. The acceleration of significant portions of the code for GPUs, combined with major restructuring of the high level structure of the LSMS code was able to maintain the excellent scalability of the code. In Fig. 3 we show the near perfect weak scaling of the LSMS code in the number of atom, while maintaining the number of atom per compute node over five orders of magnitude from 16 iron atoms to $65,536$ atoms, while achieving a speedup factor of 8.6 for the largest systems compared to using the CPUs only on the Titan system at Oak Ridge. This performance puts calculations of million atom size systems within reach for the next generation of supercomputers such as the planned Summit system at the Oak Ridge Leadership Computing Facility.

For the statistical sampling with the Wang-Landau method as described above, we tested the scaling of the code in the number of walkers. The performance tests were done for 1024 iron atoms and the energies were self-consistently calculated for a LIZ radius of $12.5a_0$ allowing us to achieve a sampling rate of nearly one Monte-Carlo sample per second on Titan. The scaling of the energy samples per wall-time is shown in Fig. 4.

To assess the improvements in the computational and power efficiency that resulted from the porting of the significant portions of the code to GPU accelerators, we have run an identical WL-LSMS calculation for 1024 iron atoms with 290 walkers on 18561 nodes on Titan at the Oak Ridge Leadership Computing Facility for 20 Monte-Carlo steps per walker. We measured the instantaneous power consumption at the power supply to the compute cabinets which includes the power for compute, memory and communication as well as line losses and the secondary cooling system inside the cabinets, but excludes the power consumption of the file system and the chilled water supply. The measurements

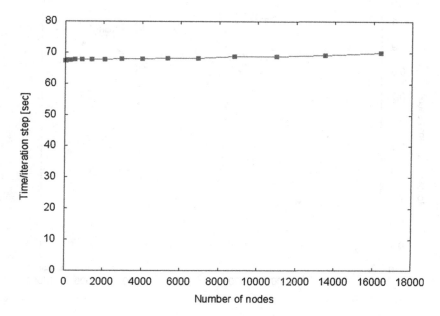

Fig. 3. Weak scaling of LSMS on Titan utilizing the GPU accelerators for a bulk iron calculation. 16 atoms on 4 nodes require 67.343 s per iteration step and 65536 atoms on 16384 nodes require 69.988 s, resulting in a parallel scaling efficiency of 96 % across Titan.

were performed both for a CPU only run as well a for a computation utilizing the GPUs. The results are shown in Fig. 5. The difference in power consumption between the compute intensive LSMS calculations, that take most of the time, and Monte-Carlo steps that are marked by a significant drop in the power consumption is obvious and this allows a clear comparison of the two runs.

4   Applications

In this section we review results that we have obtained using the method described in this paper to calculating the Curie temperatures of various materials. In particular we have applied this method to iron and cementite [1] and to Ni_2MnGa [9]. For the underlying LSMS calculations the atoms are placed on lattices with lattice parameters corresponding to the experimental room temperature values. The self-consistently converged potentials for the ferromagnetic or ferrimagnetic ground states were used for all the individual frozen-potential energy calculations in the combined Wang-Landau/LSMS algorithm. The calculations were performed by randomly choosing a site in the supercell and randomly picking a new moment direction. In the case of Fe and Fe_3C the convergence criterion for the Wang-Landau density of states was chosen to be the convergence of Curie temperature. The density of state thus obtained was used to calculate the

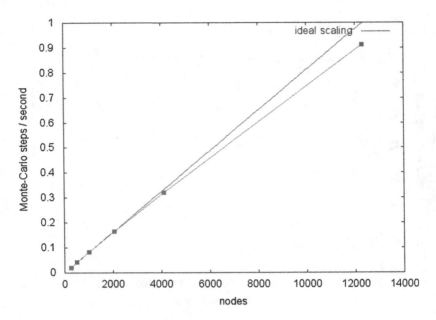

Fig. 4. Scaling of Wang Landau LSMS on Titan for 1024 Fe calculations, using 128 nodes per Monte-Carlo walker. The code shows good scaling in the number of states sampled with increasing number of walkers. With two walkers on 257 nodes the code generates 0.0208 samples/second and with 96 walkers on 12289 nodes 0.9128 samples/second are generated, thus achieving a parallel efficiency of 92 %.

specific heat. The peak in the specific heat allows us to identify the Curie temperature to be 980 K for iron, in good agreement with the experimental value of 1050 K. The Curie temperature obtained for Fe_3C is 425 K which again is in good agreement with the experimental value of 480 K. [1] For Ni_2MnGa the Curie temperature reported is 185 K, well below the experimental value of 351 K. [9] The small cell used in these calculation (144 atoms) will have resulted in a significant finite size error. Additionally, it is known that the localized moment picture that underpins our WL-LSMS calculations does miss important contributions to the fluctuations that determine the finite temperature magnetism in nickel, which will contribute to the reduction of the calculated Curie temperature from the experimental value. This was already observed by Staunton *et al.* [10] in disordered local moment calculations that underestimate the Curie temperature of pure Ni and they find 450 K as opposed to the experimental value of 631 K. We propose to include fluctuations in the magnitude of local moments. Preliminary calculations with a Heisenberg model that is extended with the inclusion of the magnitude of the local moment as a variable indicates that this can result in an increase in the Curie temperature.

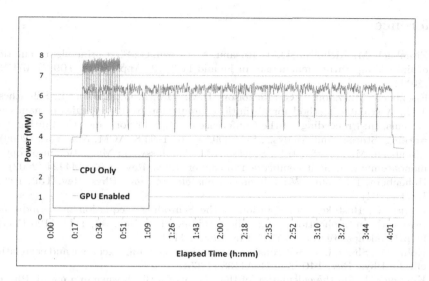

Fig. 5. Power consumption traces for identical WL-LSMS runs with 1024 Fe atoms on 18,561 Titan nodes (99 % of Titan). 14.5 PF sustained vs 1.86 PF CPU only. Runtime is 8.6X faster for the accelerated code, Energy consumed is 7.3X less. GPU accelerated code consumed 3,500 kW-hr, CPU only code consumed 25,700 kW-hr.

5   Conclusions

We have shown that for some classes of calculations, it is possible to make efficient use of GPU accelerators with a reasonable amount of code modification work. The acceleration of the code additionally results in significant energy savings while maintaining its scalability. Consequently the code and work presented in this paper enables the first principles investigation of materials at scales that were previously hard to access and pushes the possibilities for first principles statistical physics. Ongoing work involves extending the capabilities of LSMS presented in this paper to non spherical atomic potentials and to solving the Dirac equation for the electrons in solids, which will allow the first principles investigation of the coupling of magnetic and atomic degrees of freedom and other effects involving atomic displacements. These additions to the code will require additional work to accelerate the single site solvers for GPUs as in these cases a significant amount of compute resources will be needed to solve the single site equation for non spherical scatterers.

Acknowledgements. This work has been sponsored by the U.S. Department of Energy, Office of Science, Basic Energy Sciences, Material Sciences and Engineering Division (basic theory and applications) and by the Office of Advanced Scientific Computing (software optimization and performance measurements). This research used resources of the Oak Ridge Leadership Computing Facility, which is supported by the Office of Science of the U.S. Department of Energy under contract no. DE-AC05-00OR22725.

References

1. Eisenbach, M., Nicholson, D.M., Rusanu, A., Brown, G.: First principles calculation of finite temperature magnetism in Fe and Fe3C. J. Appl. Phys. **109**(7), 07E138 (2011)
2. Eisenbach, M., Zhou, C.G., Nicholson, D.M., Brown, G., Larkin, J., Schulthess, T.C.: A scalable method for ab initio computation of free energies in nanoscale systems. In: Proceedings of the Conference on High Performance Computing Networking, Storage and Analysis, SC 2009, pp. 64:1–64:8. ACM, New York (2009)
3. Eisenbach, M., Györffy, B.L., Stocks, G.M., Újfalussy, B.: Magnetic anisotropy of monoatomic iron chains embedded in copper. Phys. Rev. B **65**, 144424 (2002)
4. Hohenberg, P., Kohn, W.: Inhomogeneous electron gas. Phys. Rev. **136**, B864–B871 (1964)
5. Kohn, W., Rostoker, N.: Solution of the Schrödinger equation in periodic lattices with an application to metallic Lithium. Phys. Rev. **94**, 1111–1120 (1954). http://link.aps.org/doi/10.1103/PhysRev.94.1111
6. Kohn, W., Sham, L.J.: Self-consistent equations including exchange and correlation effects. Phys. Rev. **140**, A1133–A1138 (1965)
7. Korringa, J.: On the calculation of the energy of a Bloch wave in a metal. Physica **13**, 392–400 (1947)
8. Metropolis, N., Rosenbluth, A.W., Rosenbluth, M.N., Teller, A.H., Teller, E.: Equation of state calculations by fast computing machines. J. Chem. Phys. **21**, 1087 (1953)
9. Nicholson, D.M., Odbadrakh, K., Rusanu, A., Eisenbach, M., Brown, G., Evans III, B.M.: First principles approach to the magneto caloric effect: application to Ni2MnGa. J. Appl. Phys. **109**(7), 07A942 (2011)
10. Staunton, J., Gyorffy, B.: Onsager cavity fields in itinerant-electron paramagnets. Phys. Rev. Lett. **69**, 371–374 (1992)
11. Stocks, G.M., Eisenbach, M., Újfalussy, B., Lazarovits, B., Szunyogh, L., Weinberger, P.: On calculating the magnetic state of nanostructures. Prog. Mater. Sci. **52**(2–3), 371–387 (2007)
12. Wang, F., Landau, D.P.: Determining the density of states for classical statistical models: a random walk algorithm to produce a flat histogram. Phys. Rev. E **64**, 056101 (2001)
13. Wang, F., Landau, D.P.: Efficient, multiple-range random walk algorithm to calculate the density of states. Phys. Rev. Lett. **86**(10), 2050–2053 (2001)
14. Wang, Y., Stocks, G.M., Shelton, W.A., Nicholson, D.M.C., Temmerman, W.M., Szotek, Z.: Order-N multiple scattering approach to electronic structure calculations. Phys. Rev. Lett. **75**, 2867 (1995)

Kernel Optimization on Short-Range Potentials Computations in Molecular Dynamics Simulations

Xianmeng Wang[1](✉), Jianjiang Li[1], Jue Wang[2], Xinfu He[3], and Ningming Nie[2]

[1] University of Science and Technology Beijing, Beijing, China
wangxianmeng92@163.com
[2] Supercomputing Center, Computer Network Information Center, Chinese Academy of Science, Beijing, China
[3] China Institute of Atomic Energy, Beijing, China

Abstract. We present multi-threading and SIMD optimizations on short-range potential calculation kernel in Molecular Dynamics. For the multi-threading optimization, we design a partition-and-two-steps (PTS) method to avoid write conflicts caused by using Newton's third law. Our method eliminates serialization bottle-neck without extra memory. We implement our PTS method using OpenMP. Afterwards, we discuss the influence of the *cutoff if statement* on the performance of vectorization in MD simulations. We propose a pre-searching neighbors method, which makes about 70 % atoms meet the *cutoff* check, reducing a large amount of redundant calculation. The experiment results prove our PTS method is scalable and efficient. In double precision, our 256-bit SIMD implementation is about 3× faster than the scalar version.

Keywords: Molecular dynamics · EAM · SIMD · Vectorization · OpenMP · Short-range potential · High-performance computing

1 Introduction

Mechanical properties of materials are dedicated by its' microstructure. Let's take radiation effect of steels for an example. Structural materials in reactor systems are predominantly crystalline, metallic alloys. Neutron radiation has the capability to displace atoms from their lattice sites and formation of point defects. The migration and clustering of point defects will lead to microstructure evolution under service condition [1]. Molecular dynamics (MD) simulation provides the methodology for detailed microscopic modeling on the molecular scale; MD allows the simultaneous motion of atoms (molecules) under their mutual interactions to be simulated. The evolution of the system is governed by classical Newtonian mechanics, where the atomic forces are calculated as the negative gradient of the effective potential. The computation of the interatomic potentials and the corresponding forces dominates more than 80 % of the runtime in MD simulation, which makes the optimization on force calculation kernel very significant. Our optimization work can simulate a longer physical process in limited time.

© Springer Science+Business Media Singapore 2016
W. Chen et al. (Eds.): BDTA 2015, CCIS 590, pp. 269–281, 2016.
DOI: 10.1007/978-981-10-0457-5_25

Because Single-Instruction-Multiple-Data (SIMD) execution is an effective way to improve peak performance and multi-threading is useful to accelerate programs, both of them have been exploited to optimize MD simulations.

The main contributions of our work are:

- We propose a partition-and-two-steps (PTS) method to optimize force calculation kernel in large-scale MD simulations on shared-memory architectures. Our method eliminates write conflicts caused by using Newton's third law in a simple and efficient way. The proposed method only needs several implicit barriers. So it is unlike the traditional solving strategies, which result in too much synchronization cost, lock contentions, repeated computations or extra memory usage.
- We present a modified pre-searching neighbors strategy to optimize the meet ratio of the *cut-off if statement* in SIMD implementation. Using AVX and AVX2 to process double precision datum, we achieve up to an approximate 3× speedup over the scalar version. A higher speedup can be achieved if the variables are single precision in the simulation.
- We analyze and compare the performances of the original program and our optimized version using Intel Vtune [2, 3]. The experiment results prove that our accelerated version runs much faster than the original one. This means that in the same environment platform and limited time, our optimized version can simulate a longer physical process.

The reminder of this paper is organized as follows. Section 2 reviews related work and presents some background information. The specific multi-threading and vectorization optimization on force computation kernel are illustrated in Sect. 3. The results and evaluations of our experiments are provided in Sect. 4. The paper is concluded in Sect. 5.

2   Related Work and Background

We review some related multi-threading and SIMD optimization work in molecular dynamics simulations. Because we use Embedded Atom Method (EAM) as an example to explain our optimization strategy, we introduce the EAM potential briefly. As we have made our experiments based on Crystal MD, we then present some information about Crystal MD.

2.1   Related Work

2.1.1   Multi-threading for Molecular Dynamics

MD force computation typically employs the nature symmetry of pair forces, (i.e., $f_{ij} = -f_{ji}$) by calculating the forces of a pair of atoms only once, and then adding f_{ij} and $-f_{ji}$ to atom i and atom j separately [7]. Although exploiting Newton's third law reduces the force calculation routine by a factor of two, it creates a write conflict among threads. Different pair interactions may involve some common atoms (in Fig. 1) if Newton's third law is applied, which possibly results in a write conflict in memory (i.e., force array).

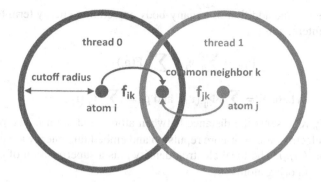

Fig. 1. The write conflict among threads caused by applying Newton's third law

Various methods have been put forward to eliminate the write conflict. Among them, the simplest way is to give up using Newton's third law. However, the solution duplicates the force calculation [8, 9]. The second solution is to use critical sections: modification of the force array need be achieved in a critical section [7]. This method has some disadvantages. The critical sections bring about a terrible synchronization cost, and the bottleneck becomes even severe with the increasing number of CPU cores. The third solution is creating thread private works areas in which store partial forces. The private data is reduced later. However, the memory cost is expensive when a large number of threads are used. Kunaseth et al. [10] have come up with some algorithms via nucleation-growth allocation to cut down the memory for privation data.

Liu et al. [11] have introduced a multi-step computation method to remove data dependence between computations of long-range forces in numbers of threads. The method firstly divides atoms into groups and then partitions the computations of real-space Ewald summation into steps based on these groups.

2.1.2 SIMD for Molecular Dynamics

Much effort has be devoted to achieving higher performance in MD using SIMD execution. Gromacs [12–14] currently has provided several SIMD acceleration options on the most compute-intensive parts. Pennycook et al. [15] have explored SIMD utilization on Sandia's miniMD benchmark using three SIMD widths (128-, 256- and 512-bit). Vectorization has been achieved by using Intel intrinsic functions as well as pragma directives on L-J force calculation in CoMD [16].

2.2 Background

Our optimization on MD force calculation kernel is applicable to various short-range potentials, such as EAM and Lennard-Jones (L-J) potential, etc. In order to elaborate our optimization strategies clearly, we choose a specific potential (EAM) as example in the following discussion. It is necessary to reiterate that our optimization strategies are applicable to all the short-range potentials. The EAM potential was suggested by Daw and Baskes [4] as a way to overcome the main problem with two-body potentials. It can be characterized as the addition of a many-body embedding energy term to a standard pair potential interaction [5]:

$$E_{tot} = \sum_{i}^{N} e_i + \sum_{i}^{N} F(\rho_i) \tag{1}$$

$$\text{Where } ei = \sum_{j} j\Phi_{ij}\left(r_{ij}\right) \text{ and } \rho_i = \sum_{j} f_{ij}\left(r_{ij}\right) \tag{2}$$

The term r_{ij} represents the distance between atom i and atom j. The pair potential term with the electrostatic core-core repulsion and embedding energy are given respectively by e_i and $F(\rho_i)$. The local electron density ρ_i is a superposition of contributions $f_{ij}(r_{ij})$ from neighboring atoms [6].

Our optimization technologies are based on Crystal MD [17]. Crystal MD is a MD package designed for BCC/FCC materials. It uses lattice neighbor list to calculate atoms' neighbor indexes according to their positions. It is necessary to note that although experimented on the basis of Crystal MD, the optimization methods we have proposed in this paper apply to other MD simulations. But the load among threads may be not perfectly balanced if PTS method is used for the material with no crystal structure.

3   Kernel Optimization of Molecular Dynamics

3.1   Efficient Multi-threading Implementation

When multi-threading is used in MD, the methods to eliminate rare condition is significant. As mentioned in Sect. 2.1.1, traditional solutions have their disadvantages. Inspired by Liu's [11] multi-step computation method, we propose a PTS computation method which partitions the simulation box (or a simulation part handled by one process) into several slabs, and then computes potential and force in two steps.

3.1.1   Principle of PTS Algorithm
The principle of the PTS method is presented as follows:

(1) The simulation area is partitioned into $2M$ (M is the number of threads used) slabs. Figures 2 and 3 demonstrate examples where M equals 2 in 2-Dimension and 3-Dimension situations separately. The simulation areas are correspondingly decomposed into 4 slabs (slab 0~slab 3).
(2) The $2M$ slabs are divided into two groups, and there are M slabs in each group. Every slab of one group need be separated from other slabs in the same group. Figures 2 and 3 illustrate that the 4 slabs are divided into a red group and a blue group. Slab 0 and 2 are in the red group and they are separated from each other. Slab 1 and 3 are in the blue group.
(3) The potentials and forces are calculated using multiple threads in two steps. In the first step, all the threads process the red group. Then the same threads continue to deal with the blue group. As our example shows, we use thread 0 and thread 1 to compute slab 0 and 2 separately in the first step. And thread 0 and 1 then process slab 1 and slab 3 respectively.

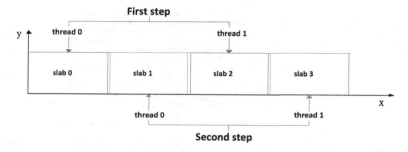

Fig. 2. Partition the 2D simulation area into 2M slabs ($M = 2$) (Color figure online).

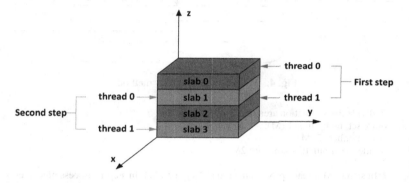

Fig. 3. Partition the 3D simulation area into 2M slabs ($M = 2$) (Color figure online).

To avoid the write conflict caused by using Newton's third law, the distance between two slabs should be larger than the *cutoff* radius. This is the limitation of our PTS method which will be discussed in detail later in Sect. 3.1.3.

The workflow of the PTS method is explained in Fig. 4. In step 0, the M threads process slabs (slab 0, slab 2…slab 2M-2) belonged to the first group. In step 1, the M threads process slabs (slab 1, slab 3…slab 2M-1) belonged to the second group. We use OpenMP [18] to implement this method.

3.1.2   Implement the PTS Method Using OpenMP

Three loops are needed in the classical calculation of EAM potential and related force. The first loop is to compute Φ (pair potential) and get ρ (electron density) from r (distance between atoms) through interpolation function. Afterwards, a loop is used for the computation of the F (embedding energy) and its derivative. The embedding energy must then be communicated as the EAM force computation requires terms from adjacent simulation areas [4]. Finally, a loop computes the embedding energy contribution to the force, and adds the final result to the two-body force [19]. Figure 5 gives the partial pseudo of the PTS method using OpenMP.

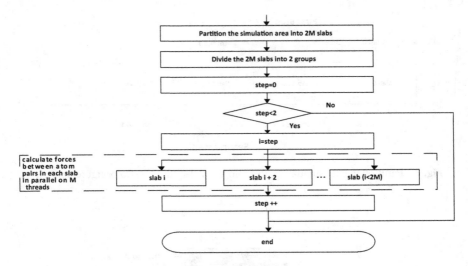

Fig. 4. Workflow of the PTS method

```
1.      /* divide the simulation area into 2M slabs*/
2.      omp_set_num_threads(M);
3.      slab_number=2*M;
4.      divide the simulation area into 2M slabs;
5.
6.      /*In step 0, M threads process the red slabs in parallel; In step 1, process blue ones*/
7.      #pragma omp parallel private(step)
8.      for(step=0; step<2; step++) {
9.          #pragma omp for
10.         for(i=step; i<2M; i+=2) {              //i is the slab index
11.         compute distance r;
12.         compute ρ from r through interpolation function in slab i;
13.             }
14.
15.         #pragma omp for
16.         for(i=step; i<2M; i+=2) {
17.         compute F and its derivative from ρ through interpolation function in slab i;
18.             }
19.
20.         #pragma omp single{
21.           Communicate F;
22.             }
23.
24.         #pragma omp for
25.         for(i=step; i<2M; i+=2) {
26.         compute the Φ in slab i;
27.         calculate force using Φ, F in slab i;
28.             }
29.  }
```

Fig. 5. Partial pseudo of the PTS method using OpenMP

3.1.3 Limitation of PTS Method

To avoid the write conflict caused by using Newton's third law, the distance d between two slabs belonged to the same group need be longer than the *cutoff* radius. Only if this condition is met, the atoms in different slabs belonged to the same group can be guaranteed to have no common neighbors. Therefore different threads will not write the same neighbor in force array simultaneously. Figure 6 demonstrates the rationale in 2D case.

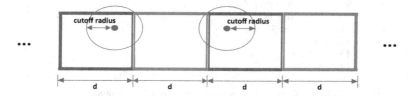

Fig. 6. If the distance *d* between two slabs (belonged to the same group) is greater than *cutoff* radius, there are no common neighbor atoms among different threads.

Our PTS method is designed for large-scale MD simulations, which can meet the limitation generally. We still support a check in our program: if the distance d is greater than *cutoff* radius, our PTS method will be used; otherwise, a traditional method will be provided. Because only half of the neighbors are calculated, the distance *d* do not need to be large than *2*cutoff*.

3.2 Improved SIMD Utilization

As discussed in Sect. 3.1.2, the computations of EAM potential and force require three loops. On account of the constrained space and the similarity of the three loops, we choose the electron density ρ calculation loop to introduce our optimization method.

3.2.1 Cut-Off if Statements

There is a *cutoff* check (indicated on line 7 in Fig. 7) to estimate if the distance *r* is shorter than *cutoff* in the short-range potential calculation. As to the neighbors whose distance between atom *i* is larger than *cutoff*, their contributions of electron density ρ should not be added to the rho array, not vice versa. Pennycook et al. [15, 20] handled this issue via blending/masking, adding value for neighbors that fail in the cut-off is set to 0. Resembling technique is used in Gromacs [13, 14].

Nevertheless, these solutions bring about redundant calculation: the neighbors which satisfy and the neighbors which fail in the *cutoff* check both execute lines 8–11 in Fig. 7. The amount of inefficiency which is caused by the redundant calculation depends upon the proportion of neighbors that fail in the cut-off check. Pennycook et al. used neighbor list in miniMD and Gromacs [18] used Verlet pair-lists to control the ratio failing in the cut-off check. So the efficiency loss is tied to neighbor list construction frequency. The more accurate the neighbor list is, the more time is spent on updating neighbor list.

```
1.   for all atoms i do
2.     for all neighbors j do
3.         delx = xi - pos[j]+[0];
4.         dely = yi - pos[j]+[1];
5.         delz = zi - pos[j]+[2];
6.         rsq = (delx * delx) + (dely * dely) + (delz *delz);
7.         if (rsq <= Rc) then              //Rc=curoff*cutoff
8.             r=sqrt(rsq);
9.             ρ= Interpolate(r);
10.            rho_i+=ρ;
11.            rho[j]-= ρ;
12.        end if
13.    end for
14. end for
```

Fig. 7. Computes electron density ρ using distance r

3.2.2 Modified Pre-searching Neighbor Method

We reference the lattice neighbor list in Crystal MD which finds neighbors according to their positions. Modification has been made on the neighbor searching method to achieve higher cut off check meet ratio. Figure 8 illustrates the principle of the lattice neighbor list: to find the neighbors of atoms i (colored red), a red rectangle whose width is N (calculated using Eq. 3) times of lattice constant is used. The yellow atoms in this rectangle are regarded as neighbors of atom i. This algorithm makes only about 30 % of the neighbors meet the *cutoff* check, which leads to a significant performance loss when SIMD is used.

$$N = \text{ceil } (cutoff \text{ radius}/\text{lattice constant}) \qquad (3)$$

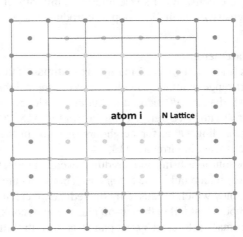

Fig. 8. Neighbor searching method in Crystal MD (Color figure online)

We modified the lattice neighbor list as follows: pre-calculate the the lattice position distance between the yellow atoms and atom i; if the distance is shorter than variable R

in Eq. (4), the yellow atoms are determined as a neighbor of atom i, otherwise the yellow atoms are removed from the lattice neighbor list. According to our application background, atoms generally do not derivate from their lattice position further than twice the lattice constant. Here we use Eq. 4 to determine variable R, and variable R can be adjusted based on other practical applications. The pre-searching work need be carried out once through the entire simulation process, so time consuming on this work can be ignored. The modified pre-searching neighbors strategy leads to about 70 % neighbors to meet the *cutoff* check. For the MD packages which use neighbor list, to achieve the same meet ratio, a large amount of time will be spent on updating neighbor list.

$$R = cutoff + 2 * (\text{lattice constant}) \tag{4}$$

The simulation variables, such as positions and forces, are double-precision in Crystal MD. And we use 256-bits SIMD [21–23] for vectorization implementation, so we process four neighbors simultaneously [15]. Figure 9 represents the AVX and AVX2 implementation outline.

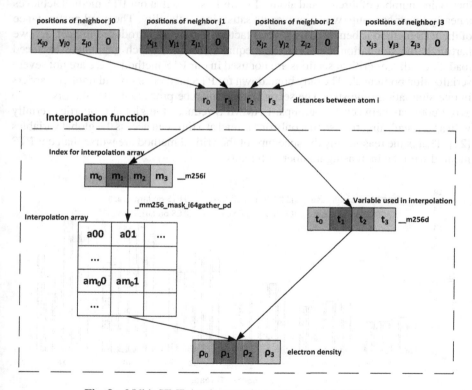

Fig. 9. 256bit SIMD implementation outline for calculating ρ.

4   Experiments Results and Analysis

We performed our experiments using Red Hat 4.8.2-16 operating system and Intel(R) Xeon(R) CPU E7-8890 v3. The original codes came from the Crystal MD.

4.1   Experiment Result and Analysis of Multi-threading Optimization

Three test cases were used in our experiments: the small case (549,250 atoms), the middle case (1,024,000 atoms), and the large case (8,192,000 atoms). Because the EAM potential and force calculation is the most compute-intensive part and this part has data dependence, the execution time in the experiment result refers to the EAM computation time.

The multi-threading speedup of LAMMPS and our improved version are presented in Fig. 10. The CS in Fig. 10 means the critical section method used in LAMMPS. It is demonstrated in Fig. 10 that the performance of our PTS version is improved with increasing number of threads and atoms. Figure 11 shows that our PTS method achieves a nearly linear speedup, which proves its satisfying scalability. The good performance of the PTS method is benefited by several factors. Firstly, as introduced in Sect. 3.1, we partitions the simulation area into some equal size slabs, which guarantees balanced load. Secondly, as critical sections are not used in our PTS method, there are not severe serialization bottleneck. The only breakdown for PTS method is several implicit barriers in one simulation time step. The second group must be processed after the first one to avoid conflicts between two groups. LAMMPS reduces the global properties serially with a "critical" directive, so that only one thread at a time can access the global variables [24]. That is the reason why the speedups of the critical method are worse than our PTS method with the increasing number of threads.

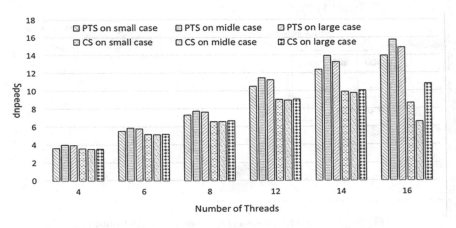

Fig. 10. Comparison of the PTS and CS methods multi-threading speedups.

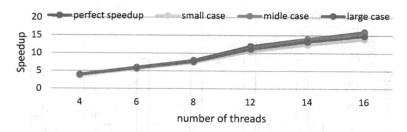

Fig. 11. Speedup of the PTS method

Unlike the method creating thread private work areas discussed in Sect. 2.2, the PTS method do not need extra memory. We observed that memory usage of our multi-threading implementation remains almost the same over the different number of threads.

4.2   Experiment Result and Analysis of Improved SIMD Implementation

The simulation parameters were fixed as follows: *cutoff* radius = 5.6, lattice constant = 2.855, time steps = 30. We use absolute performance (i.e. (atoms*time steps)/ execution time) to compare the original and AVX versions. Because our optimization focuses on force calculation kernel, we only observe execution time of EAM force calculation for both original and SIMD versions.

Table 1 presents execution time and speedups. Figure 12 compares the absolute performances of the original scalar and our SIMD versions over different atom numbers. Figure 12 also presents that the Atom-Step/s of the scalar version remain almost the same across various problem sizes. Our optimized SIMD code is consistently about 3× faster than the scalar version over different simulation scales.

Table 1. Execution time and speedup

Thousands of Atoms	54	182	250	432	1024	2000
Original execution time (s)	2.39	7.83	11.06	18.56	45.978	90.48
SIMD execution time (s)	0.83	2.57	4.26	6.61	17.48	33.53
Speedup	2.88	3.05	2.60	2.81	2.63	2.70

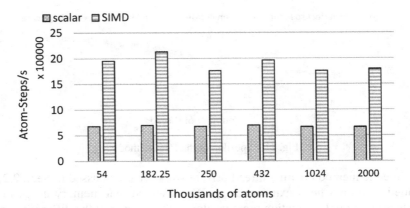

Fig. 12. Absolute performances of the original version and AVX version.

5   Conclusions

We provide multi-threading and vectorization on force calculation kernel. Our optimization method accelerates the original MD version. This means that a longer physical process can be simulated using our optimized version, in the same experimental platform and limited time.

We put forward a PTS method to avoid the rare condition when the short-range force is calculated on shared-memory multi-core platforms. Our PTS method does not need extra memory usage, redundant computation, or lead to severe serialization bottleneck with increasing threads. Our experiment results demonstrate our PTS method is scalable and efficient.

The *cut-off if statement* in short-range force calculation has great influence on the MD program performance. We modifies the lattice neighbor list in Crystal MD by adding a pre-searching process. The modified strategy leads to about 80 % atoms meet the cut-off check, which decreases numerous redundant calculations. Our vectorization version is about 3× than the scalar version.

Future research directions include auto tuning, data partitioning, and using more advanced SIMD instruction-set (such as AVX-512). Some operations required in the interpolation function are only supported by AVX-512. A better speedup is expected in the future experiment platform providing AVX-512.

Although illustrated with the EAM potential, our optimization strategies are widely applicable for other short-range potentials. Both our multi-threading and vectorization are effective and uncomplicated to implement.

Acknowledgement. The research is partially supported by Natural Science Foundation of China under Grant No. 61303050, the Hi-Tech Research and Development Program (863) of China No. 2015AA01A303 and the Youth Innovation Promotion Association, CAS (2015375).

References

1. He, X.-F., Yang, W., Fan, S.: Multi-scale modeling of radiation damage in FeCr alloy. Acta Phys. Sin. **58**(12), 8657–8669 (2009)
2. Intel. http://software.intel.com/en-us/intel-vtune/
3. Intel. http://software.intel.com/en-us/articles/using-intel-vtune-performacceanalyzer-events-ratios-optimizing-applications/
4. Daw and Baskes (M. S. Daw and M. I. Baskes, Phys. Rev. B 29, 6443 (1984); S. M. Foiles, M. I. Baskes, and M. S. Daw, Phys. Rev. B 33, 7983 (1986))
5. Daw, M.S., Baskes, M.I.: Embedded-atom method: derivation and application to impurities, surfaces, and other defects in metals. Phys. Rev. B **29**(12), 6443–6453 (1984)
6. Glosli, J.N., Caspersen, K.J.: Extending stability beyond CPU millennium– a micron-scale atomistic simulation of Kelvin-Helmholtz instability. Supercomputing (2007)
7. Tarmyshov, K.B.: Parallelizing a molecular dynamics algorithm on a multiprocessor workstation using OpenMP (2005)
8. CoMD. http://www.exmatex.org/comd.html
9. Gary, L., Masha, S., Shen, Y.Z.: Performance and energy evaluation of CoMD on Intel Xeon Phi Co-processors. Hardware-Software Co-Design for High Performance Computing (2014)
10. Kunaseth, M.: Analysis of scalable data-privatization threading algorithms hybrid MPI/OpenMP parallelization of molecular dynamics. J. Supercomput. **66**, 406–430 (2013)
11. Liu, Y.: Efficient parallel implementation of Ewald summation in molecular dynamics simulations on multi-core platforms. Comput. Phys. Commun. **182**, 1111–1119 (2011)
12. GROMACS. http://www.gromacs.org/
13. Lindahl, E., Hess, B., van der Spoel, D.: GROMACS 3.0: a package for molecular simulation and trajectory analysis. J. Mol. Model **7**, 306–317 (2001)
14. Hess, B., Kutzner, C., van der Spoel, D., Lindahl, E.: GROMACS 4: algorithms for highly efficient, load-balanced, and scalable molecular simulation. J. Chem. Theor. Comput. **4**(3), 435–447 (2008)
15. Pennycook, S.J., Hughes, C.J., Smelyanskiy M., Jarvis, S.A.: Exploring SIMD for molecular dynamics, using intel xeon processors and intel xeon phi coprocessors. In: 2013 IEEE 27th International Symposium on Parallel and Distributed Processing, pp. 1085–1097 (2013)
16. ExMatEx. CoMD proxy application (2012)
17. Baihe, Crystal MD: Molecular dynamics simulation software for metal with BCC structure. Technical report. http://scce.ustb.edu.cn/
18. OpenMP. http://www.openmp.org
19. http://imd.itap.physik.uni-stuttgart.de/userguide/eam2.html
20. Reinders, J., Jeffers, J.: High performance parallelism pearls, multicore and many-core programming approaches (2015)
21. Intel Intrinsics Guide. http://software.intel.com/sites/landingpage/IntrinsicsGuide/
22. Intel® 64 and IA-32 architectures Software Developer's Manual, Volume 2: Instruction Set Reference, A-Z
23. Intel® Architecture Instruction Set Extensions Programming Reference. https://software.intel.com/sites/default/files/managed/68/8b/319433-019.pdf
24. http://lammps.sandia.gov/

Optimizing Parallel Kinetic Monte Carlo Simulation by Communication Aggregation and Scheduling

Baodong Wu[✉], Shigang Li[✉], and Yunquan Zhang[✉]

State Key Laboratory of Computer System and Architecture, Institute of Computing Technology,
Chinese Academy of Sciences, Beijing, China
{zzubdwu,shigangli.cs,yunquan.cas}@gmail.com

Abstract. Kinetic Monte Carlo (KMC) algorithm has been widely applied for simulation of radiation damage, grain growth and chemical reactions. To simulate at a large temporal and spatial scale, domain decomposition is commonly used to parallelize the KMC algorithm. However, through experimental analysis, we find that the communication overhead is the main bottleneck which affects the overall performance and limits the scalability of parallel KMC algorithm on large-scale clusters. To alleviate the above problems, we present a communication aggregation approach to reduce the total number of messages and eliminate the communication redundancy, and further utilize neighborhood collective operations to optimize the communication scheduling. Experimental results show that the optimized KMC algorithm exhibits better performance and scalability compared with the well-known open-source library—SPPARKS. On 32-node Xeon E5-2680 cluster (total 640 cores), the optimized algorithm reduces the total execution time by 16 %, reduces the communication time by 50 % on average, and achieves 24 times speedup over the single node (20 cores) execution.

Keywords: Domain decomposition · Communication aggregation · Communication scheduling · Neighborhood collectives

1   Introduction

Kinetic Monte Carlo (KMC) [1–4] is a very popular method to simulate the dynamics of stochastic processes. It provides a simple yet powerful and flexible tool for exercising the concerted action of fundamental, stochastic, and physical mechanisms [5]. The main idea of KMC shows the evolution of a dynamical system with simulating the system transition from one state to another state. During of simulating system evolution, it is an iterative process. In every iteration, KMC chooses and performs an event by random number to update the status of the system. The iterative simulation ends until reaching the set time.

There are two parts in the KMC algorithm which can be parallelized. One part is calculating the transition probabilities, and another part is the event execution. The parallel KMC algorithm is often based on the method of domain decomposition, which is shown in Fig. 1. This algorithm arranges processes to 2d or 3d box to minimize surface area per process. The simulation domain is a 3d box, which consists of a number of

© Springer Science+Business Media Singapore 2016
W. Chen et al. (Eds.): BDTA 2015, CCIS 590, pp. 282–297, 2016.
DOI: 10.1007/978-981-10-0457-5_26

lattice sites. The key part of domain decomposition is to allocate the simulation box for every process. Figure 1 shows eight processes to be mapped into a 3d coordinate, and every process is responsible for the simulation of a sub-domain. In order to avoid data collisions in parallel simulation, the sub-domain of every process is further divided into sectors. When the sites of the process need take the data of their neighbor sites not in the same process, the process will communicate with neighbor processes.

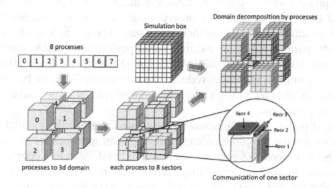

Fig. 1. Domain decomposition of KMC

The parallel KMC algorithm based on domain decomposition has good performance on a small-scale cluster. However, with the number of cluster increasing, the proportion of communication time increases sharply as shown in Fig. 2. Through experimental analysis, we find that the communication redundancy and congestion are the main factors for making the communication performance poor and the sublinear scalability of the parallel KMC algorithm, which is caused by the traditional point-to-point communication between the neighbor processes. Thus, it is very important to reduce communications overhead.

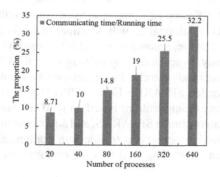

Fig. 2. The proportion of communicating time in running time

To solve the above problems, we propose a communication aggregation approach to reduce the total number of messages and eliminate the communication redundancy, and further utilize neighborhood collective operations to optimize the communication scheduling. We find there are two communications between one sector of every process and neighbor sectors of neighbor processes, it will take eight iterations to complete communications of all sectors in every process. We combine the second communication of one sector and the first communication of the next sector in the same process. The communication aggregation approach makes the total number of communications between one process and neighbor processes reduces by 40 times. In addition, using neighborhood collective operations can avoid communication congestion instead of point to point communication of neighbor processes. Experimental results show that the optimized KMC parallel algorithm exhibits better performance and scalability than the well-known open-source library—SPPARKS. On 32-node Xeon E5-2680 cluster (total 640 cores), the optimized algorithm reduces the total execution time by 16 %, reduces the communication time by 50 % on average, and achieves 24X speedup over the single node (20 cores) execution.

In the rest of paper, Sect. 2 introduces the related work of the parallel KMC algorithms. Then we discuss the communication aggregation approach for the parallel KMC algorithm in Sect. 3. Section 4 presents the neighborhood collective operations for optimization the communication scheduling. Section 5 shows experimental results and analysis. Finally, we make a conclusion of the research results and provide the direction of future research.

2 Related Work

The KMC serial algorithm exposes the problem of running time too long and is even difficult to simulate correctly in the large time scale and space scale. With the rapid development of high performance clusters, many experts and scholars are devoted to the research of parallel KMC algorithms.

Shim et al. [6] present an efficient semirigorous synchronous sublattice algorithm for parallel kinetic Monte Carlo simulations. The accuracy and parallel efficiency are studied as a function of diffusion rate, processor size, and number of processors for a variety of simple models of epitaxial growth. Martinez et al. [7] propose a synchronous parallel generalization of the rejection-free n-fold KMC method. And they test to validate the method and study its scalability by solving several well-understood diffusion problems. Sandia National Laboratories has developed an open-source KMC parallel simulation framework called SPPARKS [8, 9]. SPPARKS provides the parallel algorithm support for a variety of applications based on KMC simulation. This paper starts from the parallel KMC algorithm in SPPARKS, and adopts communication aggregation and neighborhood collective operations to alleviate communication congestion and optimize communication scheduling, and lifts the performance of parallel KMC to a higher level.

From the above references, we can see that most optimization method focus on computing parallelization, reducing synchronization overhead and optimizing sites data structures. It has been widespread concern optimizing communications between neighbor processes in the field of high performance computing. Li et al. [10, 11] utilize shared memory to optimize the point to point communication and provide excellent performance for communication of short messages between neighbor processes. O'Brien et al. [12] propose a scalable domain decomposed algorithms for Monte Carlo particle transport, and they show how to efficiently couple together adjacent domains to maintain within workgroup load balance and minimize memory usage. Hoefler et al. [13] propose the neighborhood collective operations to efficiently describe the communication patterns such as 2d and 3d Cartesian neighborhoods. In this paper, we utilize the neighborhood collective operations to schedule the communications, and find a larger performance boost when using in conjunction with communication aggregation.

3   Communication Aggregation

SPPARKS is a parallel Monte Carlo library based on MPI. The KMC simulation is an iterative process in SPPARKS. At the initialization stage, the information of simulating sites is established and assigned to the corresponding processes by domain decomposition. The messages of simulating sites to be sent and receive includes site global ID, site local ID, process ID, neighbor site ID, neighbor processes, buffer (called sendbuf) to be sent to neighbor processes and buffer (called recvbuf) to receive from neighbor processes. Each sector of every process exchanges the messages with neighbor processes. Then the process calculates the transition probabilities. Repeat this above process until reaching the set time. Figure 3 shows the process of iterative communications in SPPARKS. In every iteration, the total 8 sectors are performed in turn. The current sector communicates with neighbor sectors for the first time. Since the current sector and some of the neighbor sectors are not on the same process, point to point communication operations are used to perform data exchange between different processes. Next, the current sector randomly chooses an event to perform and update the messages of the current sector sites. Finally the current sector performs the second communication. The next sector will perform the same procedures sequentially as the previous sector. After the first iteration of simulation is done, it will continue the next iteration and exit the iteration until reaching the set time.

As can be seen from the Figs. 1 and 3, each sector of every process should perform 7 times communication with neighbor processes, the above communication operations will be performed twice. Therefore, the total number of communications is 112 (8 sectors*2*7) times in every process. We find that there is no event operation between the second communication in the current sector and the first communication in the next sector. The first time communication in sector 1 is performed at first, and then the second communication in sector 1 is merged with the first time communication in sector 2. Similarly, the second communication in the current sector is merged with the first time communication in the next sector until the last sector. The second communication in the last sector is performed normally. There are three kinds of situations in communication aggregation algorithm: (A) surface adjacent sectors situation; (B) edge adjacent sectors situation; (C) diagonally adjacent sectors situation.

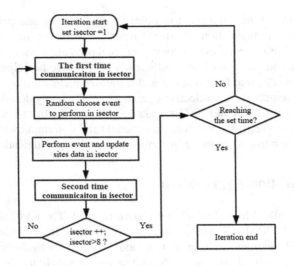

Fig. 3. The process of iterative communications

3.1 Surface Adjacent Sectors Situation

Surface adjacent sectors situation refers the two sectors location next to each other in which communication blocks will be merging. Figure 4 illustrates the situation of surface adjacent sectors and the process of communication aggregation. 8 sectors are placed in a 3d coordinate in each process, the merging of sector 1 and sector 2 is the case, also including sector 3 and sector 4, sector 5 and sector 6, sector 7 and sector 8. Each sector has 7 communication blocks due to 7 neighbor processes. The merging of sector 3 and sector 4 is shown in Fig. 4.

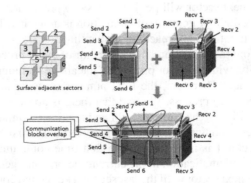

Fig. 4. The merging process of sectors of surface adjacent

However, there are some conflicts of data dependency during the merging. The reason of those problems is that these 2 sectors have some of the same neighbor processes to exchange data with each other. Thus some of the communication blocks overlap. If we don't take some measures, the order of sending and receiving messages and correctness of the data can't be ensured. We find that the data of sector 3 is always up-to-date and it will be sent to its neighbor processes, while sector 4 will receive the latest data from the neighbor processes. Therefore, we can send the latest data from sector 3 to both its neighbor processes and the process of sector 4. This method can avoid the data inconsistency problema. Before merging the twice communication operations, the number of communications is 14 times, including 7 times in sector 3 of second communication and 7 times in sector 4 of the first time communication. While after merging the number of communications reduces 13 times, including 7 times merging blocks and 6 times dealing overlapping communication blocks.

3.2 Edge Adjacent Sectors Situation

The edge is shared between two adjacent sectors which communication blocks will be merging. Figure 5 illustrates the situation of edge adjacent sectors and the process of communication aggregation. As shown in Fig. 5, the merging of sector 3 and sector 8 is the case. Each sector has 7 communication blocks due to 7 neighbor processes. As with the above surface adjacent sectors situation, there has the problem of blocks overlapping during merging the communication blocks of sector 3 and sector 8. We also take the same measure to solve the problem. The number of communications is 9 times, including 7 times merging blocks and 2 times dealing overlapping communication blocks.

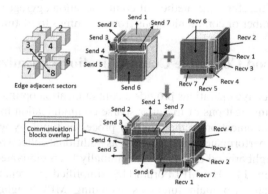

Fig. 5. The merging process of sectors of edge adjacent

3.3 Diagonally Adjacent Sectors Situation

Diagonally adjacent sectors situation refers that the two sectors are not in the same side. The position of diagonally adjacent sectors is shown in Fig. 6. The merging of sector 3 and sector 6 is the case. During merging communication blocks of sector 3 and sector 8, they don't have a same neighbor process, thus it will not exist a overlapping communication block. The number of communications is 7 times, including 7 times merging blocks.

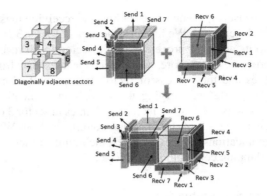

Fig. 6. The merging process of sectors of diagonally adjacent

The communication will be performed from sector 1 to sector 8 in order, there are 7 places to combine from sector 1 to sector 8 one by one. During the process, the combing of sector 1 and sector 2, sector 3 and sector 4, sector 5 and sector 6, sector 7 and sector 8 belongs to the situation of surface adjacent sectors, it will only reduce a communication. While the combing of sector 2 and sector 3, sector 6 and sector 7 belongs to the situation of edge adjacent sectors, it will reduce 5 communications. And in the situation of diagonally adjacent sectors, the combing of sector 4 and sector 5 will reduce 7 communications. Therefore, the method of communication aggregation algorithm can reduce the total number of communications from 112 times to 91 times.

4   Neighborhood Collective Communication Optimization

Neighborhood collective operations provide a communication optimization method for the collective communications of irregular sparse communication mode. The purpose is to make the communication of the arbitrary processes topology will implement the same good parallel performance and scalability as traditional collective communication. The concept of neighborhood collectives has initially been motivated under the name sparse collectives in [14]. The MPI Forum [15] simplified and renamed the proposed functions. There are two main functions including MPI_Neighbor_allgather and MPI_Neighbor_alltoall in neighborhood collective communications. The former is an operation of sending the same buffer to neighbor processes and receiving the buffer from neighbor processes in the same way. While the latter is the operation of sending different buffers to neighbor processes and receiving the buffers from neighbor processes.

Compared with point to point communication, neighborhood collective operations optimize communication topology and scheduling. The predefined collective communications and static communication patterns are used to control communication scheduling and avoid bandwidth congestion for different communication models in neighborhood collective operations.

We optimize the communication scheduling by neighborhood collective operations for parallel KMC stimulation based communication aggregation. And the optimization results can be superimposed due to optimizing different aspects of communications. Detailed optimization of this section has two parts, including graph topology processes and optimizing communication scheduling.

4.1 Graph Topology Processes

The primary condition of using neighborhood collective operations is to create the topological relations among neighbor processes. In parallel KMC algorithm, every each process communicates with every neighbor process by point-to-point communication. The communication pattern of those neighbor processes can be represented by a graph. The nodes represent processes, and the edges connect processes of communication with each other.

MPI supports three process topology types: Cartesian, graph, and distributed graph. Cartesian topologies partition a set of processes into a regular 2d or 3d Cartesian structure, every of process is equal relation. Graph and distributed graph topologies can be used to create arbitrary shape topology for processes. The function MPI_DIST_GRAPH_CREATE_ADJACENT and MPI_DIST_GRAPH_CREATE are used to create distributed graph topologies. MPI_DIST_GRAPH_CREATE provides full flexibility in order to make sure that communications will occur between every pair of the processes in the graph. MPI_DIST_GRAPH_CREATE_ADJACENT creates a distributed graph communicator with each process specifying, each of its incoming and outgoing (adjacent) edges in the logical communication graph. As a result, it will need minimal communication during initialization.

We choose MPI_DIST_GRAPH_CREATE_ADJACENT to create the neighbor processes topology for parallel KMC algorithm, because both the number of neighbor processes and sending buffer size are not fixed. This function has 7 main input arguments: comm_id, indegree, sources, sourceweights, outdegree, destinations, and destweights. The argument comm_id is current communicator, argument indegree is the size of neighbor processes for which the calling process is a destination, argument sources is ranks of above neighbor processes, argument sourceweights is weights of the edges into the calling process, argument outdegree is the size of neighbor processes for which the calling process is a source, argument destinations is the rank of above neighbor processes, argument destweights is weights of the edges out of the calling process. Running this function will return a communicator with distributed graph topology.

4.2 Optimizing Communication Scheduling

In parallel KMC algorithm, each site in simulation box is responsible for a process in 3d coordinate. Each process is further divided into 8 sectors in order to maximize the degree of parallelism. In KMC algorithm for materials science, each site represents a metal atom and has 27 neighbor atoms in body-centered cubic (BCC) structure. The ghost sites in a process interact with 27 neighbor sites. Obviously, the 27 neighbor sites may not be in the same process. Thus, each process has to communication with neighbor processes to make the ghost sites update data from neighbor sites. It may lead to communication obstruction that each process simultaneously sends messages to neighbor process due to the lack of communication scheduling in point to point communication. Thereby the communication time increases significantly.

Neighborhood collective operations use the method of messages coalescing and forwarding through proxy processes to reduce injection rate limitations, and avoid congestion and optimize communication scheduling. We compared and analyzed scheduling communication between neighborhood collective operations and point to point communication. And we set that local input bandwidth of each process is bw and network concurrency is *ncd*. Figure 7(a) shows the state of point to point communication. We use Send and Recv operation to realize MPI_Neighbor_alltoall function. From Fig. 7(a) we can see that each process receives 8 buffers from all processes simultaneously. The ideal bandwidth of each process is $bw/8$ and the total bandwidth is bw. If *ncd* is less than 8 or buffer size is so large that *ncd* can't hold all buffers from all processes, communication protocol will choose part of buffer to receive and the rest of the buffer are placed in a queue to be received. The above operations will take a lot of time for redeployment, transferring out the queue and reducing the performance and scalability of communication. The communication scheduling steps of neighborhood collective operation is as shown in Fig. 7(b). In the first step, process 0 send to process 1, process 1 send to process 2, ..., and process 7 send to process 0. In the second step, process 0 send to process 2, ..., and so on, in step j, process i send to process $(i + j)\%8$. Each process receives 1 buffer from all processes simultaneously. The ideal bandwidth of each process is bw and total bandwidth is $8*bw$.

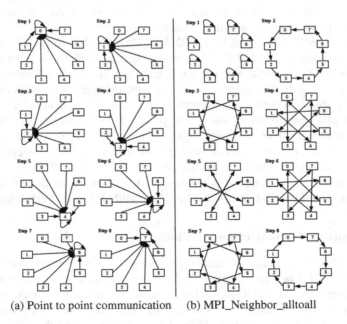

(a) Point to point communication   (b) MPI_Neighbor_alltoall

Fig. 7. The steps of communication scheduling

Then we use neighbor collective operations replace of point to point communication, and combine communication aggregation optimization in parallel KMC algorithm. Because the number of neighbor processes of each process may be different, the function MPI_Neighbor_alltoall is not applicable. We choose MPI_Neighbor_alltoallv to implement communication among processes.

MPI_Neighbor_alltoallv is the extended function of MPI_Neighbor_alltoall. It allows users to define buffer to send and to receive in vectorization. The communicator determines the neighbor relation of processes which is created by MPI_Dist_graph_create_adjacent in the initialization phase. As shown in Fig. 8, process 0 sends the different sizes of sbuf to neighbor processes.

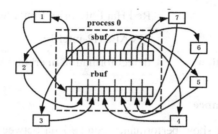

Fig. 8. Communication of MPI_Neighbor_alltoallv

5 Experimental Evaluation

We use the communication aggregation and scheduling to optimize the parallel KMC simulation. We analyze parallel performance and scalability to show the impact of each optimization method. On one hand, we show running time and communication overhead of the original parallel KMC algorithm, the optimization of communication aggregation, the optimization of communication scheduling, and using both optimization methods, respectively. On the other hand, we compare strong scaling and weak scaling to illustrate scalability.

5.1 Experimental Environment

In our experiments, we choose SPPARKS as the original parallel KMC algorithm to simulate the application of defect formation in erbium hydrides. SPPARKS is an important and popular parallel KMC simulation framework for numerous materials modeling applications. This application simulates a model of reaction and diffusion on a specialized Erbium lattice, which consists of an BCC lattice for the Erbium, additional tetrahedral and octahedral interstitial sites.

Table 1 shows our experimental environment. We use the "Era" Supercomputer located in Supercomputing Center of Chinese Academy of Sciences, which has a total of 270 sets of dawn CB60-G16 Dual Blade. The CPU overall performance reaches 120.96Tflops. Each blade is configured with 20 Intel 2.8 GHZ cores. We use the Intel

C++ Compiler version 2013_sp1 in the MPI compiler environment version MVAPCH2-intel 1.9.

Table 1. The experimental environment

Supercomputing system	"Yuan" supercomputing systems Blade computing node
CPU models	Intel E5-2680 V2
CPU info	20 Cores, 2.8GHX, 64 GB 1866 MHz DDR3 ECC memory
Operating system	Red Hat Enterprise Linux Server release 5.1
MPI version	MVAPCH2-intel 1.9
Simulating application	defect formation in erbium hydrides

5.2 Parallel Performance

In this section, we will show performance comparison between the original and the optimized algorithms. We set the simulation box in a 3d coordinate, and the maximum value in each dimension is 200. There are about 128 million sites in the box. We perform the original algorithm and optimized algorithm using different number of processes, varying from 20 to 640. Table 2 shows the running time of the original algorithm and optimization algorithm, and Fig. 9 compares communication overhead. We can see that the optimized algorithm, using communication aggregation and scheduling, can reduce the total running time and significantly minimize the communication overhead.

Table 2. Comparison of running time

Number of Processes	Time (s) of different algorithm			
	Original	Nei-coll	Aggregation	Nei-coll+ aggregation
20	1015.86	1002.02	991.36	996.48
2*20	515.94	518.654	526.169	502.617
4*20	269.76	265.054	265.982	261.593
8*20	140.00	133.538	137.96	132.883
16*20	76.06	71.8368	72.4572	70.2506
32*20	48.60	42.7146	41.34	40.9813

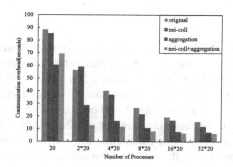

Fig. 9. Comparison of communication overhead (Color figure online)

In order to present concisely, we use the label of "original" to represent the original parallel KMC algorithm, "aggregation" to represent the optimized algorithm by communication aggregation, "nei-col" to represent the optimized algorithm by neighborhood collective communication, and "nei-col+aggregation" to represent the optimized algorithm by using both communication aggregation and scheduling.

As shown in Table 2, each optimized algorithm takes less time than the original algorithm to finish all the KMC simulation. However, there are some exceptions. "nei-coll+aggregation" takes more time than "aggregation" algorithm and less than "nei-coll" when using 20 processes. Because the 20 processes are on the same node, communication congestion is not the main bottleneck for impacting communication time. Using neighbor collective communication might take more time to create process topology, and thus may increase the communication time. What's more, the time of "nei-coll" and "aggregation" is longer than "original" using 2 * 20 processes. 2 * 20 processes use 2 nodes and each node contains 20 processes. It may be caused by the difference in bandwidth among the nodes. In conclusion, with the processes increasing, the performance improvement is more significantly when optimized by communication aggregation and scheduling.

Figure 9 analyzes the communication overheads of each algorithm on different number of processes. It takes less time than original algorithm after using neighborhood collective communication. Communication aggregation algorithm takes less time than neighborhood collective communication, while communication aggregation together with scheduling takes a little less time than communication aggregation alone. On average, communication overhead of communication aggregation and scheduling is reduced by 50 % compared with the original algorithm. It demonstrates that our two optimization methods can benefit each other and have a synergistic effect. In conclusion, our communication optimization methods using communication aggregation and scheduling are efficient.

5.3 Scalability

In this section we will show the scalability of the optimization algorithms and analyze the performance of scalability through weak scaling and strong scaling.

1. Weak Scaling

In weak scaling, the workload of each process is constant. Thus, with the number of processes increasing, we set the size of simulation box of a corresponding increase proportionally. We use 20 processes to simulate a 100 * 100 * 200 box, 2 * 20 processes to simulate a 100 * 200 * 200 box, 4 * 20 processes to simulate a 200 * 200 * 200 box, 8 * 20 processes to simulate a 200 * 200 * 400 box, and 16 * 20 processes to simulate a 400 * 400 * 400 box. Because the number of sites reaches 1.024 billion on 640 processes, the computation time for calculating the sites of potential energy occupies most of the running time. Using the running time and speedup can't reasonably explain the scalability of the algorithms optimized by the communication aggregation and scheduling. What's more, many material simulation applications get the potential energy value of sites by looking up the tables in database, which saves lots of the computation time. Thus, we can compare the communication time and communication speedup to analyze the scalability of such applications conveniently.

Figure 10 shows the performance of each algorithm under weak scaling. The using number of processes range from 20 to 640. We see that all algorithms scale is well, with the exception of original algorithm. The "original" algorithm spends up to 30 % more communication time than the optimized algorithms. Moreover, the optimized algorithm by communication aggregation and scheduling takes the minimum and the most stable time in all algorithms, and shows a good weak scaling.

2. Strong Scaling

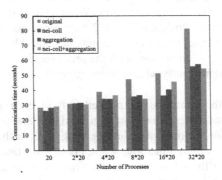

Fig. 10. The weak scaling of communication (Color figure online)

We also test the strong scaling of each algorithm on number of processes from 20 to 640. And we set the simulation box size to 200 * 200 * 200. As shown in Table 3, compared with the original algorithm, and all the optimization introduced in this paper, we have improved the speedup from 20.90 times of a single node performance to 24.32 times with 640 processes.

Table 3. Comparison of speedup performance

Number of processes	Speedup of different algorithm			
	Original	Nei-coll	Aggregation	Nei-coll+ aggregation
20	1.00	1.00	1.00	1.00
2*20	1.97	1.88	1.93	1.98
4*20	3.77	3.73	3.78	3.81
8*20	7.26	7.19	7.50	7.50
16*20	13.36	13.68	13.95	14.18
32*20	20.90	23.21	23.46	24.32

Figure 11 compares the speedup of each algorithm only considered communication time. With 640 processes, the speed of "original" is about 5 times a single node, the speed of "nei-col" is about 7 times a single node, the speed of "aggregation" is about 8 times, the speed of "nei-coll+aggregation" is about 11 times.

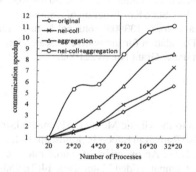

Fig. 11. The speedup of communication part

6   Conclusions and Future Work

In this paper, we showed the optimization method for parallel KMC simulation by communication aggregation and scheduling. Firstly, we introduced the KMC simulation algorithm and domain decomposition applying in the traditional parallel KMC simulation. Secondly, we analyzed the communication process of parallel KMC simulation. And we found that the communication congestion caused by point-to-point communication among the neighbor processes is the main bottleneck increasing the communication time. Thirdly, to solve the above problems, we present communication aggregation and scheduling method. Then we described each optimization in detail. It is a very efficient method to reduce the number of communication between neighbor processes using communication aggregation. Using neighborhood collective operations can implement minimum communication congestion and optimize the communication scheduling. Lastly, to evaluate the performance of optimization, we analyzed parallel performance and scalability to show the impact of each optimization in isolation. Experimental results show optimization by communication aggregation and scheduling has a good parallel performance and scalability. On 32-node Xeon E5-2680 cluster (total 640 cores), the optimized algorithm reduces the total execution time by 16 %, reduces the communication time by 60 %, and achieves 24X speedup over the single node (20 cores) execution.

The approaches of communication aggregation and scheduling proposed in this paper are not just specific to parallel KMC simulation. They can also be transplanted to those parallel applications based on domain decomposition, such as molecular dynamics.

As future work, we plan to research shared memory in MPI or hybrid MPI/OPENMP parallel programming model to reduce the size of communication buffer within the same node and overlap computing and communications. We expect that future work will further improve parallelism and scalability.

Acknowledgment. This work was supported by the National High Technology Research and Development Program of China under Grant No. 2015AA01A303; National Natural Science of China under Grant No. 61502450; State Key Program of National Natural Science of China under Grant No. 61432018 and Grant No. 61133005; National Natural Science Foundation of China under Grant No. 61272136; Foundation for Innovative Research Groups of the National Natural Science Foundation of China under Grant No. 61221062.

References

1. Voter, A.F.: Introduction to the kinetic Monte Carlo method. In: Sickafus, K.E., Kotomin, E.A., Uberuaga, B.P. (eds.) Radiation Effects in Solids, pp. 1–23. Springer, Heidelberg (2007)
2. Chatterjee, A., Vlachos, D.G.: An overview of spatial microscopic and accelerated kinetic Monte Carlo methods. J. Comput. Aided Mater. Des. **14**(2), 253–308 (2007)
3. Hastings, W.K.: Monte Carlo sampling methods using Markov chains and their applications. Biometrika **57**(1), 97–109 (1970)
4. Jansen, T.: Kinetic monte carlo. In: van Santen, R.A., Sautet, P. (eds.) Computational Methods in Catalysis and Materials Science: an Introduction for Scientists and Engineers, pp. 183–197. Wiley, Germany (2009)
5. Battaile, C.C.: The kinetic Monte Carlo method: foundation, implementation, and application. Comput. Methods Appl. Mech. Eng. **197**(41), 3386–3398 (2008)
6. Shim, Y., Amar, J.G.: Semirigorous synchronous sublattice algorithm for parallel kinetic Monte Carlo simulations of thin film growth. Phys. Rev. B **71**(12), 125432 (2005)
7. Martínez, E., Marian, J., Kalos, M.H., et al.: Synchronous parallel kinetic Monte Carlo for continuum diffusion-reaction systems. J. Comput. Phys. **227**(8), 3804–3823 (2008)
8. Plimpton, S., Thompson, A., Slepoy, A.: SPPARKS kinetic Monte Carlo simulator (2010)
9. SPPARKS. http://www.cs.sandia.gov/~sjplimp/spparks.html

10. Li, S., Hoefler, T., Snir, M.: NUMA-aware shared-memory collective communication for MPI. In: Proceedings of the 22nd International Symposium on High-Performance Parallel and Distributed Computing, pp. 85–96, ACM (2013)
11. Li, S., et al.: Improved MPI collectives for MPI processes in shared address spaces. Cluster Comput. **17**(4), 1139–1155 (2014)
12. O'Brien, M.J., Brantley, P.S.: Particle Communication and Domain Neighbor Coupling: Scalable Domain Decomposed Algorithms for Monte Carlo Particle Transport. Lawrence Livermore National Laboratory (LLNL), Livermore (2015)
13. Hoefler, T., Schneider, T.: Optimization principles for collective neighborhood communications. In: 2012 International Conference for High Performance Computing, Networking, Storage and Analysis (SC), pp. 1–10, IEEE (2012)
14. Hoefler, T., Träff, J.L.: Sparse collective operations for MPI. In: IEEE International Symposium on Parallel and Distributed Processing, IPDPS 2009, pp. 1–8, IEEE (2009)
15. MPI Forum. MPI: A Message-Passing Interface Standard. Version 3.1 (2015)

A Study on Process Model of Computing Similarity Between Product Features and Online Reviews

Xueguang Xie, Xiaoping Du, Qinghong Yang$^{(\boxtimes)}$, and Pengfei Feng

School of Software, Beihang University, Beijing 100191, China
{xgxie, xpdu, yangqh}@buaa.edu.cn, 1216862060@qq.com

Abstract. Review similarity computing is used to judge whether the content of online reviews is related to the products. It is an important prerequisite to judge the usefulness of reviews, and it is also an important basis for the classification and sorting of product reviews. This paper combines the VSM, TF-IDF algorithm and cosine similarity algorithm to build the model of similarity computing between the product online reviews and product features, and to build the process framework of review similarity computing for enterprises. Besides, this paper also verifies the model's effectiveness and correctness based on real online review data of E-business. The experiment results show that the process model can be used to quantify the similarity between reviews and product features, and the similarity results also have a good effect on the application of the review sorting.

Keywords: Reviews similarity · Process model · VSM · TF-IDF algorithm · Cosine similarity algorithm

1 Introduction

In the online shopping process, online reviews play a very important role. Consumers can judge the quality of the products according to the comments of others on products [1]. However, with the rapid development of online shopping, the number of electronic comments is also growing. It is impossible for any customer to read all online reviews. Therefore, online comment classifying, filtering, sorting and mining out valuable information are very important [2, 3] so as to improve the quality and efficiency of online shopping. For instance, consumers of books are more concerned about the content, instead of function and brand which other types of products focus on. According to the characteristics of goods, the paper builds the process model of similarity computing between the product online reviews and product features.

There are a number of representation methods of text, such as Vector Space Model (VSM), which were raised by Gerard Salton and McGill in 1969 [4]. Expressed as a mathematical model, the text can be calculated efficiently. Moreover, the three text representation models—namely, the language model, the suffix tree model [6] and the ontological model— are integrated with the semantic knowledge, and can more accurately reflect the contents of the text features [6–9]. However, VSM with the simplicity,

© Springer Science+Business Media Singapore 2016
W. Chen et al. (Eds.): BDTA 2015, CCIS 590, pp. 298–308, 2016.
DOI: 10.1007/978-981-10-0457-5_27

low computational complexity and good effect of representation becomes the most widely used text representation model.

Feature extraction is picking out from all features the characteristics which have a really great contribution to characterize text [10]. Feature extraction can be divided into supervised and unsupervised. Supervised feature extraction methods [5] include mutual information (MI), χ^2 - test (CHI) and information gain (IG); Feature extraction algorithm for the unsupervised have term strength (TS), document frequency (DF) and feature contributions (TC).

TF-IDF algorithm, improved DF algorithm based on vector space model, considers the word's frequency in the text and the word's ability to distinguish between different texts. Therefore, the algorithm is widely used in various fields.

Differences in text representation model lead to different ways in similarity calculation. Taking VSM as an example, the text is represented as a point in a vector space. Thus, the distance between the point and the other point represent the degree of similarity between the text and the other text. The degree of similarity and distance is inversely proportional to the value. Widely used methods of distance calculating in mathematics are Euclid distance, Manhattan distance and Minkowski distance [11].

In addition to distance, inner-product, Jaccard similarity coefficient, Dice coefficient and cosine coefficient can also represent the text similarity. These are all called similarity coefficient. The larger similarity coefficient is, the more similar the texts are. Among them, the calculating method of the cosine coefficient is the cosine similarity algorithm, which is actually an improvement of the vector inner-product. Using this algorithm, all results can be controlled in the interval (0, 1). Thus, there is a relatively clear threshold in judging. Besides, in the case of a large number of calculations, computation effort can be reduced effectively.

Thus, the paper put out a process model of similarity computing between the product online reviews and product features, based on VSM model, TF-IDF algorithm and the cosine similarity algorithm.

2 Research Design

This paper studies the similarity calculation algorithm between online reviews and product features and the process model applying this algorithm. Based on the real data of a certain website, the experiment is performed to test and verify the process model. The research process consists of three phases:

Phase I: Research theory and algorithm. This article covers text mining knowledge including text representation model, feature extraction method and similarity calculation algorithm. This section is introduced in the section of Introduction.

Phase II: Based on the theory, build the process model on similarity computation of product reviews.

According to the general process of data mining, and combined with the objectives of this paper, the process model is built, which includes the three main parts: data preprocessing, feature selection and weighting, and similarity calculation. The 3rd section of this paper will focus on the process model.

Phase III: Experiments and result analysis. Collect data from some E-commerce sites, and divide them into two classes. One is real books' reviews data and the other is the description of the goods from sellers. The description from sellers can represent the product features. Then, based on those data, this paper studies review correlation, and performs the experiment step by step according to the process model. Finally, experimental results of each step will be displayed and analyzed. This phase is introduced in the 4th quarter.

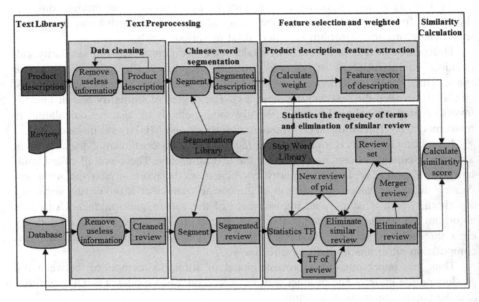

Fig. 1. The framework of process model of review similarity computing

3 Framework of Process Model of Review Similarity Computing

In the area of text mining, product review mining becomes a rising hot spot in recent years. It uses natural language processing technology to mine and find useful information from large amounts of the product reviews on the Web [16]. Mining product reviews are divided into four main processes, i.e. comment text pre-processing, feature extraction, useful information mining and show of mining results. Similarity computing is an important part of the review mining.

Combining the general process of review mining with specific review mining's objective (mining the degree of similarity between the goods comments and the description of the goods), this paper constructs the correlation calculation process model. Figure 1 is the framework of process model of review similarity computing. The review texts and descriptions are the model's input, and the model includes three processes: text preprocessing (data cleansing, Chinese word segmentation), feature vector representation and feature extraction, and review text similarity calculation. Further, the model can be further divided into the following 7 modules:

(1) Text library module: This part is used to store the texts of reviews and product descriptions;

(2) Data cleaning module: Clean the texts of reviews and product descriptions, such as removing numbers and punctuation;

(3) Library module: Administration and maintenance of segmentation library and stop word library, which store a large number of sub-words and stop words to support Chinese automatic segmentation;

(4) Chinese word segmentation module: Segment Chinese text and regulate the ambiguity of words;

(5) Product description feature extraction module: According to the segmentation and statistic results, this module calculates feature weight and extracts text features;

(6) Review feature extraction module: According to the segmentation and statistic results, this module represents the review texts as feature vectors, calculates feature weight, removes the duplicate comments and merges the new comments with the original;

(7) Similarity computing module: According to feature vectors of reviews and product descriptions, calculate the similarity of each comment and description of the goods.

3.1 Text Preprocessing

Data Cleaning. Since product description in the text library is published at the E-commerce platform by sellers, the texts are standard and clear. However, the product reviews are messages from different users, and contain a large amount of spam information [10], such as advertisement and useless information, which need to be filtered in the module. Moreover, data cleaning needs to remove punctuation, symbols, and numbers. Then, the text will become the format standard and valuable, and lay the foundation for further work.

Chinese Word Segmentation. In this article, the research objects, reviews and product descriptions are texts written in Chinese. In Chinese, a single word is the basic unit of writing. There are no visible marks to distinguish the terms, and thus the computer cannot understand the meaning of those words [17]. Therefore, dividing the text into appropriate terms is the base and key to dealing with Chinese information. Currently, word segmentation algorithms can be divided into four major categories: word segmentation method based on dictionaries and thesaurus matches, word segmentation method based on word frequency statistics, word segmentation method based on semantic and word segmentation method based on knowledge [12].

In this model, this module use the word segmentation method based on dictionaries and thesaurus matches to segment the product description and reviews. This method is in accordance with certain policies to match specific strings with the terms in the thesaurus. If the string is in the thesaurus, the match is successful.

3.2 Feature Selection and Weighting

Product Description Feature Extraction. After segmenting a series of terms from the text of product descriptions, the model uses the stop thesaurus to extract feature terms from the segmented terms.

Stop words are translated as "Computer Retrieval function words or Non-retrieval words" in the dictionary. It can be divided into the following two categories: One is the words used very widely or even too often. Such words as "我(I)" and "就(on)" that appear in almost every document are useless for similarity calculation, and cannot improve the accuracy of the results, and even reduces the efficiency. The other is meaningless words. This category includes modal particles, adverbs, prepositions, and conjunctions, such as "的(of)", "在(on)", and "和(and)".

After removing those stop words, the features of goods descriptions can be obtained. Then according to the TF-IDF algorithm [13] the weights of each feature are calculated.

Step 1. Calculate term frequency (TF). Suppose D_j represents the ith text, and T_i is the ith word. Then in D_j, the frequency of T_i is represented as TF_{ij} and computed as follows:

$$TF_{ij} = \frac{N_{i,j}}{\sum_k N_{k,j}}. \tag{1}$$

In the above formula, $N_{i,j}$ refers to the frequency the word T_i appears in D_j, and $\sum_k N_{k,j}$ is the total number of all words in the text D_j.

Step 2. Calculate inverse document frequency (IDF). IDF is a measure of general importance of words [18], which can be computed by taking logarithm of the ratio between the total amount of text and the number of text containing T_i:

$$IDF_i = log\frac{|D|}{|\{j : T_i \in D_j\}|}. \tag{2}$$

In this formula, $|D|$ stands for the total amount of text in the library, and $|\{j : T_i \in D_i\}|$ is the number of text containing T_i (the number of text whose $N_{i,j}$ are not equal to zero). Due to the fact that the words may not be in the library, the dividend may be zero. Therefore, $1 + |\{j : T_i \in D_j\}|$ is used;

Step 3. Calculate the weight W_{ij}. W_{ij} is proportional to the frequency of T_i in the text D_j, and is inversely proportional to the frequency of T_i in the other text from the text library. The equation for W_{ij} is:

$$W_{ij} = TF_{ij} \times IDT_i = \frac{N_{i,j}}{\sum_k N_{k,j}} \times log\frac{|D|}{1 + |\{j : T_i \in D_j\}|}. \tag{3}$$

Step 4. Normalize the weight. In order to be clear about the threshold value in calculation, the document vectors need to be normalized, as shown in the formula:

$$W_{ij} = \frac{TF_{ij} \times IDT_i}{\sqrt{\sum_{T_i \in D_j}[TF_{ij} \times IDT_i]^2}}. \tag{4}$$

According to the above processes, all the feature weight values can be computed. The feature with low weight value should be removed. Then, by using the VSM model [13] we can put the text into the mathematical model. The basic idea of the process is simplifying the text into certain features. Feature weights as components form a vector. Thus the vector in n-dimensional space is the quantitative form of the text.

Features: The document can be divided into multiple independent terms. Remove the stop word, and the remaining entries are called the features T_i. The document can be expressed as $(T_1, T_2, T_3, \cdots, T_n)$.

Feature weight: In representing documents, the importance of every feature T_i is different. The degree of feature importance can be quantized into weight W_i. Then the document can be expressed as $(T_1 * W_1, T_2 * W_2, T_3 * W_3, \cdots, T_n * W_n)$, which are simplified as $(W_1, W_2, W_3, \cdots, W_n)$.

Vector Space Model: Use $(T_1, T_2, T_3, \cdots, T_n)$ as coordinates and an n-dimensional vector space can be formed. The document $(W_1, W_2, W_3, \cdots, W_n)$ is a vector in the n-dimensional space [14].

Through the three processes, remove the stop words, calculate the feature weight and form the feature vector, and the segmentation results are represented in a feature vector form, which lay the foundation for computing similarity.

Statistics and Frequency of Terms. Like the process of product description, the added reviews in the database also need to be clean and segmented and the stop words need removing, and then the comment features can be obtained. However, the calculating of weights is different from the processing of goods description. Review weights should be calculated by counting up the frequency of feature items because most reviews are too short and the frequency of features can sufficiently represent the important degree of words. In this module, similar reviews should be eliminated.

3.3 Similarity Calculation [14, 15]

In the above process, the feature vector about product descriptions and reviews can be obtained. Taking the two vectors as input and using cosine similarity algorithm, the similarity between the product descriptions and reviews can be calculated. The detail is as follows [15]:

First, represent product description D_1 in feature vector $X = \{X_1, X_2, \cdots, X_n\}$, and express the text of product review D_2 in $Y = \{Y_1, Y_2, \cdots, Y_n\}$. n means the dimension of feature vectors, and X_k, Y_k is the kth dimension of the vector X and Y.

Second, use the vector cosine formula, the similarity between X and Y is calculated. The cosine formula is as follows:

$$sim(X, Y) = \frac{X \cdot Y}{|X| \cdot |Y|} = \frac{\sum_{k=1}^n X_k \times Y_k}{\sqrt{(\sum_{k=1}^n X_k^2)(\sum_{k=1}^n Y_k^2)}}. \tag{5}$$

4   Model Application and Effect Analysis

The objects of the experiment are book reviews and descriptions from some e-commerce sites. In this experiment, 10 product details and the corresponding 1000 comments are selected. Each book corresponds to 100 comments. Take the book of *Les Aventure De TinTin* as an example, the paper introduces in detail the proceedings of applying the model to the similarity calculation. Figure 2 is the input data of the model. Based on the above model of similarity calculating, the data were preprocessed and the experimental results were obtained.

4.1   Treatment Effect of Product Descriptions

First, in order to obtain the standard data, this paper preprocessed the descriptions and reviews of *Les Aventure De TinTin*. Next, the standard data were divided into some words using the Chinese word segmentation. Finally, all the features of the descriptions were extracted by matching up with the library of stop words. The feature items are 242 in total. Figure 3 presents the detailed process of this part of experiment.

Having obtained the features of details, the weight of each feature was calculated by the two methods of TF and TF-IDF. Tables 1 and 2 are sorted features and the weight of features by TF and TF-IDF respectively.

Comparing the two results above, it can be seen that those characteristic words such as "世界 (world)", "创作(creation)", "成(into)" and "读者(readers)" appear in the top ten sorted according to TF, but not in the top ten sorted according to TF-IDF. This shows that the TF-IDF algorithm weakens the importance of such words as "世界

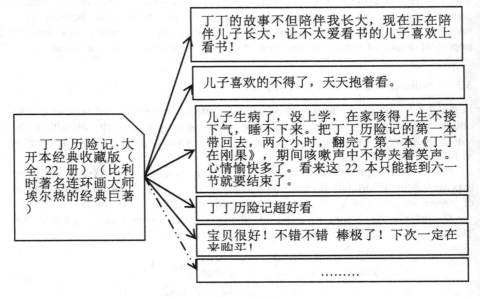

Fig. 2. The model input—the product description and corresponded comments

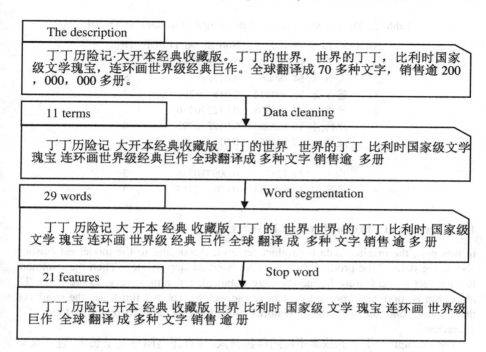

Fig. 3. The treatment process of product descriptions

(world)", "创作(creation)", "成(into)" and "读者(readers)", and strengthens the weight of such words as "米卢(Milutinovic)", 托托(Toto)" and "狗(dog)" simultaneously. It is known based on experience that such words as "米卢(Milutinovic)" and "托托(Toto)" can characterize the book more precisely compared with "世界 (world)" and "读者 (readers)".

Experimental data show that in this case, TF-IDF algorithm can characterize the text better than TF algorithm and is more appropriate to the actual situation. Therefore, the TF-IDF algorithm is used to calculate the feature weight of descriptions of the book.

Table 1. The sorted features and the weight of features by TF

Num	Word	TF-IDF	Normalized TF-IDF	TF
1	丁丁	122.627449	0.079276244	16
2	埃尔热	64.789063	0.04188486	8
3	世界	9.051818	0.005851823	5
4	比利时	19.153393	0.012382293	4
5	创作	10.681673	0.006905492	4
6	成	8.812729	0.005697257	4
7	读者	5.571782	0.003602048	4
8	一个	5.397255	0.00348922	4
9	蓝莲花	28.469944	0.018405261	3

Table 2. The sorted features and the weight of features by TF-IDF

Num	Word	TF-IDF	Normalized TF-IDF	TF
1	丁丁	122.627449	0.079276244	16
2	埃尔热	64.789063	0.04188486	8
3	蓝莲花	28.469944	0.018405261	3
4	米卢	19.753508	0.012770256	2
5	比利时	19.153393	0.012382293	4
6	托托	17.124866	0.01107089	2
7	历险记	16.912283	0.01093346	3
8	狗	15.1287	0.009780408	3
9	名字	13.676311	0.008841467	3

4.2 Treatment Effect of Product Reviews

In this part, the original data is product reviews. According to the model established above, the data of the product reviews were handled step by step. Then the comment feature vector and feature frequency were obtained. Since most reviews are short, TF algorithm is more appropriate to calculate the weight. Table 3 illustrates the results of the features and the weights of features on comments, and there are a total of 8 characters.

For example: "丁丁的故事不但陪伴我长大，现在正在陪伴儿子长大，让不太爱看书的儿子喜欢上看书 [19]!"

Table 3. The feature weight of reviews by TF

Num	Word	TF	TF归一化	isInTitle
1	丁丁	1	0.083333	1
2	故事	1	0.083333	1
3	陪伴	2	0.166667	0
4	长大	2	0.166667	0
5	书	2	0.166667	1
6	儿子	2	0.166667	0
7	爱	1	0.083333	0
8	喜欢	1	0.083333	0

4.3 Results of Similarity Calculation and Effects

In the above process, the feature vectors of product descriptions and reviews were obtained. The feature vector of product descriptions has 242 vector elements and the feature vector of product reviews has 8 elements. These two vectors have three same elements, which are {丁丁(Tintin), 故事(story), 书(book)}, and the total feature vector includes 247 elements, {丁丁(Tintin), 故事(story), 书(book) … …}. Use 247 elements as coordinate to form a vector space. Then the feature vector of product descriptions is {0.079, 0.0013, 0.00089, ……} and the feature vector of product reviews is

{0.083, 0.083, 0.083, ...}. Therefore, according to the cosine similarity algorithm the similarity between product descriptions and reviews could be obtained. Table 4 lists the top three most relevant reviews which correspond to the book *Occasion Not Allowed to Cry*.

The model of similarity calculating can be applied to many cases of text mining; for example, apply the similarity model to comment sorting. On the E-commerce website, showing the product reviews sorted by relevance has more advantages than showing the disordered ones. And it can boost book sales [19]. This also indicates that the proposed process model of the similarity computing between product features and reviews is effective.

Table 4. The similarity score between book reviews and product features

Num	The context of reviews	Similarity score
1	成长这场孤独的旅行, 只有真正走过人才会在磕磕碰碰中明白, 这世上, 还有不允许哭泣的场合.	0.08086834
2	有时候觉得"情"是一团乱麻, 复杂无解, 本能的会去逃避。看了《不允许哭泣的场合》后, 觉得情的是与非关键是看你如何去处理, 越来越觉得情商是个很重要的玩意!	0.07839214
3	绿色的封面预示着生命以及成长中的抚慰, 即使在路上摔倒也不允许哭泣拔去残留在身体上的荆棘刺继续奔跑, 跟不畏惧电脑辐射的绿萝一样。	0.07709725

5 Conclusion

This paper established a similarity computing process model based on VSM. The first step of the process is text pre-processing for the text. Then, through feature extraction and weight calculation, the feature vector can be obtained. Finally, according to the cosine algorithm, the score of similarity can be computed. In this paper, the experiments applying the model are performed, based on the books' electronic comments from the E-commerce websites. The experiment shows that the process model can support the review text mining such as comment usefulness analysis, comment sorting and comment categorizing. How to strengthen some features' impacts on similarity will be the next focus.

References

1. Duan, W., Gu, B., Whinston, A.B.: Do online reviews matter? - an empirical investigation of panel data. Decis. Support Syst. **45**, 1007–1016 (2008)
2. Huang, T., Zeng, G., Xiong, H.: Trustworthy sort method for shopping customer reviews based on correlation degree with product features. J. Comput. Appl. **08**, 2322–2327+2341 (2014)

3. Huang, A.H., Yen, D.C.: Predicting the helpfulness of online reviews—a replication. Int. J. Hum. Comput. Interact. **29**(2), 129–138 (2013)
4. Fang, J., Pu, D., Bai, S.: Research and implementation of text categorization system based on VSM. Appl. Res. Comput. **09**, 23–26 (2001)
5. Guo, Q., Li, Y., Tang, Q.: Similarity computing of documents based on VSM. Appl. Res. Comput. **11**, 3256–3258 (2008)
6. Zhang, J.: The study of streaming text representation method based on suffix tree model (STM) and application. In: Graduate School of the Chinese Academy of Sciences (Institute of Computing Technology) (2005)
7. Peng, J., Yang, D., Tang, S., Wang, T., Gao, J.: A new similarity computing method based on concept similarity in Chinese text processing. Sci. China Ser. F: Inf. Sci. **51**(9), 1215–1230 (2008)
8. Liu, X., Zheng, Q., Ma, Q., Lin, G.: Text similarity computing based on thematic term set. EN **4**(6) (2012)
9. Wu, S., Cheng, Y., Zheng, Y., Pan, Y.: A survey on text representation and similarity calculation in text clustering. Inf. Sci. **04**, 622–627 (2012)
10. Li, P.: Research on opinion extraction and classification technologies for product review mining. ChongQing University (2009)
11. Peng, J., Yang, D., Tang, S., Wang, T., Gao, J.: Text similarity computing based on concept similarity. Sci. China **05**, 534–544 (2009)
12. Long, S., Zhao, Z., Tang, H.: Overview on Chinese segmentation algorithm. Comput. Knowl. Technol. **10**, 2605–2607 (2009)
13. TF-IDF algorithm [EB/OL]. http://www.cnblogs.com/biyeymyhjob/archive/2012/07/17/2595249.html. 17 July 2012/04 June 2015
14. Guo, T.: Text similarity computing based on semantic field of vector space model. Yunnan University (2013)
15. Peng, J., Tang, C.J., Yang, D.Q., Zhang, J., Hu, J.J.: Similarity computing model of high dimension data for symptom classification of Chinese traditional medicine. Appl. Soft Comput. **9**(1), 209–218 (2009)
16. Wu, X., He, Z., Huang, Y.-W.: Product review mining: a survey. Comput. Eng. Appl. **44**(36), 37–40 (2008)
17. Wang, F., Wang, Z.: Research on Chinese automatic segmentation. Comput. Digit. Eng. **11**, 57–59 (2008)
18. Liu, J.: Chinese webpages feature extraction in learning to rank algorithm. Harbin Institute of Technology (2009)
19. Ju, Q.: Book reviews sorted for users purchasing influence. Programmer, pp. 122–125 (2014)

Research on the Tendency of Consumer Online Shopping Based on Improved TOPSIS Method

Yuekun Ma$^{(\boxtimes)}$, Pengfei Liu, and Donghuan Huang

College of Information Engineer, North China University of Science
and Technology, Tangshan 063009, China
mayuekun@163.com

Abstract. Analyzing and predicting the tendency of consumers online shopping is the precondition of providing personalized recommendation service, and has attracted more and more attentions. Most of e-commerce platform shave various types of products, and there exists tremendous difference in consumers' occupation, education background and other personalized features. This Paper realizes a TOPSIS Method which is based on entropy and fuzzy numbers. Compared to the traditional TOPSIS method, with the Association Rules mining method of data mining, the improved TOPSIS solves the problem in traditional TOPSIS method which requires manual intervention during execution. In this study, to implement intelligent tendency predicting and analysis of consumers online shopping based on data driven, three steps is carried out. Firstly, the data mining method is leveraged to obtain the fuzzy weights of evaluation indicator through analyzing the electric business transaction data, and then a fuzzy decision-making matrix is established between product and consumer's attribute; finally, a product category sequence which can indicate the tendency of consumer online shopping is established through calculating.

Keywords: Purchase tendency · Entropy · Fuzzy TOPSIS · Data mining

1 Introduction

With the development of computer science and internet, electronic commerce develops rapidly, and online shopping has become the most important way for consumers purchasing. According to the 36 times report of Internet development status statistics of china [1], which is published by CNNIC (China Internet Network Information Center) in July 2015, the amount of Chinese online shopping users is delivering up to 374 million, which take up 55.99 % of the total number of Internet users, by the end of June 2015. And the total turnover of electronic commerce is increasing rapidly, according to the data provided by Alibaba, he total turnover of commodity transaction through "Alipay" settlement merchandise is delivering up to ￥ 571 billion, only in November 11, 2014. Under this background, to analyze and predict the tendency of consumers online shopping has an important significance for the enterprise to carry out business of online marketing and advertisement.

Most of e-commerce platforms have various types of products, and every customer's personalized features are tremendous different from each other, including

© Springer Science+Business Media Singapore 2016
W. Chen et al. (Eds.): BDTA 2015, CCIS 590, pp. 309–322, 2016.
DOI: 10.1007/978-981-10-0457-5_28

gender, age, educational background, occupation, income, region, etc. This paper aims to realize the goal that analyzing and predicting the tendency of consumer online shopping through analyze and study the relationship between the personalized attribute feature set and the set of possible purchased product.

The traditional TOPSIS method can solve the above analysis and predicting problem, and has made some progress in the domain that analyzing and predicting tendency of consumer online shopping. But the traditional TOPSIS method exist some problems which are listed as follows.

(1) When we predict the tendency of consumer online shopping, the customer's different personalized features have different effects on the decision that whether they will purchase certain kind of product. The traditional TOPSIS method generally makes use of the expert investigation method or Analytic Hierarchy Process to determine weights of evaluation indicator which is along with a greater subjective factor, because everyone is different from each other on the degree of importance evaluation for every indicator.

The Entropy-weighted Method is a method which is to determine weight of indicator based on the information content of the observed values of the indicators. The data distribution is more dispersed, the method's accuracy is more higher [2]. This paper uses the Entropy-weighted Method to determine the weight for each indicator of consumer's personal attribute. And then, we can use the weights to predict the tendency of consumer online shopping. Through analyzing the entropy of each indicator, we can determine the degree of important for each indicator which have an influence on whether the consumer will buy a certain product, so as to determine the weights, that is entropy weight. Entropy weight method is an objective weighting method, which determines the weight of indicator according to the amount of information contained in each indicator, and overcomes the defects of the subjective weighting method [3, 4].

(2) Using Entropy-weighted Method to determine the weights avoids the deviation caused by human factors, but exists the shortcomings of heavy bias, and in the real application process, the weights themselves often can't be expressed by an accurate numerical. Therefore, we take advantage of the fuzzy number to express the degree of importance for indicator.

(3) The traditional TOPSIS method is a static multi-objective decision-making algorithm. The weights used in the forecasting process can't be automatically updated in time. In this paper, we gain and update the weights dynamically in the algorithm by using the method of machine learning.

In summary, some improvement is made on the TOPSIS method, which is described as follows. We get the degree of importance for indicators by using big data analysis method to mine e-commerce transaction data, and then make use of the information entropy method based on fuzzy number to determine the weights of indicators. It not only overcomes the shortcomings that the decision-makers' decisions may be inaccuracy, but also gains the degree of importance for evaluation objects based on decision matrix we make. The method described in this paper reduces the difficulty of the decision makers and make their views expressed better, also eliminates the influence of human factors and increases the real-time performance of the dynamical prediction.

2 Improvement on the TOPSIS

2.1 Traditional TOPSIS

TOPSIS (Technique for Order Preference by Similarity to an Ideal Solution) algorithm was first proposed by C.L.Hwang and K.Yoon in 1981 [5]. TOPSIS is usually used as an evaluation method which executes the process of multi-objective decision-making analysis with finite alternatives. It can establish a sorted sequence by calculating the degree of closeness between the evaluated object and the ideal target, and gives the relative merits of evaluation.

In the basic idea of the TOPSIS, some steps are carried out. Firstly, it will find out the optimal and the worst scheme using cosine method based on the normalized data matrix, and then calculate the distance between the evaluation object and the optimal scheme or the worst scheme, finally obtain the relative closeness, which is considered as the Basis for evaluating merits, between each evaluation object and the optimal scheme.

Assume that the number of evaluated objects is n, and the evaluation indicators is m, the original data is shown in the Table 1, X_{ij} indicates that the value of indicator j for the evaluation object i, where $i = 1, 2, 3..n$ and $j = 1, 2, ..., m$.

Table 1. The original data of TOPSIS

Evaluated object	Indicator 1	Indicator 2	$\cdots$	Indicator m
1	X_{11}	X_{12}	$\cdots$	X_{1m}
2	X_{21}	X_{22}	$\cdots$	X_{2m}
$\cdots$	$\cdots$	$\cdots$	$\cdots$	$\cdots$
n	X_{n1}	X_{n2}	$\cdots$	X_{nm}

The steps of using the TOPSIS algorithm to solve the problem are as follows:

Step 1: Establish the original matrix.
According to the original data can establish the corresponding matrix A

$$A = \begin{bmatrix} X_{11} & X_{12} & \cdots & X_{1m} \\ X_{21} & X_{22} & \cdots & X_{2m} \\ \vdots & \vdots & \ddots & \vdots \\ X_{n1} & X_{n2} & \cdots & X_{nm} \end{bmatrix} \tag{1}$$

Step 2: Preprocessing of matrix data
When using the TOPSIS as an evaluation method, all the indicators should be changed congruously, also is called communality, which can transform upper-quality indicator (benefit indicator) into low-quality Indicator (cost indicator), or low-quality indicator into upper-quality indicator, we usually use the latter. Through leveraging the reciprocal method shown in Eq. (2), the low-quality indicator can be transferred

into upper-quality indicator, so as to establish the original data table after data communality. The value after data communality processing is shown in the Eq. (3).

$$X'_{ij} = 1/X_{ij} \tag{2}$$

$$X'_{ij} = \begin{cases} X_{ij} & \text{lowquality} \\ 1/X_{ij} & \text{upperquality} \end{cases} \tag{3}$$

X'_{ij} represents the value of the indicator j for the evaluation object after transformation with reciprocal method, as shown in Table 2.

Table 2. The value after data communality

Evaluated object	Indicator 1	Indicator 2	$\cdots$	Indicator m
1	$1/x_{11}$	$1/x_{12}$	$\cdots$	$1/x_{1m}$
2	$1/x_{21}$	$1/x_{22}$	$\cdots$	$1/x_{2m}$
...	...	...	...	...
n	$1/x_{n1}$	$1/x_{n2}$	$\cdots$	$1/x_{nm}$

Step 3: Calculating the original data matrix after data communality, the calculation method is shown in the formula (4).

$$Z_{ij} = \begin{cases} \dfrac{X_{ij}}{\sqrt{\sum_{i=1}^{n}(X_{ij})^2}} & \begin{array}{l} \text{original upper} \\ \text{quality indicator} \end{array} \\ \dfrac{X'_{ij}}{\sqrt{\sum_{i=1}^{n}(X'_{ij})^2}} & \begin{array}{l} \text{original low} \\ \text{quality indicator} \end{array} \end{cases} \tag{4}$$

The matrix after the normalized processing is shown as follows.

$$Z = \begin{bmatrix} Z_{11} & Z_{12} & \cdots & Z_{1m} \\ Z_{21} & Z_{22} & \cdots & Z_{2m} \\ \vdots & \vdots & \ddots & \vdots \\ Z_{n1} & Z_{n2} & \cdots & Z_{nm} \end{bmatrix} \tag{5}$$

Step 4: Find out the optimal value vector the worst value vector according to the matrix Z that is to find the optimal and the worst scheme in the finite solution. Among them, the optimal scheme Z^+ is constituted by the maximum value of each column in the matrix Z, as shown in the formula (6).

$$Z^+ = (\max Z_{i1}, \max Z_{i2}, \ldots, \max Z_{im})\, 1 \leq i \leq n \tag{6}$$

The worst scheme Z^- is made up of the minimum value of each column in the matrix Z, as shown in the formula (7).

$$Z^- = (\min Z_{i1}, \min Z_{i2}, \ldots, \min Z_{im}) \, 1 \leq i \leq n \tag{7}$$

Step 5: Calculate the distance D_i^+ between each evaluated objects and the optimal scheme Z^+, D_i^- between each evaluated objects and the optimal scheme worst scheme Z^-. As shown in formula (8).

$$D_i^+ = \sqrt{\sum_{j=1}^{m} \left(\max Z_{ij} - Z_{ij}\right)^2}$$
$$D_i^- = \sqrt{\sum_{j=1}^{m} \left(\min Z_{ij} - Z_{ij}\right)^2} \tag{8}$$

If each indicator has weight W_{ij}, the distance formula is Eq. (9).

$$D_i^+ = \sqrt{\sum_{j=1}^{m} w_{ij} \left(\max Z_{ij} - Z_{ij}\right)^2}$$
$$D_i^- = \sqrt{\sum_{j=1}^{m} w_{ij} \left(\min Z_{ij} - Z_{ij}\right)^2} \tag{9}$$

W_{ij} is the weight coefficient of the indicator j.

Step 6: Calculate the closeness C_i between the evaluated object and the optimal scheme, the formula is shown in formula (10).

$$C_i = \frac{D_i^+}{D_i^+ + D_i^-} \quad 0 \leq C_i \leq 1 \tag{10}$$

C_i Get a value between 0 and 1, we should know that C_i is more closer to 1, the evaluated object is more closer to the optimal scheme. On the contrary, C_i is more closer to 0, the evaluated object is more closer to the worse scheme.

Step 7: Get the optimal scheme among the evaluated object by sorting the evaluate object according to the numeric value of C_i.

2.2 Improve on TOPSIS

2.2.1 Determine the Weight Using Entropy Method

In this paper, the weight of each indicator is calculated by using the entropy method [6]. Thus the subjective deviation caused by the artificial intervention in forecast is eliminated. The rules which are used to calculate the value of weight by using the entropy method are as follows:

There is a matrix A (as shown in Formula 1) in which the number of evaluation objects is m and the number of indicators is n. X_{ij}, which represents the evaluation value of the attribute j for the evaluated object i, is defined as the Formula (11).

$$X_{ij} = -\frac{X_{ij}}{\sum_{i=1}^{m} X_{ij}} \tag{11}$$

E_j which is shown as Formula (12) represents the entropy of the evaluated object about attribute j.

$$E_j = -k \sum_{i=1}^{m} X_{ij} \ln X_{ij} \tag{12}$$

Here $k = 1/\ln m$.

D_j is the information deviation degree, whose definition is shown as formula (13).

$$D_j = 1 - E_j \tag{13}$$

The weight of attribute for the indicator j is W_j.

$$W_j = \frac{D_j}{\sum_{j=1}^{n} D_j} \tag{14}$$

If the indicator has attribute preference, the weight of attribute is:

$$W_j = \frac{\lambda_j D_j}{\sum_{j=1}^{n} \lambda_j D_j} \tag{15}$$

λ_j is the eigenvalue, the formula for calculating the maximum eigenvalue is:

$$\lambda_{max} = \sum_i^n \frac{(AW)_i}{nW_i} \tag{16}$$

2.2.2 The TOPSIS Algorithm Based on Fuzzy Number

The assigning method by using Entropy avoids the deviation caused by human factors, but the decision object itself still exists fuzziness and uncertainty in the problem of multi-attribute decision-making, the value of each attribute factors is difficult to get an accurate result [7, 8]. So in this paper, the degree of importance for the indicator is expressed by the form of interval number, which aims to improve the TOPSIS algorithm further.

Interval numbers are defined as follows:

Definition 2.1. Assuming that

$$\tilde{a} = [a^L, a^U] = \left\{ x | a^L \leq x \leq a^U, a^L, a^U, a^U \in R \right\},$$

Here call $\tilde{a}$ an interval number. Particularly, if $a^L = a^U$, $\tilde{a}$ is degenerated to a real number.

The comparison rules of interval numbers are defined as follows (refer to the Definitions 2.2 and 2.3):

Definition 2.2 [9]

$$p(\tilde{a} > \tilde{b}) = \begin{cases} 1, & \text{when } \tilde{a} > \tilde{b} \\ \frac{1}{2}, & \text{when } \tilde{a} = \tilde{b} \\ 0, & \text{when } \tilde{a} < \tilde{b} \end{cases}$$

here $p(\tilde{a} > \tilde{b})$ is defined as the possible degree of $\tilde{a} > \tilde{b}$, and $\tilde{a}$, $\tilde{b}$ are real number.

Definition 2.3 [10]. When there exists at least one interval between $\tilde{a}$ *and* $\tilde{b}$, the possible degree of $\tilde{a} \geq \tilde{b}$ is defined as follow:

$$p(\tilde{a} \geq \tilde{b}) = \frac{\min\{l_{\tilde{a}} + l_{\tilde{b}}, \max(a^U - b^L, 0)\}}{l_{\tilde{a}} + l_{\tilde{b}}}$$

here,

$$\tilde{a} = [a^L, a^U], \tilde{b} = [b^L, b^U],$$

$$l_{\tilde{a}} = a^U - a^L, l_{\tilde{b}} = b^U - b^L,$$

For a given set of interval number $\tilde{a}_i = [a_i^L, a_i^U]$, $i \in N$, all of them will be compared with each other, and use the above probability degree definition to obtain the possible degree $P(\tilde{a}_i \geq \tilde{a}_j), i, j \in N$ between each other. And then build the matrix of possible degree $p = (p_{ij})_{n \times n}$. The matrix contains all the probability degree information between each interval number of the set. Therefore the problem of ranking the interval number is transformed into the problem of ranking the vector of the probability degree matrix.

In the past research, subjective analysis method such as expert analysis methods were used to determine the interval weights in some applications, but the subjective analysis is easily influenced by human factors. The indicator weight assignment problem belongs to multi-attribute decision-making problem in which the related attributes have different preference information. In this paper, fuzzy number theory and a kind of interval weight obtain method, which is based on interval fuzzy preference relation, is used to implement the indicator weight assignment with non-subjective factor.

For example, the indicator is divided into five kinds of levels according to the important degree: unimportant, slightly unimportant, important, slightly important and very important, and is expressed by fuzzy interval, as shown in Table 3.

Therefore, for the indicator index by j, a kind of interval weight obtain method, which is based on interval fuzzy preference relation, is used to implement the indicator

Table 3. Fuzzy interval

Important degree	Fuzzy interval
not important	(0.0, 0.2)
slightly not important	(0.2, 0.4)
important	(0.4, 0.6)
slightly important	(0.6, 0.8)
very important	(0.8, 1.0)

weight assignment in which the weight is expressed by the form of fuzzy interval number, then the weight λ_j is transformed into the formation as follows:

$$\left(\lambda_{j1}, \lambda_{j2}\right).$$

Here λ_{j1} is the upper bound of the fuzzy interval, and λ_{j2} is the lower limit of the interval.

Assume that λ, shown in the formula (17), is the weight vector of all the indicators,

$$\lambda = \left((\lambda_{11}, \lambda_{12}), (\lambda_{21}, \lambda_{22}) \cdots (\lambda_{j1}, \lambda_{j2})\right) \qquad 1 \leq j \leq n \qquad (17)$$

Then, the lower and upper bound entropies can be calculated according to the formula (12). The form of weight vector is shown as formula (18).

$$W = \left((w_{11}, w_{12}), (w_{21}, w_{22}) \cdots (w_{j1}, w_{j2})\right) \qquad 1 \leq j \leq n \qquad (18)$$

Finally, the weight vector W is substituted into the TOPSIS algorithm.

$$A = \begin{bmatrix} X_{11} & X_{12} & \cdots & X_{1m} \\ X_{21} & X_{22} & \cdots & X_{2m} \\ \vdots & \vdots & \ddots & \vdots \\ X_{n1} & X_{n2} & \cdots & X_{nm} \end{bmatrix}$$

At this time, each value of the matrix A is multiplied by the corresponding lower and upper bound weight, the result is shown as formula (19):

$$A = \begin{bmatrix} \left(X_{11}^-, X_{11}^+\right) & \left(X_{12}^-, X_{12}^+\right) & \cdots & \left(X_{1m}^-, X_{1m}^+\right) \\ \left(X_{21}^-, X_{21}^+\right) & \left(X_{22}^-, X_{22}^+\right) & \cdots & \left(X_{2m}^-, X_{2m}^+\right) \\ \vdots & \vdots & \ddots & \vdots \\ \left(X_{n1}^-, X_{n1}^+\right) & \left(X_{n2}^-, X_{n2}^+\right) & \cdots & \left(X_{nm}^-, X_{nm}^+\right) \end{bmatrix} \qquad (19)$$

In order to obtain the optimal and the worst scheme, further improvement on the TOPSIS algorithm is carried out.

The optimal scheme:

$$Z^+ = \left(\left(\max(X_{i1}^-), \max(X_{i1}^+) \right), \left(\max(X_{i2}^-), \max(X_{i2}^+) \right), \cdots, \right.$$
$$\left. \left(\max(X_{im}^-), \max(X_{im}^+) \right) \right) \; 1 \le i \le n \tag{20}$$

The worst scheme:

$$Z^- = \left(\left(\min(X_{i1}^-), \min(X_{i1}^+) \right), \left(\min(X_{i2}^-), \min(X_{i2}^+) \right), \cdots, \right.$$
$$\left. \left(\min(X_{im}^-), \min(X_{im}^+) \right) \right) \; 1 \le i \le n \tag{21}$$

Then using the formulas (20) and (21) to get the distance between the evaluation object and the ideal optimal or the ideal worst scheme:

$$D_i^+ = \sqrt{\sum_{j=1}^{m} \left[\left(X_{ij}^- - \max(X_{ij}^-) \right)^2 + \left(X_{ij}^+ - \max(X_{ij}^+) \right)^2 \right]} \tag{22}$$

$$D_i^- = \sqrt{\sum_{j=1}^{m} \left[\left(X_{ij}^- - \min(X_{ij}^-) \right)^2 + \left(X_{ij}^+ - \min(X_{ij}^+) \right)^2 \right]} \tag{23}$$

Finally, according to the formula (24) we can obtain the closeness of distance between the evaluation object and the optimal scheme in order.

$$D_i = \frac{D_i^+}{D_i^+ + D_i^-} \tag{24}$$

3 Algorithm Application

This work is based on the e-commerce platform of Bone China supply chain. We analyze and forecast the tendency of different kinds of products for different consumers.

3.1 Product Category

The e-commerce platform of Bone China supply chain involves all product categories of China's Bone Porcelain. In this paper, we choose the ceramics for daily use product category for analysis and predicting, as shown in Table 4.

Table 4. Ceramics for daily use product category

Class number	Class name	Code
0355602090001	Magnesia porcelain tableware	MC
0352002130010	Magnesia strengthen porcelain tableware	MQC
0355602090001	Strengthen porcelain tableware	QC
0353602070015	Shell porcelain tableware	BC
0352802810002	Color glazed tableware	SC
0735604180002	Elegant light porcelain tea set	YC
0355600213008	Bone china tableware	GC

3.2 Determination of the Consumers' Attributes Value and the Corresponding Fuzzy Weights

In this paper, the consumer's attribute including gender, age, education background, occupation, income, region and so on are mentioned.

According to requirement of the method used to calculate the weights of attributes parameter, which is involved in the prediction process, through fuzzy calculation, the fuzzy weight of each consumers' attribute indicator is obtained by analyzing the historical transaction data with the data mining method on association rules. The method is described as follows.

In order to obtain the fuzzy interval data (w_i^-, w_i^+) of the parameter indicators, the upper and lower bound value of the corresponding fuzzy interval w_i^- and w_i^+ are required to obtain respectively.

We mine the transaction data to obtain the association rules between product categories and each indicator of consumers' attributes by using A priori algorithm. The results of mining will be assigned to the upper and lower bound of each indicator corresponding fuzzy interval. Detail method is as follows:

The calculation method of w_i^-:

The confidence degree of association rule mining algorithm is set to 1, and then the support degree, which is assigned to lower bound of the fuzzy weights of the indicators, is calculated under this confidence degree.

The calculation method of w_i^+:

The confidence degree of association rule mining algorithm is set to 0.5, and then the support degree, which is assigned to upper bound of the fuzzy weights of the indicators, is calculated under this confidence degree.

Divides about value of each consumers' attribute and the corresponding fuzzy weight are obtained by using the above method to analyze the Transaction data about Shell porcelain tableware. Part of analysis results above is shown as between Tables 5, 6, 7, 8, 9 and 10. Due to the number of product category is too large, only the parameters about Shell porcelain tableware are listed.

Table 5. Fuzzy weight of sex

Sex	Weight
Male	(0.4, 1.0)
Female	(0.0, 0.4)

Table 6. Fuzzy weight of age

Age	Weight
≤19	(0.0, 0.1)
20–29	(0.1, 0.2)
30–39	(0.2, 0.5)
≥40	(0.5, 1.0)

Table 7. Fuzzy weight of degree

Degree	Weight
Junior middle school and below	(0.0, 0.1)
Senior middle school, Technical secondary school, Technical school	(0.1, 0.2)
Junior college	(0.2, 0.4)
Undergraduate	(0.4, 0.7)
Graduate student	(0.7, 1.0)

Table 8. Fuzzy weight of occupation

Occupation	Weight
State organs, Party organizations, Enterprise, leaders of enterprise unit	(0.5, 0.6)
Professional and technical personnel	(0.4, 0.5)
The staff and related personnel	(0.6, 0.8)
Business and service personnel	(0.8, 1.0)
Agricultural, forestry, animal husbandry, fishery, water conservancy industry production personnel	(0.2, 0.3)
Production, transport equipment operators and related personnel	(0.3, 0.4)
Soldier	(0.1, 0.2)
Other practitioners of inconvenience	(0.0, 0.1)

The vector (42, male, 06, 01, The coastal region) is an instance of a consumer A's attribute corresponding to the indicator. At first, the values in above vector are mapping into the Table 11, and then middle result is substituted into formulas (16) and (17), the data in Table 12 are final decision-making matrix.

According to formulas (17) and (18), the optimal scheme and the worst scheme are shown as follows:

The optimal scheme:

$$Z^+ = ((0.248, 0.354), (0.177, 0.314), (0.26, 0.458), (0.333, 0.481), (0.219, 0.294))$$

Table 9. Fuzzy weight of income

Income (¥)	Weight
No income	(0.0, 0.1)
<1000	(0.1, 0.2)
1001–3000	(0.2, 0.3)
3001–5000	(0.3, 0.5)
5001–8000	(0.5, 0.6)
8000–10000	(0.6, 0.9)
More than 10000	(0.9, 1.0)

Table 10. Fuzzy weight of region

Region	Weight
The coastal region	(0.7, 1.0)
The central region	(0.5, 0.7)
Northeast region	(0.2, 0.5)
Western Region	(0.0, 0.2)

Notes: We select and analyze 100 thousand trading records related to several kinds of products involved in Table 4 in our study. And then obtain the weight value of indicator about each kind of tableware, as shown in Table 11.

Table 11. Decision-making matrix

P_C	Age	Sex	Deg.	Occu.	Rigion
BC	0.063,(0.0,0.1) 0.483,(0.1,0.2) 0.635,(0.2,0.5) 0.981,(0.5,1.0)	1.234,(0.4,1) 0.795,(0,0.4)	...	......	......
MCQ	0.098,(0.0,0.1) 0.123,(0.1,0.2) 0.342,(0.2,0.6) 0.323,(0.6,1.0)	......	...	......	......
QC	......	......	...	......	......
MC	......	......	...	......	......
SC	......	......	...	......	......
YC	......	......	...	......	......
GC	......	......	...	......	......

The worst scheme:

$$Z^- = ((0.107, 0.229), (0.084, 0.196), (0.124, 0.266), (0.202, 0.343), (0.151, 0.215))$$

Table 12. Final decision-making matrix

P_C	Age	Sex	Deg.	Occu.	Rigion
BC	(0.133,0.266)	(0.084,0.212)	(0.241,0.385)	(0.331,0.397)	(0.204,0.291)
MCQ	(0.163,0.271)	(0.123,0.246)	(0.154,0.307)	(0.291,0.436)	(0.219,0.274)
QC	(0.134,0.334)	(0.177,0.295)	(0.26,0.364)	(0.333,0.466)	(0.172,0.294)
MC	(0.115,0.229)	(0.137,0.196)	(0.229,0.458)	(0.202,0.354)	(0.185,0.264)
SC	(0.172,0.287)	(0.115,0.287)	(0.219,0.383)	(0.274,0.343)	(0.205,0.256)
YC	(0.107,0.268)	(0.129,0.258)	(0.124,0.289)	(0.321,0.481)	(0.151,0.215)
GC	(0.248,0.354)	(0.126,0.314)	(0.157,0.266)	(0.262,0.459)	(0.165,0.274)

Finally, the distance D^+ between each candidate scheme and the optimal scheme can be obtained by using formula (22); the distance D^- between each candidate scheme and the worse scheme can be obtained by using formula (23). The value of D^+ and D^- in our study in shown as follows:

$$D^+ = (0.229, 0.241, 0.196, 0.292, 0.559, 0.301, 0.243)$$

$$D^- = (0.241, 0.19, 0.312, 0.229, 0.222, 0.223, 0.27)$$

According to the formula (24), the closeness between the evaluated object and positive scheme can be obtained which is included in the vector (0.487, 0.559, 0.386, 0.561, 0.715, 0.574, 0.274).

And then the priority about the purchasing tendency among the selected seven kinds of product for consumer A is SC, YC, MC, MCQ, BC, QC, GC.

The method above can used to implement analysis and predicting of tendency about consumer online shopping in some other e-commerce platform.

4 Conclusion

This paper puts forward a method TOPSIS based on entropy and fuzzy numbers which can implement the tendency predicting analysis of consumer online shopping. This method offsets the impacts of human factor in the subjective analysis method and the deviation which is caused by many factors in objective method, such as too many evaluation indicators. It should be pointed out that the predicting process of the method is dynamical. In this process, data mining method is leveraged to dynamically obtain the fuzzy weights of evaluation indicator through analyzing the electric business transaction data, and the fuzzy data which can reflect the weights of evaluated indicator will dynamically updated with the electric business transaction data renewal.

Acknowledgement. This paper is Acknowledged by the key technologies R&D program project of Hebei province (15210110D), and the key technologies R&D program project of Tangshan (14130233B).

References

1. The 36 times report of Internet development status statistics of china. China Internet Network Information Center, pp. 3–11 (2015)
2. Harremoes, P., Topsoe, F.: Maximum entropy fundamentals. Entropy **3**, 191–226 (2001)
3. Yang, H., Lin, L., Yu, Z.: A class of fuzzy multiple attributes TOPSIS decision making based on exponential type fuzzy numbers. Comput. Eng. Appl. **48**(34), 120–124 (2012)
4. Wu, C., Fan, X.: Fuzzy entropy based human resource's structure optimal configuration. Chin. J. Manage. Sci. **10**(4), 43–47 (2002)
5. Li, C-Y., Xu, M.-Q.: The importance analysis of equipment based on the improved TOPSIS. J. Vib. Shock **28**(6), 19–27 (2009)
6. Deng, X., Li, J., Zeng, H., Chen, J.: Research on computation methods of AHP wight vector and its applications. Math. Pract. Theory **42**(7), 93–100 (2012)
7. Abbattista, F., Degemmis, M., Fanizzi, N., et al.: Learning User Profiles for Content-Based Filtering in E-Commerce. SSRN
8. Zhang, X., Wu, Q.: Research on the personalized recommendation based on TOPSIS method. J. Intell. **12**(2), 23–28 (2009)
9. Xu, Z.: New method for uncertain multi-attribute decision making problems. J. Syst. Eng. **17**(2), 176–181 (2002)
10. Xu, Z., Da, Q.: Possibility degree method for ranking interval numbers and its application. J. Syst. Eng. **18**(1), 67–70 (2003)

Author Index

Printed in the United States
By Bookmasters